# Alcoholism in America

## From Reconstruction to Prohibition

## Sarah W. Tracy

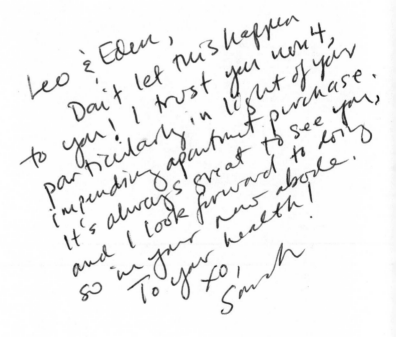

The Johns Hopkins University Press

*Baltimore and London*

© 2005 The Johns Hopkins University Press
All rights reserved. Published 2005
Printed in the United States of America on acid-free paper

(Johns Hopkins Paperbacks edition, 2007)

9  8  7  6  5  4  3  2

The Johns Hopkins University Press
2715 North Charles Street
Baltimore, Maryland 21218-4363
www.press.jhu.edu

*The Library of Congress has cataloged the hardcover
edition of this book as follows:*

Tracy, Sarah W., 1963–
  Alcoholism in America : from reconstruction to prohibition /
Sarah W. Tracy.
      p.   cm.
  Includes bibliographical references and index.
  ISBN 0-8018-8119-6 (hardcover : alk. paper)
    1. Alcoholism—United States—History. 2. Alcoholism—
Treatment—United States—History. 3. Alcoholism—
Hospitals—United States—History.
1. Title.
  RC565.7.T73 2005
  362.292′0973—dc22          2004025484

ISBN 10: 0-8018-8620-1 (pbk.: alk. paper)
ISBN 13: 978-0-8018-8620-1

A catalog record for this book is available from the British
Library.

# Alcoholism in America

*To the memory of my grandfather,*
*F. Erwin Tracy, M.D.*

# Contents

# Preface

There is a personal story behind my choosing to study the history of alcoholism, but it does not involve alcohol. At times I have wished that I had a more direct connection, rationalizing that I might have finished the book earlier had I been propelled by the immediacy of the problem in my own life. But the truth is that the more I have learned about the history of alcoholism—and its present state—the more grateful I am that my life has not been directly affected by it. Addiction to alcohol is a complex and devastating phenomenon: medically, personally, and socially. My work on this book over the past decade has given me the chance to meet dozens of people who have had problems with alcohol. I have benefited from their willingness to share their experiences and insights, and from their enthusiasm for the project.

Although I have developed a profound respect for individuals who contend with addiction on a daily basis, as a historian my primary interest is in the past. This book is an attempt to understand how and why many physicians, social reformers, and heavy drinkers in an earlier day came to regard habitual drunkenness as a disease, and what they chose to do about it. I hope, however, that my historical study will engage those with interests in contemporary addiction as well as historians of medicine, alcohol, and social welfare policy in America.

I was drawn to the study of alcoholism while studying for my comprehensive examinations in graduate school. Having defined the history of American medicine, American social history, and the sociology of knowledge as my three areas of interest, I made the commitment to finding a topic that would link these different subjects. Ever patient as I searched for a topic, Charles Rosenberg, my advisor at the University of Pennsylvania, dispatched me to the Historical Collections of the College of Physicians for inspiration. There, on a wintry Philadelphia day, I discovered that many late-nineteenth-century physicians believed that "inebriety" and "dipsomania" were constitutional diseases in which heredity played a heavy hand. My research antennae began to twitch. Here was a topic that allowed me to probe the connections among reform movements, social welfare policy, the professionalization and special-ization of medicine, and the social construction of disease. When I learned how little attention historians had paid this episode in the history of disease, my commitment grew more resolute.

Writing a dissertation is a humbling experience. When I selected "the med-icalization of alcoholism" as my topic, I imagined that I would produce what one of my professors, Henrika Kuklick, dubbed "The Great American Mono-graph," titled, in my case, "From Vice to Disease: the History of Alcoholism in America from Benjamin Rush to Bill Wilson." While both Kuklick and Rosen-berg will attest to the fact that I may have collected enough archival data to fulfill my ambition, the practical task of completing the dissertation meant that I narrowed my scope considerably. I wrote my thesis on the efforts of the Commonwealth of Massachusetts to build a state system for the medical man-agement of inebriates between 1833 and 1919.[1] It was an interesting and im-portant story, for the rest of the nation watched what went on in the Bay State, anticipating that success in Massachusetts might be incorporated into their own states' public health and social welfare programs.

Early in the dissertation writing process, however, I realized that the narra-tive I was telling concerned contested or incomplete medicalization.[2] I had defined medicalization quite simply as the process by which society comes to regard specific conditions and particular forms of social deviance (e.g., preg-nancy, homosexuality, opiate addiction) as diseases. In the case of Massachu-setts, it was clear that nineteenth-century social reformers, physicians, and politicians were promoting new medical understandings of a resilient social problem: habitual drunkenness. They were also putting a new disease frame-work into practice through the creation of specialized state and private insti-tutions and the revision of laws concerning public drunkenness. Still, there remained resistance at a number of levels, and ultimately Prohibition ap-peared to bring an end to these remedial efforts. I had not anticipated the complex dynamics of inebriate reform in the Bay State: a dynamics that linked medical experts, social reformers, state institution builders, politicians, the general public, and the inebriates themselves. This was no simple case of "From Vice to Disease," and hardly a case of medical "imperialism." Did a sim-ilar story play out across the country? I wondered. *Alcoholism in America* re-sults from my desire to understand the national story of which Massachusetts was a part.

As I wrote this book, I realized that two questions steered my research. First, how do diseases reflect social values, economic imperatives, and political and professional priorities? And second, what consequences do new diagnostic categories have on social institutions, public opinion, and ultimately on the lives of ordinary individuals?[3] These questions place my work squarely within the social history of medicine and the social construction of disease, but they are also important questions about professional power, individual rights, and social justice. To declare that alcoholism is a disease, for example, may shift attention away from social and economic circumstances such as unemployment and poverty that can precipitate alcohol abuse. Instead, individual pathology commands our attention. Likewise, to discuss children's unruly behavior in terms of attention deficit–hyperactivity disorder may deflect attention from problems in the schools—teacher shortages, large class sizes, and the underfunding of our educational system. Should we not address the larger infrastructural problems, too? The act of diagnosis frequently allows us to recognize and liberate sick individuals from their plight, but we must also remember that it may prove a powerful instrument of social control. In this sense, disease is not just clinical pathology; it is social commentary.

Along the road to completing this book, I also realized that alcoholism's status as a hybrid sociomedical pathology meant that I had my work cut out for me. The consumption of alcohol is intricately woven into the fabric of American society; its use and its abuse have figured literally and symbolically in politics, religion, and culture, from colonial times to the present.[4] For this reason, alcoholism has provided an occasion not only for pathological speculation but also for the projection of myriad social concerns. To understand what alcoholism meant between the eras of Reconstruction and Prohibition, it is important to see how the chronic over-consumption of alcohol resonated not only with ideas about disease and disability but also with issues of gender, class, race, ethnicity, and social order. Thus, I have addressed alcoholism's social significance as I have discussed its clinical and professional importance.

As one might expect from contemporary debates over alcoholism's status as a disease, a symptom of a disease, a vice, or a crime (in the case of pregnant women), the issue of individual agency has reverberated throughout the history of physicians' and reformers' thinking on habitual drunkenness. Caught between biological determinism on the one hand and individual free will on the other, physicians of the late nineteenth and early twentieth centuries tended to emphasize both in their explanations for habitual drunkenness.

This did little to quell controversy. Aberrant behavior such as chronic drunkenness was often explained through recourse to vague somatic causes such as
neurasthenia and hereditary degeneration, but overindulgence, overwork,
poverty, and disappointment remained oft cited causes of alcoholism. Remedies emphasized self-vigilance, new habits of living, and other environmental
interventions for which the individual was ultimately responsible. As we shall
see, there was no united intellectual front for the advancement of the disease
concept of inebriety. Instead, a full range of explanation for the condition obtained within medicine and social science, changing gradually, but not uniformly, between 1870 and 1920. In writing this book, I have tried not to evade
the complexity of Gilded Age and Progressive Era thought about inebriety but
to probe the interesting inconsistencies and different emphases of medical
and popular views of the condition.

As a social and cultural historian of medicine, however, I have felt compelled to compare theory with practice. From the start, I wanted to find out if
the varieties of inebriety proposed by physicians held specific consequences
for the treatment of inebriates. Accordingly, *Alcoholism in America* pays particular attention to the history of state hospitals and asylums for inebriates.
From the founding of the American Association for the Cure of Inebriates
(AACI) in 1870 to the post–Volstead Act closing of facilities for alcoholics,
ideas about inebriety were brought to life in the management of patients at
inebriate homes, asylums, and hospitals. Few issues were as important to the
medical entrepreneurs who promoted the disease concept as the creation of
inebriate institutions, particularly public ones. In 1876, for example, members of the Section on Public Hygiene and State Medicine of the American
Medical Association (AMA) declared that inebriates required special institutions and that "it is the imperative duty of each Commonwealth to establish
and maintain public institutions for the treatment of inebriety."[5] Reformers
envisioned grand returns for the state, for a public hospital for inebriates was
"a means of preventing insanity, diseased and degenerate offspring, and dependency and crime. Aside from its incalculable value as a saver of men and
women, it would be a great financial gain in the end."[6] Thus, these institutions were both an essential element of medicalization efforts and a reflection
of an important development in the history of American social welfare policy:
the rationalization and centralization of state authority.

In many ways, this book's progression from ideas to institutions parallels
the trajectory of my own education in the history of science and medicine. As

an undergraduate at Harvard, I was drawn to the history of ideas and their cultural significance. As a graduate student at Penn, I began to investigate the institutional and professional dimensions of science, technology, and medicine. If the structure of this book mirrors this intellectual journey, it also reflects my desire to understand how professions and institutions change the way people think about themselves. Thus, my final concern is whether or not the disease concept and the institutions based on it mattered to people with drinking problems.

Here I have not been as concerned about the efficacy of institutional treatment as about its perception by those who took it. Did inebriates adopt a medical perspective on their drinking problems? Did they find the care that they received at medical institutions helpful or oppressive? What role did they or those close to them have in shaping treatment? In many respects, these have been the toughest questions to answer. Correspondence among patients, their families, and hospital physicians yielded important insights into the utility, futility, and reception of various medical interventions; the published autobiographical accounts of inebriates who sought medical care were also helpful. In the end, however, *Alcoholism in America* is a bigger book than I had originally envisioned because I adopted this inclusive approach. I hope that readers will share my commitment to interrogating the relations among theory, practice, and perception.

As I have shared particular facets of my research with different audiences over the years, I am always asked: "So, what do you really think about alcoholism? Is it a disease?" Generally, I have retreated to my position as an historian, observing that I am not a medical expert, but a student of medicine's past, and that although alcoholism is a tremendous social and medical problem the solution to this problem will not come from my work. My focus is not on solving the persistent clinical or cultural puzzles of alcoholism but on understanding how people conceived of them in the late nineteenth and early twentieth centuries.

That said, I believe that I do owe readers some explanation of what I think about alcoholism. I believe it is useful to treat chronic drunkenness as a disease. Many may disagree with me. Dissenters emphasize scientists' failure to find a discrete mechanism for the condition apart from the consumption of alcohol.[7] Critics also point out the inadequacies of E. M. Jellinek's typology of alcoholism and the biased sample of alcoholics on which it was based.[8] Still others take objection to the disease label, feeling that it defines the sick person as an invalid, a person unable and (it is often implied) unwilling to take responsibility for his or her own recovery.[9] Frequently, in the same breath that alcoholism is discussed as a disease, total abstinence is referenced as its only cure. Many disagree with this formulation, for there is significant evidence that some heavy, self-destructive drinkers are able to attenuate their alcohol habits over time, learning to drink moderately.[10] Other critics of the disease concept, especially those within the medical field, look at alcoholism as a symptom rather than a genuine disease. Just as some early twentieth-century physicians argued that neurasthenia underlay each case of alcoholism, so many contemporary physicians see the condition as one particular expression of an underlying personality disorder.

Each of these objections has some merit, of course. The truth is that we do not know what causes people to embark on a course of increasingly heavy, self-destructive drinking. Although genetics may play an important role in an individual's tolerance to alcohol and susceptibility to alcohol addiction, genes alone cannot explain the wide range of individuals whose lives are endangered by chronic problem drinking. The ungainly truth about alcoholism, and addiction generally, is that its etiology lies in both nature and nurture, in our genes and our environment. Just as our parents share their heredity with us, so too they transmit their behavioral repertoires for socializing, celebrating, and coping with loss. Anthropologists who have studied the consumption of alcohol and other psychoactive substances have demonstrated that a person's relationship to a drug depends not only upon pharmacology and the user's physiological response but also on the mind-set of the consumer and the setting in which the substance is taken.[11] Alcoholism is a product of both constitution and culture.

My conviction that chronic drunkenness is best placed within a disease framework stems from a pragmatic desire to provide effective and sympathetic care for those in clear physiological and psychological distress. If this can be accomplished within the larger, more current therapeutic paradigm of "alcohol problems," which recognizes the diversity of peoples' relationships with beverage alcohol, I would prefer such an approach.[12] Treatment experts have proven persuasively that individually tailored care is needed. Echoing William Osler's writing on syphilis, one could easily say, "To know alcohol problems is to know medicine," precisely because drinking problems present to clinicians in such variable forms. Indeed, George Vaillant has suggested that "it is the variety of alcohol-related problems, not a unique criterion, that captures what clinicians really mean when they label a person alcoholic."[13] But at the dawn of the new millennium, to label something a disease still seems—at least to me—to garner more public sympathy for the affected individual than to simply say that they have a "problem." For all of the ambiguity that surrounds alcoholism's etiology and symptoms, the diagnosis still affords people access to the health-care system—a reasonable starting point for acquiring help, even if the strengths of modern medicine do not lie in the management of chronic disease.

I would add that the failure to identify a single, discrete disease mechanism is of little consequence to me (although I will argue that the want of one proved increasingly problematic for early disease concept advocates, much as it bothers alcoholism's critics today.) There are other less controversial diseases where multiple or discrete mechanisms are assumed but not proven. The disease of hypertension, for example, stems from unknown origins. There are very likely multiple pathophysiological mechanisms at work that ultimately end in hypertension, yet we do not question its legitimacy as a disease or the necessity of medically treating those affected by high blood pressure. Alcoholism is different in obvious ways: one must consume a legally controlled psychoactive substance to acquire the condition, and volition is at work, at least initially. The strict regulations regarding the sale of alcohol testify to its potential dangers—dangers that adults are entrusted to manage. When they fail to do so, the risk of alcoholism increases significantly, and likewise the risk of moral and legal censure.

Hypertension is similar in that it may be caused in genetically susceptible people by the overconsumption of potentially dangerous but (unlike alcohol) dietetically necessary substances: foods rich in sodium and fat. In hypertension, volition is similarly involved: too many super-sized servings of french fries or grilled ham-and-cheese sandwiches will meet with swift disapproval from an individual's physician. Yet where the alcoholic routinely runs the risk of endangering those around him or her, this is rarely the case with the hypertensive, unless of course one has a stroke and loses consciousness or the ability to function. Alcohol's disinhibiting effects and its associations with partying, sex, and drug use make it anything but "the neutral spirit" Berton Rouche once labeled it.[14] In the United States alcohol is, and has always been, a complicated cultural signifier, and this has tended to make its excessive use and abuse more stigmatizing (and more interesting) than the overconsumption of french fries.

*Alcoholism in America* is not a tidy tale of transformation in medical and social thought and policy. Nor do I think that the history of alcoholism will ever be characterized this way. Today, as one hundred years ago, we are faced with a series of interesting contradictions about alcoholism. While the American Medical Association "endorses the proposition that drug dependencies, including alcoholism, are diseases and that their treatment is a legitimate part of medical practice,"[15] most physicians receive little training in managing addiction of any variety. Indeed, the strongest advocates for the disease concept of alcoholism are not physicians but the lay members of Alcoholics Anonymous (AA), who also remain the largest group devoted to the treatment of the condition. Likewise, although AA adheres to a disease concept of alcoholism, it organizes recovery through a series of 12 steps that emphasize spiritual awakening and individual agency, matters traditionally outside the purview of Western biomedicine. Does this history of the early campaign to provide medical care for the inebriate shed any light on these contemporary contradictions? I think the answer is yes.

My story highlights the failure of the medical profession to discover a specific mechanism that caused alcoholism (other than the excessive consumption of ethanol). Yet it also reveals that in spite of this failure—one that still haunts research on alcoholism—physicians and lay reformers implicated a wide range of environmental and constitutional factors in alcoholism's origins and a diversity of interventions for its treatment—many of which, for better or worse, resemble those in place in today's "treatment industry." They were able to do this because they adhered to a holistic model of disease that was more characteristic of the early nineteenth century than of the early twentieth. Their model emphasized factors like patient attitude, social and economic stresses, nutrition, and heredity. To quote turn-of-the-century psychiatrist William Alanson White, "The causes of drinking are infinitely varied and intimately bound up in the heart of man—at once an expression of his strength and weakness, his successes and his failures."[16] Then as now, alcoholism was a chronic condition with "overdetermined" or multiple origins that required an integrated approach emphasizing the biological, the psychological, and the social.

In the early twenty-first century it is largely the result of biomedical success that we are living in an age of chronic disease. Diabetes, arthritis, hypertension, and obesity have affected our population to an unprecedented degree. These are chronic conditions with multifactorial etiologies that are also out of step with the reductionism and specificity of the reigning acute disease model. It may be time to reconsider our understanding of what constitutes a disease and how best to treat it. It may be time to embrace a more holistic or comprehensive view of disease and disability. If this is the case, then an appreciation of the multiplicity of factors—from germs to gender roles, from class to constitution—that affect what we consider "healthy" and "diseased" may prove valuable. History, as well as anthropology and sociology, are quite useful in this regard. It would be ironic indeed if alcoholism, long a stepchild of modern medicine, might lead us in developing this new understanding.

# Acknowledgments

This book would not have been possible without the support of many friends, colleagues, and institutions. First, I would like to thank Charles E. Rosenberg, who introduced me to the history of disease and the ways it can illuminate cultural and societal change. As a scholar and mentor, he has always set the highest standards. I am grateful for his continuing support of both this project and my interests in constitutional and chronic disease. Henrika Kuklick has likewise been an unstinting source of insight and inspiration. Her friendship and mentorship have proved invaluable. The ties I formed with scholars during my graduate years in the History and Sociology of Science Department at the University of Pennsylvania have continued to be important. For their various and sundry contributions to this book, I would like to thank Jeff Brosco, Chris Feudtner, Janet Golden, Jennifer Gunn, Julie Johnson, Michael B. Katz, Maneesha Lal, Donna Mehos, Steve Peitzman, the late Jack Pressman, Eric Schneider, Janet Tighe, Nancy Tomes, and Elizabeth Toon.

I am grateful, too, for a National Institute of Mental Health postdoctoral fellowship at the Institute for Health, Health Care Policy and Aging Research, Rutgers University. There Gerald N. Grob, Allan Horwitz, David Mechanic, Jerry Wakefield, Jamie Walkup, and Rod Wallace encouraged me to think about the connections between past and present and to take a systemic view of mental health care in the United States, while bearing in mind the individual burdens of mental illness. DeeDee Davis, Ruth Kranes, and Ruth Peters provided valuable computing and administrative assistance as well. I benefited immensely from my time in this stimulating research environment. I am especially grateful for Gerald Grob's encouragement and continued scholarly advice over the years.

My thanks to the Francis C. Wood Institute of the College of Physicians of Philadelphia for giving me the post of scholar-in-residence. Through this fellowship I gained daily access to one of the nation's greatest collections of historical and archival materials in medicine. The Wood Institute seminar series also gave me a collegial group of scholars and physicians with whom I could share my research. Tom Horrocks, Charles Greifenstein, and Monique Bourque offered me constant support. This opportunity also brought me the chance to meet William H. Helfand, whose research and writing in and support for the history of medicine, pharmacy, and addiction has proved invaluable to my work.

Former colleagues in the history department at the University of Delaware have also been enduring in their support for me and this project, in particular Anne Boylan and Peter Kolchin. I continue to appreciate their generosity as historians and as friends.

I feel fortunate to have participated in the exciting intellectual atmosphere of the History of Medicine Department at the University of Wisconsin. Though my stay there was short, Judith Leavitt, Ronald Numbers, and Hal Cook made me feel at home. They engaged me and my research, and they gave me teaching opportunities that encouraged me to place alcoholism within the larger context of Progressive Era public health.

The Section of the History of Medicine at Yale University offered me another teaching opportunity before I took a position in Oklahoma. There Toby Appel, the late Larry Holmes, Joan Jackson, the late Stanley Jackson, David Musto, Naomi Rogers, and John Warner provided an excellent environment in which to pursue research and teaching. I learned much from sharing my work with these scholars and this book is the better for it.

The support I have received since arriving at the University of Oklahoma Honors College in late summer 1999 has been impressive. The faculty with whom I have worked at the Honors College—Ben Alpers, Jim Bennett, Julia Ehrhardt, Steve Gillon, Robert Griswold, Catherine Gudis, Rich Hamerla, Randy Lewis, Carolyn Morgan, Jane Park, James Treat, and Andy Wood—are gifted scholars, teachers, and administrators. Several graciously read and commented on drafts of the early chapters of this book. In addition, they allowed me to take a year's leave from teaching and service when I received a fellowship from the National Endowment for the Humanities. I am grateful for the intellectual and moral support of these faculty members and several colleagues outside the Honors College: Julia Abramson, José Canoy, Hunter Crowther-Heyck, Elyssa Faison, Ellen Greene, Susan Kates, Catherine Kelly, Steve Livesey, Ron Schleifer, Zoe Sherinian, and Jerry Vannatta. Marty Thompson, the director of the Robert M. Bird Health Sciences Library, has gone beyond the call of duty to facilitate my research. And my research assistant Kevin Epperson worked long hours to help me explore the microfilmed archives of the Woman's Christian Temperance Union.

Over the years, I have had the good fortune to collaborate with several scholars in the history of alcohol and drug use. No one has been more stimulating and encouraging than Caroline Acker. Her insights from the history of narcotics addiction have been helpful and have contributed to an energizing and ongoing dialogue. I also thank Allan Brandt, Jim Baumohl, Jack Blocker, David Courtwright, W. Scott Haine, Harry Gene Levine, Michelle McClellan, Ron Roizen, Robin Room, and Bill White for piquing my interest and engaging many others in the history of addiction and its treatment.

Other historians who share my interest in the management of chronic disease have provided words of wisdom and support. I would like to thank in particular Chris Crenner, Ellen Dwyer, Barron Lerner, and Chris Warren.

Chapter six has appeared in two earlier versions: the first, as "Contesting Habitual Drunkenness: State Medical Reform for Iowa's Inebriates, 1902–1920," pp. 241–85, in the summer 2002 issue of *Annals of Iowa*; the second, as "Building a Boozatorium: State Medical Reform for Iowa's Inebriates, 1902–1920" in Sarah W. Tracy and Caroline Jean Acker, eds., *Altering American Consciousness: The History of Alcohol and Drug Use in the United States, 1800–2000* (Amherst: University of Massachusetts Press, 2004), pp. 124-64. My thanks to Marv Bergman, editor of the *Annals*, and Clark Dougan, senior editor at the University of Massachusetts Press, for allowing me to include this material in my book and for providing excellent suggestions for its revision.

Historical research depends upon the extensive use of archives, and this is made possible through the time, expertise, and generosity of the archivists who manage them. I have many to thank in this regard: the staffs of the Bizzell Library of the University of Oklahoma, the Boston Athenæum, the Boston Public Library, the Jan Eric Cartwright Memorial Library of the State of Oklahoma, the Connecticut Historical Society, the Connecticut State Library, the Countway Medical Library, the Hartford Medical Society, the Harvey Cushing / John Hay Whitney Medical Library at Yale University, the Historical Collections of the College of Physicians of Philadelphia, the Historical Society of the Town of Warwick, New York, the Knoxville, Iowa Public Library, the Library Company of Philadelphia, the Marion County, Iowa Historical Society, the Massachusetts Archives at Columbia Point, the Massachusetts Historical Society, the Minnesota History Center Library, the Library of the State Historical Society of Iowa in Des Moines, the State Library of Iowa, the State

Law Library of Iowa, the State Library of Massachusetts, and Widener Library at Harvard University.

Several individuals at these institutions were of special help to me. Bill Milhomme at the Massachusetts Archives provided guidance and good humor as I spent months immersed in the Norfolk State Hospital patient case files. Besieged by an army of genealogists, Sharon Avery and Ellen Sulser nevertheless made sure that I always had access to the documents I needed at the State Historical Society of Iowa. Sue Gardner, the archivist of the Town of Warwick, New York, helped me track down key details of the New York City Inebriate Colony; and New York Academy of Medicine archivists Arlene Shaner and Miriam Mandelbaum patiently shared their expertise as I explored the recent history of alcoholism. I would also like to thank Bhasker J. Dave, M.D., superintendent of the Independence Mental Health Institute; Thomas Deiker, M.D., superintendent of the Cherokee Mental Health Institute; and Dave Scurr, superintendent of the Mount Pleasant Correctional Facility. They and their staff guided me through the proper channels to gain access to inebriate patient records that had long been buried in basements and stowed away in attics. I am grateful that they shared my curiosity about their institutions' histories.

This project has received the generous support of fellowships and grants from the College of Physicians of Philadelphia; the National Endowment for the Humanities and the Agency for Healthcare Research and Quality (Grant no. FB-37732-02); the National Institute of Mental Health and the Institute for Health, Health Care Policy, and Aging Research, Rutgers University (Grant nos. MH 16242-12, MH 16242-13, and MH 16242-14); the National Library of Medicine (Grant no. R01-LM005993-01), the National Science Foundation (Grant no. DIR-91-12092), the New York Academy of Medicine, the State Historical Society of Iowa, and the University of Oklahoma. The contents of this book are my responsibility alone and do not necessarily represent the views of these institutions or agencies. Without their assistance, however, this project would not have been possible.

I would be remiss if I did not mention the ongoing support that I have received from the Johns Hopkins University Press. I have benefited especially from the skill, wisdom, and patience of editor Jacqueline Wehmueller. I am grateful, too, for the excellent comments of the anonymous reader of my manuscript. Thanks also to Julie McCarthy and Susan Lantz, who expertly guided the book to completion.

Finally, I must thank close friends and family—people who routinely offered me their unconditional support, occasionally furnished their editorial eye, and always provided good cheer. Indeed, Brian and Susan Boggs, Fran Dalton, Judith Freilich, Emily Haddad, Isabella Halstead, Ken and Sara Hieke, Kristin Kennedy, Scott Kluksdahl, Tim Lyons, Candace McCaffrey, John Nicholson, Kim Pelkey, Elizabeth and Iwan Praton, Christy Seaton, Bob and Mimi Teghtsoonian, and Michael Willis have helped me in ways too numerous to mention. For their love and their moral and material support, I am grateful to my family, especially Anne Carter, Chris and Brian Courtney, Amy W. Ellis, Mel Ellis, Anne Ellis Ruzzante, the late Elizabeth Anne Ellis Tracy, Fred Tracy, and the late Sarah M. Tracy.

# Alcoholism in America

At noontime on 29 November 1870, sixteen men convened in the New York City Young Men's Christian Association to change the way society understood and treated habitual drunkenness. This group of physicians, clergy, and businessmen from New York City, Boston, Chicago, Philadelphia, and Baltimore called themselves the American Association for the Cure of Inebriates, and they declared that "intemperance" was an illness caused by an inherited or acquired constitutional susceptibility to alcohol and could be cured "as other diseases are."[1] The AACI hoped that through an educational campaign about the nature of alcohol, its physiological and psychological effects, and the disease of inebriety, the public might come to "demand protection against a disease, infinitely more destructive than cholera, yellow fever, small pox, or typhus, which are now so carefully quarantined."[2] But how to confine a disease the public regarded as a moral contagion? The last three principles of the AACI's charter spoke to this issue:

6. All methods hitherto employed having proved insufficient for the cure of inebriates, the establishment of asylums for such a purpose, is the great demand of the age.

7. Every large city should have its local or temporary home for inebriates, and every state, one or more asylums for the treatment and care of such persons.

8. The law should recognize intemperance as a disease, and provide other means for its management, than fines, stationhouses and jails.[3]

According to the charter, inebriety was not just a disease. It was a public health problem that required municipal and state facilities for its containment and treatment.

The act of defining intemperance as a disease was not original to the AACI. Colonial physician and statesman Benjamin Rush, often regarded as the "father" of American psychiatry, had asserted the same in his 1784 *An Enquiry into the Effects of Spiritous Liquors Upon the Human Body, and Their Influence Upon the Happiness of Society.*[4] Rush later urged the construction of "sober houses" where drunkards could find refuge and medical assistance for their condition.[5] Likewise, Scottish physician Thomas Trotter made the case for considering intemperance a disease twenty years later in *An Essay, Medical Philosophical, and Chemical on Drunkenness, and its Effects on the Human Body.* By 1849 Swedish physician Magnus Huss had written *Alcoholismus Chronicus, or Chronic Alcoholic Illness. A Contribution to the Study of Dyscrasias Based on My Personal Experience and the Experience of Others,* introducing the term *alcoholism* to the medical world for the first time. What distinguished the AACI's efforts from its predecessors' was its collective commitment to creating a new medical specialty that would bring medical science to bear on a resilient social problem that the church, the courts, insane asylums, and the temperance movement had so far failed to conquer.

Spearheading the efforts of this new organization were Willard Parker and Joseph Parrish. Parker, a renowned surgeon and medical author, was president of both the AACI and the New York State Inebriate Asylum in Binghamton. Parrish, a well-regarded Philadelphia physician and a reformer by temperament, was editor of several medical journals and superintendent of the Pennsylvania Sanitarium, a private facility treating inebriates in Media, Pennsylvania, just outside Philadelphia. Like their confreres, Parker and Parrish had extensive experience managing patients with drinking problems, and both were convinced that science held the key to discovering what they called "the natural laws of inebriety." Parker's opening address made the AACI's goals plain: "The purpose of this meeting is the discussion of the subject of inebriety, and its proper treatment. It is not a *temperance,* but a scientific gathering, made up of men having charge of the asylums and homes already established in the United States, for the cure of the unfortunate victims of alcoholism."[6] Medical science, continued Parker, was to do for inebriates what it had accomplished for insanity's lost souls: reveal that they were victims of a curable condition, worthy of public sympathy and medical care rather than punishment.

Thus, at the end of November 1870, the first organized campaign to understand and treat habitual drunkenness as a disease—what I have called the attempt to "medicalize" intemperance—began. This book is about that campaign and its mixed success, and it takes the AACI's founding as its starting point. The group's inaugural meeting suggests some of the most important dimensions of this episode in the history of disease, professionalization, and social welfare policy. The AACI's use of the terms *intemperance, inebriety, dipsomania,* and *alcoholism* suggests multiple medical meanings for habitual drunkenness and presages the ongoing renegotiation of these meanings among physicians and the lay public. The AACI's emphasis on securing state sponsorship of medical treatment for inebriates, as was available for the insane, signaled its belief not only that inebriety was a form of mental disease that could lead to insanity but also that the creation of new state-run institutions for inebriates was an essential step in the campaign to medicalize habitual drunkenness. This form of state recognition gave the new disease concept validity while reinforcing the medical profession's social and cultural authority and conferring legitimacy upon the nascent specialty of inebriety medicine. Even more elemental, the founding of the AACI highlighted the professional ambitions of its members. They identified themselves as scientists, not temperance leaders, and they hoped that rational data collection and analysis in the "laboratory" of the inebriate asylum would unlock the secrets of a disabling condition that moral and penal authorities had failed to cure. Later in the meeting the new secretary of the AACI, Joseph Parrish, even read a paper from the inebriates at his Pennsylvania Sanitarium, an act that highlighted the important role alcoholics might play in this sociomedical campaign.

Members of the AACI envisioned a revolution in medical and social theory and practice with regard to the inebriate, but in the end their revolution failed. Their calls and those of their allies for state action on behalf of the inebriate, although at times successful, were eclipsed by federal legislation that was billed as a prophylactic measure of the first order: Prohibition. The ban on the manufacture, sale, and distribution of alcoholic beverages, it was said, would not only eliminate the inebriate but would reduce the larger social costs associated with alcohol consumption. As we will see, it was not so much the failure of treatment interventions as the relative failure and success of political and professional ones that spelled the end to the "early alcoholism movement."[7]

Yet if the "Noble Experiment" inaugurated in 1920 was the death knell for early efforts to medicalize habitual drunkenness, "Prohibition failed to prohibit!"[8] as popular wisdom proclaims. The Eighteenth Amendment was repealed in 1933. Within a decade, different versions of the disease concept were revived: as an "allergy" within the framework of Alcoholics Anonymous and in the multiphase alcoholism model of physiologist Elvin M. Jellinek at the Yale Center of Alcohol Studies. The birth of the "modern alcoholism movement" in the 1940s and 50s reminds us that within a generation medical and social attitudes can come full circle even if history does not repeat itself.

It is not surprising that the first collective efforts to promote the disease concept of habitual drunkenness occurred in the late nineteenth century. The Gilded Age and Progressive Era constituted a time of intense scrutiny of deviant behavior, fueled by the arrival of millions of imbibing newcomers and the expanding poverty and disarray of America's immigrant-swollen cities. At about the same time as the Commonwealth of Massachusetts was opening its state hospital for dipsomaniacs and inebriates, legendary cultural critic and writer Mark Twain wryly observed "Nothing so needs reforming as other people's habits."[9] Sexual promiscuity, prostitution, drunkenness, petty theft, gambling, and vices of all varieties met the interrogating gazes of urban vice commissions, settlement house reformers, state legislatures, and adjudicators such as Louis Brandeis, who argued in favor of limiting working hours for women because he believed undue exhaustion posed a threat to women's moral fiber and led many to seek relief in narcotics and stimulants.[10]

At the same time, flexible medical concepts such as degeneration theory and neurasthenia offered biological explanations for the transmission of deviant behavior or ingrained "bad habits" from one generation to the next; but their reliance on the notion of inherited "tendencies" meant that there was room for potentially effective somatic and environmental intervention. In the late nineteenth century, diagnoses such as homosexuality, hypersexuality, inebriety, kleptomania, and feeble-mindedness offered new medical frameworks for viewing deviant behavior. And these ideas increasingly met with a warm public reception, for medicine was but one form of expert authority that experienced a renaissance in the golden era of professionalization between 1880 and 1920. State policymakers and bureaucrats enlisted the advice of experts to preserve public health and moral order and to establish new specialized institutions that reformers claimed would deal more rationally and humanely with society's misfits. Thus, the turn-of-the-century attempt to medicalize habitual drunkenness is an example of synergy among medical ideas, cultural values, professional authority, and policymaking.

It is this synergistic relationship that I have attempted to capture in this book. Some will regard this as a case study in the social negotiation of disease among all varieties of actors: problem drinkers and their families, physicians, politicians, court officers, and temperance workers. After all, each of these parties had some say in the construction of professional and public definitions of habitual drunkenness. Others will see a failed case of medicalization, identifying problems in the translation of deviance definitions into diagnoses, in the professional and popular acceptance of these diagnoses, and in their institutionalization through the creation of new inebriate laws and new public and private asylums and hospitals for chronic drunkards. Others may see a story about the Gilded Age and Progressive Era state: the growing dissatisfaction with the local administration of welfare relief (the almshouse); the subsequent rise of supposedly more rational, more modern, more humanitarian forms of state-centered social reform and social control (the public inebriate asylum); and the eclipse of state-based reforms by federal measures (Prohibition). A subset of this group may find in the inebriate hospital a breakdown of the long-touted polarity between "conscience and convenience"—widely regarded as one of the guiding imperatives of the Progressive asylum at the turn of the century since David Rothman's monograph of the same name appeared in 1980. Still others will see that it is not just in the construction of saloons and restaurants, in the fashioning of cocktails, and in the development of drinking and temperance cultures that gender, ethnicity, and class played influential roles, but also in the construction of alcoholism and its treatment. My goal has been to write a book that speaks to all of these audiences in a way that engages readers with each of these interwoven stories. My narrative begins with the place of alcohol in American society.

## Alcohol in American Society

Alcohol consumption has been a part of American life from the earliest days of the republic. For sixteenth- and seventeenth-century British and Europeans, drinking was a way of life that was essential in societies where most sources of water were contaminated. Colonists imported this way of life to America, where alcohol was regarded as the "Good Creature of God." Whether rum distilled from West Indian sugar, home-brewed beer, or imported wines from the Continent, alcohol was a staple of colonial life, and wine and opium were mainstays of colonial medical practice. Colonial Americans drank early and often—at pubs, barn and house raisings, weddings, elections—and simply to fortify their constitutions against the harsh elements of their everyday existence. These early settlers, however, drew a distinction between drinking and drunkenness; they considered the latter the work of the devil.

Although the first temperance reformers may have been American Indians attempting to curtail the damage colonials introduced to their people through alcohol, the American temperance movement is said to have begun with physician Benjamin Rush's temperance tract in 1784. Of course, Rush was also the first to describe intemperance as a disease. Fearful for his new nation's future, Rush recoiled at the prospect of intoxicated voters shaping the country's destiny—no small consideration at a time when elections often featured heavy drinking and when annual per capita consumption of absolute alcohol figured between four and six gallons (approximately twice the rate in 2000). Yet like many in the earliest days of the American temperance movement, Rush distinguished between fermented beverages, which he regarded as healthful, and distilled alcohol, which he deemed dangerous. Evidence suggests that Americans consumed even more alcohol between 1800 and 1830. The efficiency and profitability of turning corn into whiskey, the heavy drinking on the frontier, the spread of urban saloons, and the arrival of beer-drinking Germans and whiskey-swilling Irish all encouraged the nation's bibulous tendencies. These tendencies, however, elicited a reaction within the Protestant churches, which linked salvation to temperance and other reforms. The American Society for the Promotion of Temperance (ASPT), founded by evangelical clergymen in 1826, gained support from farmers, industrialists, and homemakers. The temperance campaign—really a series of reform drives—constituted the nineteenth century's largest social reform movement.[11]

Members of the temperance cause viewed alcohol as imperiling capitalist enterprise, domestic tranquility, and national virtue. As the industrial revolution made its way through the Northeast and Midwest, a drunken workforce was increasingly out of step with a routinized and mechanized workplace.[12] By 1836 the ASPT, renamed the American Temperance Society, advocated total abstinence. In the early 1840s, Americans thronged to temperance rallies, "took the pledge" for sobriety, and in record numbers lobbied to end the licensing of saloons. The Washingtonian movement, a grassroots total-abstinence campaign, sponsored parades and speeches, offered recruits financial and moral assistance, and established sober boardinghouses for drunkards. The Washingtonian enthusiasm soon gave way to better-organized temperance fellowships such as the Good Templars and the Blue, Red, and White Ribbon societies. By 1840, per capita alcohol consumption had dropped to 3.1 gallons; ten years later, it was a gallon lower.[13] The late Antebellum Era also saw renewed middle-class drives for local and state prohibition. Eleven states passed prohibitory legislation, although most of these laws were soon repealed.

After the Civil War the brewing and distilling industries expanded, and alcohol consumption, especially in the immigrant-rich cities, remained high. But the temperance movement revived as well, linking "demon rum" to concerns about newcomers, workplace efficiency, social welfare, and urban political corruption. Frances Willard's Woman's Christian Temperance Union redefined temperance as a women's issue involving home protection.[14] Reformers were quick to identify the saloon as the nation's most threatening fount of social evil—a Pandora's box of poverty, prostitution, disease, and social chaos that permeated the urban landscape. At the WCTU's prompting, the U.S. Congress mandated the inclusion of "scientific" temperance instruction in high school physiology texts.[15] Increasingly, experts in sanitation, public health, urban planning, and scientific charity were enlisted to protect the commonweal, promote the efficient coordination of human resources, and maintain social order. It was in this time of social reorganization and professionalization, extending from roughly 1880 to 1920, that the first widespread attempt to medicalize habitual drunkenness took place. Public concerns about alcohol consumption increased in the last few decades of the nineteenth century and first two decades of the twentieth—a time, ironically, when annual per capita consumption of absolute alcohol had stabilized at just over two gallons, where it remains today.[16]

This illogical timing points to the importance of assessing the context of any social response to heavy drinking, including the effort to medicalize habitual drunkenness.[17] This attempt was but one part of the "moral panic" over alcohol and habitual drunkenness that took place during the Gilded Age and Progressive Era and culminated in Prohibition.[18] As Morton Keller has observed, "Modern America nourished both a burgeoning variety of forms of social behavior and a rising reaction against that diversity. . . . The conservative desire for social control, the religious quest for moral conformity, and the reformist search for social uplift joined to turn private vices into threats to public mores."[19]

To Keller's mix of factors controlling social deviance, I would add the rapidly professionalizing discipline of medicine, for physicians postulated and promoted new clinical entities such as neurasthenia and degeneration theory that tied individual behavior to environmental stress and defective heredity. The physicians who promoted the disease concept of inebriety were medico-moral entrepreneurs who wished to engineer new solutions for a resilient social problem that had stimulated heightened concern at the approach of the new millennium. It is important to observe, however, that authorities such as the church and court system—institutions that had wrestled with the vicious and criminal aspects of inebriety for years without great success—were frequently willing to cede responsibility to the doctors, or at the very least to cooperate with physicians' new plans for reform. Initiatives for the construction of inebriate hospitals frequently came from those outside the medical field, as was the case in the states of Massachusetts and Iowa.

Why did alcohol problems command the public's attention between 1870 and 1920? For late-nineteenth-century Americans—especially community leaders, captains of industry, and public health reformers—several factors made alcohol use increasingly alarming. First, alcohol was inextricably tied to the saloon and therefore to the world of vice. "Gambling and the social evil [prostitution] are closely allied with the perils of drunkenness," Raymond Calkins wrote in his 1903 summary of *Substitutes for the Saloon*.[20] Likewise, Perry Duis reported that in 1893 a passenger on Chicago's "Alley El" (the elevated train that passed the backsides of various liquor establishments on its way to the World's Columbian Exposition) submitted a scathing letter to the *Chicago Herald* to complain that "loose women" employed at Appleton's saloon were "exposing themselves" to train riders.[21] In his 1890 *How the Other Half Lives*, reform journalist Jacob Ries observed that "gaming" and "dice playing" were endemic in working-class saloons, reducing the laboring man's wage further and moving him a few steps closer to penury. "The saloon," commented Ries, "projects its colossal shadow, omen of evil wherever it falls into the lives of the poor."[22] As tens of thousands of single young men and women flooded the cities in pursuit of employment, reformers feared the corruption of their morals. Boarding houses for the cities' newcomers frequently encroached upon the entertainment and tenderloin districts, where the ubiquitous saloon seductively offered patrons free lunches and booze-filled conviviality.[23]

In the meantime, manufacturers and captains of industry had their own concerns about alcohol. As Roy Rosenzweig has argued, the large factories of the late nineteenth and early twentieth centuries did not allow "the gambling, storytelling, singing, debating, and especially drinking" that were common in the smaller workshops of the early nineteenth century.[24] Factory workers at the dawn of the twentieth century faced a far more regulated, routinized, and mechanized workplace than had their predecessors. Indeed, Lillian Brandt observed in 1910 that in New York, "employers prefer men who do not drink, for there is a consensus that . . . alcohol interferes with efficiency. It is probable that this handicap is much greater now than formerly . . . because of the changed nature of the work which we have to do. Factory work, work on railways, office work, intellectual work, are all incompatible with drinking habits."[25]

It was not only the workplace that had changed: the automobile offers an excellent example of a modern technology that intensified the dangers associated with drunkenness. To have an accident in a car traveling thirty-five miles per hour was more dangerous than colliding with something at the speed limit set by horse and carriage. The first Driving Under the Influence (DUI) laws were passed in 1910 in New York State. California followed the next year, reaffirming the dangers of drinking on the road. Automobile manufacturers were also concerned. Henry Ford, who pioneered mass production techniques in his automobile factory at River Rouge, offered workers a $5.00 daily wage if they refrained from drinking and lived temperate, family-oriented lives. Alcohol and an efficient workforce did not mix in Ford's eyes. Yet in other ways the technology-driven, efficiency-guided workplace facilitated drinking. Ford also offered his workers an eight-hour day, giving them more leisure time in addition to a large income. Together, these factors frequently conspired against sobriety. If workers could not drink on the job, alcohol consumption became a powerful symbol of leisure time, and workers had more of that with a shortened work day.[26]

The same cities that were teeming with single men and women were also crowded with immigrants. Much to the consternation of a largely middle- and upper-class community of civic leaders and reformers, the immigrants drank. Nowhere was this more apparent than in Boston, where three Brahmin graduates from Harvard University founded the Immigration Restriction League in 1894. Nativist sentiment ran high among the city's establishment, with Congressman Henry Cabot Lodge forecasting that the unfettered immigration of "alien or lower races of less social efficiency and less moral force" would not only bring about the decline of "a great race," but "of civilization."[27]

Fears of the bibulous Irish, German, and Italian newcomers were rampant, and the saloon, often run by an Irish political boss, was portrayed not only as a source of vice but also as a conduit of corrupt politics. If 1894 was a watershed year for immigration restriction in Boston, it was also a pivotal year for Irish politics: for the first time, the Brahmin-led City Committee "had been forced to consult with the ward chairmen about the choice of a mayoral candidate."[28] When Josiah Quincy ascended to the office of city mayor the following year, he supported public bathhouses, reasoning that if baths were available on a daily basis the saloon goers might be more inclined to spend their evenings at home. As Thomas O'Connor reminds us, for Republican reformers throughout New England and into the Middle West, prohibition was but "one more way of reducing the disastrous impact of foreign immigration on American institutions."[29]

Without question, alcohol appeared to pose the greatest threat to civil order within the urban jungle of the Gilded Age and Progressive Era, or what settlement house pioneer Robert Woods termed "the city wilderness."[30] The metropolitan middle and upper classes were forced to view alcohol's social damage firsthand as they lived and worked in the ever-crowded city. New professional groups—social workers, settlement house reformers, sanitary engineers, industrial physicians, and public health advocates—busily explored the world of their less-fortunate fellow citizens and publicly reported their findings. Poverty, prostitution, unemployment, violence, and ill health became tied to alcohol on an unprecedented scale. "No one fact, other than the fact of poverty itself, confronts social workers, in whatever particular field they may be engaged, so constantly as alcoholism," declared Homer Folks, secretary of the New York State Charities Aid Association, in 1910.[31] Jane Addams, Mary Richmond, Florence Kelley, Mary Van Kleeck, Josephine Baker, and Robert Woods all made visible the human wreckage of rapid urbanization, immigration, and unchecked industrial capitalism. They were aided by muckraking journalists such as Upton Sinclair, Ida Tarbell, Lincoln Steffans, and Samuel Hopkins Adams.

Mixing social justice with social control, and social science with moral uplift, settlement reformers and social workers conducted social surveys of the city's neighborhoods, organized well-baby clinics, taught English classes, gave cooking lessons, erected pure-milk stations, and promoted education in American citizenship.[32] In all these activities, as Homer Folks indicated, they met alcoholism head-on. In her landmark text, *Social Diagnosis*, social worker Mary Richmond stressed the important role that social workers played in treating inebriety, initially through their "gathering the pertinent social data, and later in rallying to the patient's aid every tonic influence which can supplement the medical means employed."[33] As the public health community mounted a successful campaign against tuberculosis, Folks reminded readers of *The Survey* that the campaign against the White Plague would not be complete until a war had been waged on alcoholism, too. Habitual drunkenness undermined the constitution of the drinker, making him or her more susceptible to infectious disease.[34]

It is not surprising that alcoholism would consume the attention of social and settlement house workers, for their ranks were dominated by middle-class women in pursuit of careers that combined the public lives of policymakers and social scientists with the private, and traditionally female, realms of domesticity and family. Most importantly, by the late nineteenth century, the alcohol issue had become a woman's issue, thanks largely to the efforts of the Woman's Christian Temperance Union, which campaigned against the saloon and its denizens under the banner of "home protection." Indeed, when we think about the level of public agitation over excessive alcohol consumption, it would be hard to overestimate the role of the WCTU. Frances Willard's "do everything" policy meant women crusaders were present in a wide range of social-improvement campaigns—child welfare, kindergartens, the Americanization of foreign-speaking peoples, health education, the suppression of impure literature, international peace, women's suffrage—and the WCTU delivered its anti-alcohol message in all of these arenas. Moreover, the WCTU won congressional approval for mandated alcohol physiology education in high schools across the country, pressing the temperance cause through scientific instruction.[35] Almost as ubiquitous as alcohol itself was the image of the temperance woman in white, campaigning for the prohibition of alcohol.

One more factor that placed alcohol squarely in the public's eye was its relationship to crime. In 1883, drunkenness accounted for 62 percent of the total number of criminal prosecutions in Boston. In 1893, the year that the Massachusetts Hospital for Dipsomaniacs and Inebriates opened its doors, prosecutions for drunkenness in Boston had dropped to 48 percent of all criminal prosecutions.[36] Yet the number of simple arrests for drunkenness in Boston for the year ending September 30, 1892, had climbed by 8,029 from the preceding year, according to the *Boston Daily Globe*.[37] The situation was little different in other major cities, where arrest figures for drunkenness and disorderly conduct (presumed to be linked to alcohol consumption) dwarfed other offenses. Court systems were routinely clogged with alcoholic "rounders," or recidivists. The State Charities Aid Association of New York laid out the situation neatly in 1909: "The habitual drunkard is a menace to society, a burden to his family, and an economic loss to the community. He is the cause of the largest single item of expenditure of police departments, police courts and jails and workhouses."[38] Nor did these concerns address the use of alcohol by individuals who committed acts against persons or property. The Committee of Fifty's 1899 investigation of 13,402 convicts in seventeen state penitentiaries scattered throughout the United States revealed that intemperance was linked to 50 percent of the crimes committed, and it was listed as the "first cause" in 31 percent.[39] Figures such as these promoted the perception that alcohol was a growing social plague.

Here, then, was the context in which physicians began to promote the disease concepts of intemperance, inebriety, dipsomania, and alcoholism. The "liquor problem," as it was called, commanded the attentions of both the general public and professionals of all types. The physicians who began to diagnose and treat the habitual drunkard as a sick person were well aware of his problematic status in American life.

## The Medical Profession and Habitual Drunkenness

Benjamin Rush's 1784 essay on the effects of ardent spirits suggests that members of the medical profession had long been involved in treating the intoxicated patient. But the physicians forming the American Association for the Cure of Inebriates in 1870 had something else in mind as well: the creation of a new medical specialty. Establishing a new medical specialty that dealt with a tremendous social problem was one way to "do good" while doing well professionally. And the period between 1830 and 1880 was a difficult time for professionals. As historians have pointed out, an overall "disestablishment and humbling of the professions" took place in the middle third of the nineteenth century. Consider, for example, that three-quarters of all states had educational requirements for lawyers in 1800; by 1860, only 25 percent of states mandated specific education for the practice of law. Likewise, while just about every state in the union enforced medical licensing laws in 1800, only a handful of states did so in 1860. Following the Civil War, however, the professions were given new life. Industrialization, urbanization, and the accumulation of capital created new needs for expert knowledge and new infrastructures for obtaining it. The years between 1880 and 1920 amounted to a professional renaissance in the United States.[40]

Medicine was no exception to this trend. Chemist and Harvard University president Charles Eliot initiated a "revolution of the system of medical education" in 1871 by attempting to bring Harvard's medical school on par with its other graduate programs. He lengthened the academic year from four months to nine, increased the duration of the program from two to three years, and gave a central spot to the laboratory sciences of chemistry, pathological anatomy, and physiology in the curriculum.[41] Harvard students were also required to pass their courses to receive their diplomas. The most revolutionary development in medical education took place south of Cambridge, however, at the Johns Hopkins University in Baltimore. In 1893 Hopkins embarked on a new system of medical education by requiring entering students to have undergraduate degrees prior to matriculation and by mandating four years of graduate medical education grounded in both the basic sciences and hospital medicine. A new standard was set. Scientific competence became a criterion for medical practice. Between 1875 and 1900, every state in the union passed some kind of medical licensing law.

With a more rigorous medical education system, one tied to clinical research and hospital care, medical specialties began to proliferate. Ambitious physicians quickly realized that specialists experienced the satisfaction of mastering a particular area of an ever-expanding medical knowledge base. They also commanded higher prestige and better pay. Two specialties in particular guided the physicians who organized the AACI: public health, which became a discipline in its own right in the early decades of the twentieth century; and psychiatry, whose practitioners were still called "alienists" in 1870. Inebriety was a key problem within each of these fields.

Urban sanitarians working to make the city a cleaner, healthier, safer environment viewed drunkenness as a proverbial thorn in their sides. Not only did the habitual drunkard's health suffer, but so did his family's. "What shall we do with the inebriate?" was a question that required public discussion "until the people succeed in obtaining laws and police regulations that shall take adequate cognizance of the cause of murderous frenzy, the delirious riot, the hideous brutalism and pauperized and starved wife and children," argued New York City public health advocate and physician Elisha Harris, who joined the AACI two years after its founding.[42] Twenty-three years later, J. W. Grosvenor, M.D., of Buffalo, New York, concurred: "For the purpose of conserving the public health and of promoting the elevation of the race, the medical profession should take an active and abiding interest in this subject.... State care and control of the alcoholic inebriate should be the persistent policy of every State in the Union."[43] For Grosvenor, alcohol was "antagonistic to any plans whose object was the promotion of health."[44] Public health reformers thus saw treatment for inebriates as a sound investment in the prevention of a host of social and physical evils: poverty, mental and physical illness, and crime might all be alleviated by helping the drunkard change his ways.[45] The mental hygiene movement of the early twentieth century capitalized on these views, encouraging psychiatrists to see the treatment of alcoholics as prophylactic mental health work.

Through their promotion of the new specialty of inebriety medicine, physicians attempted to gain more cultural authority: they alone possessed the necessary training to distinguish between the vicious and the diseased. They also hoped to acquire more social authority as the gatekeepers of a public system for managing inebriates.[46] This distinction between social and cultural authority is an important one, for as Paul Starr argued in *The Social Transformation of American Medicine,* institutions and agencies in society possessing social authority—the police, the judiciary, the state legislature—rely on the cultural authority of medicine to validate their policies. "In this regard," observed Starr, "medical authority is a resource for social order as well as for the profession and its clients."[47] The physician-reformers who promoted the disease concept of inebriety pursued both social and cultural authority as they urged state and municipal governments to adopt new policies for the management of the habitual drunkard—policies that recognized both the disease aspects of inebriety and the authority of physicians to diagnose inebriates and commit them to new institutions for their medical care.

Of course, physicians working within mental medicine and the care of the insane already had the power to commit mentally ill patients to institutions. These physicians found the suggestion of separate facilities for the treatment of inebriates attractive on a number of different levels. First, there was a strong consensus that inebriates did not belong in an asylum for the insane. Once sober, habitual drunkards were quick to protest their confinement and to demand release, creating havoc within mental hospitals.[48] As early as 1876, Charles Nichols, president of the Association of Medical Superintendents for American Institutions for the Insane, observed that insane and inebriate patients did not get along well: "I think a sharp lookout is necessary to prevent the inebriate from interfering with the rights and privileges of the insane, when both occupy the same wards."[49] Separate institutions for the inebriate would remove this obstacle to institutional order. Second, it was widely assumed that a substantial portion of the insane got that way through alcohol consumption. Nichols likewise noted that drunkenness was "sometimes both the cause and principal manifestation of real insanity—of real disease."[50] As a preventable psychosis, alcohol-related insanity held obvious appeal for physicians who worked with patients with notoriously pessimistic prognoses. An estimated one-eighth of all admissions to hospitals for the insane were due to excessive drinking. To stem the tide of inebriety, therefore, was to reduce the caseloads of asylums across the United States.[51] As hospitals for the insane grew increasingly custodial in the late nineteenth and early twentieth centuries, treatment for the inebriate had especial appeal, for it appeared to be a simple intervention with a great reward.

Finally, there was "dynamic" psychiatry. During the first decades of the twentieth century, the psychiatric profession moved outside the asylum and into the realm of everyday life, away from the institutional world of psychosis. In growing numbers, psychiatrists treated the neurotic, whose milder behavioral symptoms they regarded as symbolic expressions of inner conflicts preventing a patient's normal adjustment to social roles. Few conditions exemplified individual "maladjustment" better than alcoholism. Psychiatrists envisioned themselves, in the words of Elizabeth Lunbeck, as "men of action, bringing their disciplinary perspective to bear on 'practically the entire human world.'"[52] In selecting alcoholism as a suitable clinical problem, psychiatrists chose a subject that reformers believed permeated every nook and cranny of modern society. In many ways, then, institutional and out-patient treatment for the inebriate represented a stepping stone for the psychiatric profession from the management of the institutionalized mentally ill to the outpatient treatment of all manner of neurotics. In approaching inebriety as a legitimate medical condition, alienists and psychiatrists extended their reach to more pervasive and theoretically more tractable forms of social deviance than severe mental illness; yet as we shall see, their ties to the asylum model remained strong.

## The Inebriate Asylum as a Progressive Era Institution

In the creation of new inebriate asylums—specifically, public asylums—we see another influential factor in the inebriate reform campaign: the growth of state-centered mental health, social welfare, and public health reform. The creation of state asylums for inebriates was a priority for the American Association for the Cure of Inebriates from the time of its founding in 1870. The connection between these specialized institutions and the medicalization program was clear: new medical institutions, if reasonably successful, could enhance the cultural and social authority of physicians and validate the disease interpretation of particular forms of social deviance: in this case, habitual drunkenness.[53] To gain the state's imprimatur—whether through new laws that recognized that inebriates were in need of some form of medical rather than penal care (an initial move toward the creation of specialized inebriate asylums that was enacted in many states), or through the creation of new state hospitals for alcoholics (a less common development)—constituted a clear step forward for advocates of the disease concept.

California, Connecticut, Iowa, Massachusetts, Minnesota, and New York constructed public inebriate hospitals as, at the dawn of the twentieth century, the climate of state welfare reform nurtured institutional growth. Indeed, at few times in American history has the state been regarded as such a welcome partner in reform. As Gerald Grob has observed, the economic depressions of 1873–79, 1882–85, and 1893–96; the generalized discontent of labor and the waves of strikes; the immigration of Eastern and Southeastern Europeans; and the endemic poverty of the cities—all cast into relief the failures of locally managed care for the poor, the sick, and the mentally ill. Late Gilded Age and Progressive Era reformers reasoned that uniform standards of care determined by scientific experts could be rationally and economically applied in centrally administered state institutions and that this would constitute a decided improvement of matters. Indeed, state governments, much like the corporate world, sought monopolies as they consolidated and rationalized their authority within welfare, public health, and education. The State Care Act of New York, passed in 1890, was a shining example of this trend, placing authority for the care of the insane in state hands and reducing the county asylums to poorhouse status.[54]

The irony is that the therapeutic ideology that obtained within mental health, welfare, and penal reform stipulated that there was no standard, or rather that the standard was individually tailored rehabilitation. In *Conscience and Convenience: The Asylum and Its Alternatives in Progressive America*, David Rothman characterized the general orientation of turn-of-the-century scientific charity workers: "Coming from the world of the college, the settlement house, and the medical school, the Progressive reformers shared optimistic theories that at once clarified the origins of deviant behavior and shaped their efforts to control it.... Progressives aimed to understand and to cure crime, delinquency, and insanity through a case-by-case approach."[55] Thus, reformers of the mental hospital and the prison alike emphasized the need to focus on the particularities of each individual; institutional rehabilitation and moral regeneration would take as long as was necessary, given the idiosyncrasies of the case. With regard to the mentally ill, difficult or severe cases might be best served through prolonged periods of institutionalization; others less afflicted might be better restored by living in the community and availing themselves of outpatient care or acute short-term treatment. For the criminal, indeterminate sentencing to the penitentiary with probation eligibility dependent upon the behavior of the inmate represented the strong arm of the law; in lesser cases, probation with or without fines might prove sufficient. The upshot of this general policy was that state institutions and their managers received unprecedented levels of discretionary power, but their status as experts—as professionals—appeared to justify this to their contemporaries. So the growth of state authority, the expansion of state facilities for the diseased and the delinquent, and the rise of professional authority went hand in hand: few institutions embody these trends better than the public inebriate hospital.

The inebriate hospital provides an opportunity to view medical ideas about alcoholism put into action—to compare rhetoric with reality, inasmuch as historians can gauge "what really happened." As we do so, we gain some insight into the nature of the distinction David Rothman drew some twenty-five years ago between "conscience" and "convenience" as the governing dyad of the Progressive Era asylum.[56] He argued that the best intentions of reformers were at work in the formulation of policies, but when those policies were put to work in specific settings, the outcome was less influenced by high ideals than by the everyday administrative demands of the asylum.[57] Rothman's assessment of the turn-of-the-century asylum and penitentiary was remarkably apt, and it set the stage for decades of scholarship. The histories of specific institutions such as the state hospitals for inebriates in Massachusetts and Iowa allow us to extend this analysis and to see that conscience and convenience were at times more complicated than Rothman suggested. The Massachusetts and Iowa stories suggest that "conscience" varied between institutions and even between institutional regimes, influenced by a host of local, medical, and nonmedical factors. "Convenience," likewise, was frequently measured by what the patients, their families, and their employers considered appropriate, not just the criteria of the hospital administration. Indeed, the responses of patients to the care they received suggest that the politics of the asylum was more than the politics of resistance and accommodation. Patients at the inebriate hospital might embrace or reject the rhetoric of treatment, but they were quite capable of turning it to their own ends in empowering ways. Thus, understanding what went on inside the inebriate hospital affords insight into both alcoholism's incomplete medicalization process and the nature of the Progressive Era state-run asylum.

## Three Narratives

I have organized this study around three concurrent narratives concerning the negotiation of social and scientific knowledge, professional power, public policy, and the changing meaning of alcoholism. The first story is the aforementioned rise of concerns about alcohol problems in the late nineteenth century and the development of the medical profession's claims of the scientific expertise and authority to address them. The second narrative focuses on physicians' foray into the arena of social reform. And the third story considers the popular negotiation of disease.

The temperance campaigns of the nineteenth and twentieth centuries comprised America's biggest social reform movement. Doctors hoped to capitalize on this anti-alcohol sentiment and to develop new approaches to a tremendous social problem; they wished to advance more effective medico-moral solutions in the face of judicial and welfare systems that had failed to ameliorate either the drunkard's condition or that of his dependent family.[58] This was not, however, a case of medical imperialism, for these other social authorities were often as eager to cede responsibility for managing the habitual drunkard as doctors were to embrace him.

Instead, it was a case of physicians working with reformers of all stripes to reconstruct the social, moral, and political context in which new sociomedical approaches might operate. Yet as they did this, physicians did not jettison their Judeo-Christian interpretations of behavior, free will, and appropriate social roles. Instead of promoting exclusively medical interpretations of inebriety, grounded in biology, physiology, and psychology, physicians continued to address the moral and social aspects of the condition. They recognized that any disease concept that ignored the moral dimensions of drunkenness would face both public and professional resistance. Medico-moral entrepreneurs, physicians worked within the cultural thought styles of their day. Thus, their "rational" and "effective" clinical solutions to alcoholism also constituted a campaign to redefine manhood, womanhood, and citizenship in America in sober, responsible, and fiscally independent ways. Physicians saw their curative efforts in role-restoring terms, helping their inebriate patients develop the skills they deemed necessary for life in a modern, patriarchal, industrial-capitalist democracy. The cultural and moral priorities of the professional classes made their imprint on the science of inebriety.

Throughout the book, I also analyze the popular perception of inebriety as seen in turn-of-the-century periodical literature, novels, newspaper reportage, and legislative documents. The inebriate asylum, whether public or private, was the crucible in and around which the popular negotiation of disease took place. I look at the exchanges that resulted in the creation, management, and closing of inebriate asylums—negotiations involving physicians, legislators, state administrators, patients and their families, the media, and others. I also focus on the published narratives of inebriates, stories of medically assisted recovery that became the new "temperance tales" of the early twentieth century. Finally, I reconstruct the stories of individual inebriates from a large cache of patient records and patient-physician correspondence from Foxborough and Norfolk State Hospitals, the two state facilities that treated inebriates in Massachusetts between 1893 and 1920. Through these sources, we see the impact of the disease concept(s) of inebriety on the lives of drinkers and their circles of friends and family. Moving from the actions of physicians, policymakers, and reformers to the experiences of those they sought to reform provides a multilayered understanding of the dynamics of the medicalization—or "incomplete medicalization"—of alcoholism and the alternately facilitating and antagonistic roles that inebriates themselves played in this process.

Chapter 1 focuses on the ways in which physicians and reformers defined habitual drunkenness as a disease. I argue that there was a general terminological movement from *intemperance* to *dipsomania* and *inebriety,* and finally, to *alcoholism.* Each of these terms represented a particular constellation of cultural assumptions about gender, class, ethnicity, heredity, and personal responsibility. Of course, each diagnostic label was also a reflection of the state of medical thinking on habitual drunkenness. The two are intimately related: terms such as *dipsomania,* which stressed the connection between drinking and insanity, also comprised a rhetorical strategy for the promotion of the disease concept and the authority of the medical profession.

Scholars within history, alcohol studies, and sociology have argued that one of the reasons that this early movement to medicalize habitual drunkenness failed was the lack of consensus on the definition of inebriety. More interesting to my mind, however, is the general agreement among inebriety experts that there was both a hereditary and a volitional element, both a medical and a moral aspect, to the disease. Although the labels changed, reflecting different developments in medicine and culture, there was a remarkable resistance to monocausal explanations of inebriety. Inebriety specialists consistently embraced a complex model of their disease. Indeed, regardless of the term used to describe chronic drunkenness, treatment routinely aspired to restore the inebriate's weakened moral fiber as well as his physical constitution and mental stability. Inebriety's status as a chronic medico-moral affliction, however, was increasingly out of step with the ascending model of disease that linked specific pathogenic mechanisms or causes to specific diseases. This posed a significant obstacle to the condition's medicalization.

Furthermore, the distinctions that medical experts drew among varieties of inebriety mirrored those established by social reformers who separated the "worthy" from the "unworthy" poor. Class, gender, and ethnic biases, too, were built into the disease concept. Drawing distinctions among the types of inebriates was not a sign of muddled or naive thinking on the part of medical experts. Rather, it was both an acknowledgment of the complexity of inebriety and an attempt to define the disease in ways that would garner support for treatment. Physicians wanted to treat people that they believed they could help: people who wanted to be cured and whom the public would find deserving of treatment.

The act of defining inebriety was much more than a description of pathology. It was an explanation of individual and societal illness and a prescription for the nation's health. Chapter 2 focuses on the larger social and cultural issues that framed inebriety in the Gilded Age and Progressive Era. Debates over the nature of inebriety brought to the surface a range of nineteenth- and early-twentieth-century concerns about the status of women in American society, competing definitions of manhood, the perils of modern civilization and the urban metropolis, the responsibilities of the state to its citizens, and the individual's rights within the modern state.

The definition and redefinition of inebriety also offered physicians a venue for the expression of their own professional aspirations and insecurities. A significant patent medicine trade surrounded the treatment of alcoholism. The most popular proprietary cure, Leslie E. Keeley's Bi-Chloride of Gold cure, arguably helped popularize the disease concept more than the efforts of the AACI. Yet the regular physicians and reformers who promoted the disease concept opposed the efforts of Keeley and others in the patent medicine trade, whom they regarded as quacks. Inebriety specialists were stuck agreeing with Keeley and the WCTU that inebriety was a disease, but disagreeing with both organizations about how to cure it. At least initially, the WCTU supported medical treatment for the inebriate, but care for inebriates always took a backseat to its primary goal of prohibition. Medical advocates for the disease concept could not ignore the influential roles these two groups played in shaping the public discourse on inebriety.

Chapter 3 explores the range of private institutions for inebriates that reformers developed between 1870 and 1920. Within the walls of the inebriate asylum, the disease concept helped script the actions and interactions of physicians and patients. From moral instruction to physical culture, from drug therapy to psychotherapy, the different therapeutic regimens revealed the extent to which new medical and traditional moral perspectives clashed and coexisted. Just as there was no single disease concept of inebriety, so too there was no one therapeutic approach, although the different types of institutions available to habitual drunkards at the turn of the century had more in common than historians have acknowledged.

Chapter 4 introduces readers to the world of the public inebriate asylum and hospital farm, looking not only at the forms of treatment offered in these institutions but also at the public discussions that framed the construction of state and municipal inebriate facilities in Connecticut, Minnesota, and New York City. The world surrounding the inebriate asylum offers as much insight into the popular reception of the disease concept as the world inside. The legislative battles that were fought to create and maintain public institutions, the local community response to inebriate asylums, and the relations between the court systems and the inebriate hospitals all suggest that the public was willing to embrace a new vision of an old problem, if reluctant to abandon the traditional moral latticework that had framed habitual drunkenness. Individuals and agencies supported the public inebriate asylum because it met their particular needs, although their needs often opposed the therapeutic course charted by hospital physicians. Court systems, for example, readily embraced state inebriate facilities as a promising alternative to repeated fining and sentencing of alcoholic recidivists. Yet these were the least hopeful cases in the eyes of physicians—the last patient group doctors wished to treat. Likewise, disgruntled wives committed their besotted husbands to the asylum not just to restore their husbands' health, but also to legitimate their requests for divorce and to punish their spouses. In short, support for inebriate asylums came from many parties, but it did not necessarily signal acceptance of the disease concept.

Chapters 5 and 6 are in-depth case studies of two states that were able to build medical systems for the treatment of inebriates: Massachusetts and Iowa, respectively. Massachusetts was the first state to establish a treatment system for its inebriates, and of all the states that experimented with inebriate hospitals, the Bay State was by far the most successful. The story of the Massachusetts Hospital for Dipsomaniacs and Inebriates highlights the political capital, inter-institutional relationships, and medical innovation necessary to make medical care for habitual drunkards a reality. The disease concept received support from the state's large and integrated health, education, and social welfare infrastructure; a century-old tradition of activism among state officials, social reformers, educational leaders, and medical luminaries—especially within Boston Brahmin society; a large Irish population, whose drink habits had long been the focus of reform efforts; and the Commonwealth's prioritization of innovative penal and welfare policies.

Chapter 6 moves from the Northeast to the Midwest. Iowa, a largely agricultural state in the nineteenth century, began to send habitual drunkards to special inebriate wards within the state hospital system in 1902. Four years later, Iowa's first and only state hospital for inebriates opened its doors in Knoxville. The Iowa story differs in a number of regards from that of Massachusetts. Whereas the Irish population was an important motivating force for reformers in the Bay State, Iowa's inebriates were mostly second- and third-generation Americans. Learning that inebriety was "an American disease," rather than an immigrant affliction, helped propel Iowa's medical community forward in its campaign to create a new state hospital for alcoholics. Medical reform of the state's habitual drunkards was but one of many innovative strategies the state employed to curtail the liquor "traffic" between 1850 and 1920, a period when prohibition dominated state politics.

What did people with drinking problems think about their condition? How were the daily lives of inebriate patients and their families and friends affected by the disease concept of inebriety? Did the thoughts and experiences of inebriates and their families and friends help shape the disease concept? What role did patients and their loved ones play in the process of medicalization? If the disease of alcoholism is socially negotiated, we need to hear from the inebriates themselves and from those close to them, not just from the medical experts and social reformers. Chapter 7 completes the book's trajectory from ideas to institutions to individuals. It relies primarily on patient correspondence from the case files of the Massachusetts State Hospital for Dipsomaniacs and Inebriates and published patient narratives.

These records of the micro-negotiations of medicalization suggest that many inebriates adopted the disease concept without much difficulty. They also indicate that treatment at the inebriate hospital did not lead them to abandon the Judeo-Christian morals that had framed their condition from the start. If the disease concept freed the chronic drunkard from blame for his compromised condition, it did not remove his duty to vigilantly guard his health, avoid stressful situations at home and work, and develop wholesome habits such as exercise and churchgoing—all steps that physicians thought essential to their patients' continued improvement. The normative language inebriates used to describe their "besotted sin," the "drink evil," or their "intemperate habits" suggests that they were able to integrate both the medical and moral dimensions of their disease. Some inebriates, however, internalized the medical perspective more than others and became their own nurses, monitoring their food intake, exercise, weight gains and losses, and general mental state; they reported this data to hospital physicians by mail or in person.

Some individuals did more than make their own medical observations: they observed and diagnosed other drinkers, referring them to the inebriate hospital for treatment. Exceptional cases published their own heroic autobiographical tales of recovery. These stories championed the steadfast, expert aid of the psychiatrists and neurologists who ran special inebriate sanitaria and hospitals, and reminded the public that inebriety was a nervous disease. By looking at the way inebriates and their immediate families, friends, and others thought about their affliction, and by examining their responses to medical treatment, this chapter reveals the significant roles played by lay people in shaping the disease concept of inebriety, the identity of the inebriate hospital, and ultimately, the process of medicalization.

*Alcoholism in America* comes to an end with the passage of the Volstead Act in 1920. With this sea change in national policy, the problems of the alcoholic did not vanish, but they were obscured. Most institutions for inebriates closed their doors with Prohibition, and the alcoholic patient once again awoke from his stupor within the confines of the state mental hospital, the local jailhouse and work farm, or more likely, his own abode. In retrospect, it seems odd that a half-century of research and treatment on inebriety would be forgotten or dismissed. Yet this was the case when the disease concept was resurrected in the modern alcoholism movement. Indeed, Elvin M. Jellinek sought to distance his own typology of the alcoholic from the work of his turn-of-the century predecessors, which he impugned as particularly vague and "not even first approaches to definitions."[59] Readers, of course, must come to their own conclusions about Jellinek's assessment. Yet, in their embrace of the complexity and many manifestations of the disease of inebriety and in their development of treatment options that addressed the social, economic, psychological, and physical manifestations of the condition, the early inebriety specialists may appear more prescient than quaint. Today the disease concept of alcoholism is hotly contested, and public health officials appear more enamored with the "alcohol problems" approach precisely because it addresses the diversity of medical and social problems associated with excessive drinking.

# CHAPTER ONE

# Disease Concept(s) of Inebriety

On the eve of the U.S. Centennial, the American Medical Association convened in Philadelphia, where founder of the American temperance movement Benjamin Rush had signed the Declaration of Independence. It was 6 June 1876, and members of the AMA Section on Public Hygiene and State Medicine gathered at the Horticultural Hall of the Centennial Exposition in Fairmount Park to discuss, among other concerns, the disease status of inebriates and the necessity of inebriate hospitals. Only five months earlier, at its own annual meeting, the Association of Medical Superintendents of American Institutions for the Insane (AMSAII) had passed several resolutions recommending: first, that each state establish an inebriate asylum; second, that laws be modified to commit inebriates to these institutions just as the insane are committed to state hospitals; and third, that the treatment of the inebriate in hospitals for the insane was not beneficial to either population.[1] It was now time for the AMA to declare their own views on the issue.

At the meeting of the Section on Public Hygiene and State Medicine that same day, Dr. B. N. Comings, a Civil War surgeon from Connecticut, introduced two resolutions that brought the AMA in line with the official opinion of the AMSAII. The first declared that inebriates should not be treated at public asylums for the insane. The second observed that "inebriety being both a disease and a vice—a vice as related to man's normal nature, and a disease of his physical organization—special treatment in institutions adapted to the purpose is required for the inebriate, and it is the imperative duty of each Commonwealth to establish and maintain public institutions for the treatment of inebriety."[2] Comings's interest in inebriety derived from his recent charge to explore the topic for the Connecticut State Medical Society and from his unease about committing alcoholics to the state hospital for the insane at Middletown.

The Section on Public Hygiene and State Medicine adopted Comings's resolutions, but not without some heated discussion. John Shaw Billings, who found Comings's explanation that inebriety was both vice and disease taxing, declared: "The vice is the disease. They are the same thing." Comings responded to Billings by clarifying his position. In doing so, he revealed the holistic, moral-medical perspective on inebriety that many of his contemporaries shared. The clergy, Comings insisted, recognized inebriety as a vice only, and not a disease, while the medical profession treated inebriety as a disease alone, not a vice. "The consequence is, both fail," he concluded, adding "I desire therefore to recognize it as both a vice and a disease, that the combined influence of the clerical and medical professions may be brought to bear upon its treatment." Hardly the position of a medical imperialist, Comings's interpretation of inebriety won the acceptance of his colleagues, who passed the resolution forthwith. Thus, the AMA recognized inebriety as a hybrid medico-moral affliction.[3]

Defining intemperance as a disease was an essential step in the early campaign to medicalize the condition, for most people were accustomed to viewing habitual drunkenness as a moral failing or petty crime. Indeed, describing the disease of inebriety was an ongoing process during the fifty years that followed the AACI's formation, a process that reflected much more than the state of medical knowledge about the condition. As physicians and reformers proposed new frameworks for understanding this very old problem, their terminology revealed their opinions about a host of other important issues of the day.

This chapter examines the evolution of the terms used to describe habitual drunkenness between 1870 and 1920: a linguistic continuum that began with *intemperance,* moved from there to *dipsomania* and *inebriety,* and ended with a

diagnosis that is familiar to twenty-first-century readers: *alcoholism.* Although
historians have emphasized the inconsistency of the terms used by inebriety
physicians and the ways that this lack of consensus may have hampered the
disease concept's acceptance both within and outside the medical profession,
these terms also reveal much about the particular cultural and medical mo-
ments in which they were employed. *Dipsomania,* for example, emphasized
the connections between habitual drunkenness and hereditary forms of insanity, at once releasing drunkards from responsibility for their condition and
implicating physicians as their natural caretakers.

Admittedly, members of the medical profession and the general public at
times used the four terms—*intemperance, dipsomania, inebriety,* and *alcoholism*—interchangeably. Yet a reading of over two hundred medical, social reform, and popular articles and monographs on habitual drunkenness
published between 1870 and 1920 suggests that there was a general progression from one diagnostic label to the next. Although they may have shared a
core meaning, the terms differed from one another in specific ways. They
served different political functions, and they reflected various developments
in medicine's and the public's vision of the alcohol problem in American
society.

## Varieties of Inebriety: Evolution of the Disease Concept

### Intemperance

Of the four terms, the oldest and most morally suffused was *intemperance,* a
term that suggested a deviation from the golden rule of moderation championed by church and health evangelists alike—a term that physicians hoped to
replace with *inebriety.* When the American Association for the Cure of Inebriates made a formal statement of its principles at its inaugural meeting in 1870,
the first was "Intemperance is a disease." The choice of the word *intemperance*
was an important one for an organization devoted to the promotion of the disease concept of inebriety. *Intemperance* was the term most commonly used at
midcentury to describe habitual drunkenness, and it was recognizable to a
public accustomed to temperance rhetoric and conditioned by seventy years
of anti-drink agitation. Perhaps more to the point, several of the AACI's prominent members, including cofounder Nathan Smith Davis, were temperance
advocates who ran inebriate homes that sought to rehabilitate the drunkard's

moral character through enlightened Christian fellowship as well as medical care. Chicago Washingtonian Home superintendent Thomas Vancourt and others referred to the guiding principle in these places as "the law of kindness."[4] The nascent professional organization could not afford to lose any treatment allies this early in the battle by insisting on terminology that might alienate reformers.

What were the characteristics of intemperance as a disease? The AACI's "Preamble and Declaration of Principles" asserted: "Its primary cause is a constitutional susceptibility to the alcoholic impression."[5] This vulnerability to alcohol could be either inherited or acquired, but medical reformers such as Joseph Parrish claimed that even in cases where the disease of intemperance was acquired through repeated indulgence, there was a defective constitution at fault: "We have primarily, a defective condition of the body or mind, and an impaired will, among its earliest evidences; then an appetite, and lastly drunkenness with all its resulting evils."[6] For Parrish, habitual drunkenness was both a disease *and* a symptom of a compromised constitution. One could damage one's constitution—deplete it of nervous energy—through a variety of means, but Parrish, in his "Philosophy of Intemperance," pointed to one means in particular: the frenetic pace of American life. The "haste with which we live," claimed the Pennsylvania physician, was not a short-lived, impetuous haste; it was a way of life that governed business and education and permeated the very fabric of the nation: "The pressure upon the brain of children by a forcing system of education, the subsequent tax upon the supreme nervous centre, in the struggle for wealth, power or position; the unhealthy rivalry for display, and all the excitements which produce the 'wear and tear' in our life, are so many means of exhausting nervous energy, and producing a condition that demands relief."[7] Words such as these make plain the cultural ties that bound intemperance, and later inebriety and alcoholism, to the social body.

The search for the cause of intemperance was ongoing. Those who attended the AACI's first meeting would point to a variety of factors. John Willett, the superintendent of the Inebriates' Home of Kings County, Fort Hamilton, Long Island, assured his audience that most of the intemperate men he treated "first began to drink because their religious training had been overlooked, and there was no fear of God before their eyes."[8] Willett's moral perspective was not unlike that of Vermont physician William Sweetser, the author of an early prize-winning essay, "A Dissertation on Intemperance." Sweetser spent more than half of his 1829 essay elaborating on the physiological changes brought about by the chronic consumption of alcohol; he understood intemperance as a habit that might evolve into a disease. He claimed that it could be acquired as early as infancy from a nursing mother's consumption of alcohol. Sweetser singled out several social practices implicated in the etiology of intemperance: the army's habit of giving soldiers grog, farmers' distribution of alcohol to their weary laborers, the consumption of overstimulating foods, the despair resulting from personal or economic hardship, and political candidates' serving of liquor to their constituents on election day. Yet even if Sweetser could find a multitude of factors conspiring to promote intemperance—the habit *and* the disease—he warned that "it is a disease produced and maintained by voluntary acts, which is a very different thing in my view from a disease with which providence inflicts us." "Ignominy and disgrace," he continued, "should ever be associated with intemperance, no matter how much, there is not yet enough to prevent the spreading evil. . . . I feel convinced that should the opinion ever prevail that intemperance is a disease like fever, mania, &c., and no more moral turpitude be affixed to it, drunkenness, if possible, will spread itself even to a more alarming extent."[9] For Sweetser, moral censure and infamy were to befall the weak-willed. Although he wrote several decades before the formation of the AACI, his notion of intemperance provides a sense of the disease's moral trappings.

A central issue in the acceptance of the disease concept highlighted by Sweetser and acknowledged by Joseph Parrish in 1870 was accounting for the role of individual volition in inebriety. Embodied within the older concept of intemperance was a Christian notion of individual potency and free will. In 1870, as in 1829, many people—including physicians—believed that all persons were divinely endowed with the power to resist temptation and to find salvation through the exercise of the will; only the vicious and the weak failed to exert the necessary self-discipline to remain sober, or at least temperate. If, as Sweetser and others suggested, intemperance was self-inflicted, shouldn't the inebriate be made to pay the price for his violation of social custom and law? Shouldn't he take responsibility for his illness? Nowhere was public skepticism of the disease concept more apparent than in the words of police-court judge Nathan Crosby of Lowell, Massachusetts, who spoke before the State Legislature's Committee on Charitable Institutions in 1871: "The great general fact remains that the *drunkard is self-made,* progressively self-taught, and obstinately self-immolated. I regard the doctrine of 'disease' and 'insanity' a new incentive to intemperance, the waiver of imprisonment, of prosecutions, a removal of criminality, and an asylum a bounty for drunkenness."[10] Crosby deemed the state's philanthropic practices overindulgent. Sympathy and largesse were not to be extended to those he thought had spoiled their own lives.

Parrish responded to such criticism in his 1872 presidential address to the AACI: "Nothing could be further from sound reason and common experience than this statement. When a person knows that he has a disease, he applies himself to its relief or cure. He takes counsel, changes his mode of life, and does what his physician may prescribe."[11] As Parrish intimated, intemperance in its early stages might only be a habit, and an individual might then be able to contend with it on his own, but when the craving for alcohol became acute and the demand "so imperious as to overpower the judgement and will," it was time for medical treatment and support. Thus, Parrish hoped to reassure his audience that seeking medical help *was* taking responsibility for one's condition. Were an individual aware of his "inherited tendency" to alcoholic excess, Parrish reasoned that he should be held more, not less accountable for acquiring the habit.[12]

As its *Proceedings* for the first five years make plain, the AACI was determined to supplant the morally tainted vision of *intemperance* with the disease of the same name, or better still, with the disease of *inebriety*. And yet neither the AACI nor the majority of physicians who wrote about habitual drunkenness in the 1870s were willing or able to abandon the moral aspects of this medical condition. Not only would the public not permit it, but physicians themselves also realized the importance of the individual inebriate's moral conscience, both in the origins of habitual drunkenness and in its treatment. At its third annual meeting in 1872, the AACI appointed a committee to respond to the religious and temperance press's attacks on the association's principles. The AACI's "Report of the Special Committee on Principles" offered opponents of the disease concept three major concessions. It acknowledged "the criminality of the acts and habit of drunkenness"; it agreed with moralists "that inebriety will neither be induced nor kept up except by the use of intoxicants, which is in every case its exciting cause"; and finally, it stated that the AACI's opinions did not "excuse the crime of intemperance, or promote indulgence in the use of intoxicants."[13]

How could calling alcohol a poison encourage its use? asked Association physicians. How could declaring habitual drunkenness a potentially lethal disease help promote the condition? No one doubted "the sinfulness of the habit," they assured their critics. No one doubted the results of excessive indulgence. To quell lay concerns about the moral nature of treatment, they reminded readers that daily religious services were held in most of the institutions for habitual drunkards. They also insisted that medical treatment for intemperance entailed much more than administering drugs; it included "the encouragement derived from intercourse with others who have been partially or wholly restored, and the higher moral influences which are derived from revealed truth, its admonitions and its invitations."[14] It was no accident that upon the opening of the Massachusetts Hospital for Dipsomaniacs and Inebriates in 1893, patients were admitted only if they were certified in a court of law as "not of bad repute or of bad character, apart from [their] habits of inebriety."[15] Over the intervening two decades, the term *intemperance* had fallen from grace, but its moral valence had persisted even as physicians employed the two more medicalized terms, *dipsomania* and *inebriety*.

## Dipsomania

Literally a mad (Greek *mania*) thirst (Greek *dipsa*) for alcohol, *dipsomania* was perhaps the most colorful of terms for habitual drunkenness. "The craving for drink in real dipsomaniacs, or for opium or chloral in those subjugated, is of a strength of which normal persons can form no conception," wrote psychologist William James in his 1890 discussion of human will.[16] Of course, a poor understanding of something rarely prevents any "normal" person from venturing an opinion. Like all the other terms and frames for habitual drunkenness, dipsomania's disease status was contested. Still, James's specific use of the term *dipsomaniac* suggests that it had attained a fairly wide currency in the United States at the turn of the century, regardless of its contested nature. As political scientist Mariana Valverde has argued in *Diseases of the Will*, *dipsomania* was the "most medical" of the terms used to describe habitual drunkenness in the late nineteenth and early twentieth centuries. But while Valverde sees *dipsomania* as a term whose meaning changed over time and one that "never caught on outside of biomedical circles," the use of *dipsomania* in fact signified an important step in physicians' efforts to promote the disease concept in the United States.[17] Its meaning evolved over time, as did every term used to describe habitual drunkenness, but initially, *dipsomania* was used by physicians and reformers simply to describe habitual drunkenness that was a disease rather than a vice.[18] In the closing years of the nineteenth century, however, *dipsomania* more often denoted a special form of inebriety, one characterized by its periodicity, its hereditary nature, its middle- and upper-class victims, and above all, its relationship to insanity.[19]

Dipsomania's nosological connection to insanity (a disease that was clearly within the physician's domain) and its status as a special form of monomania appear to have captured the public's imagination. In other words, contrary to Valverde's estimation, the term did catch on. The popularity of dipsomania may be seen in a wide variety of periodical literature. A series of newspaper headlines from turn-of-the-century Iowa announced different features of the state's experiment in providing medical care for its inebriates: "Dipsomaniac Law Put Into Effect," "Dipsomaniacs Provided For," "Dislike Dipsomaniacs: Insane Patients Incensed at Being Confined with Bestial Drunks," "Disregarded the Rules: Inebriates at Dipsomaniac Asylums Make Easy Escapes," "Recapture Dipsos: All Escaped Inebriates will be Placed in Knoxville 'Jag' Asylum," and finally, "'Dips' Have Gone: Cherokee Contingent Were Taken to Knoxville on Friday."[20] These sensational headlines testify not only to the popularity of the term but also to the difficulties encountered in treating inebriates at state institutions (see chaps. 4 and 5). Likewise, in a 1907 act of retrospective diagnosis, the *Boston Evening Transcript* optimistically offered the following commentary on author Edgar Allan Poe, quoting from Charles H. Goudiss, M.D.:

> Had Poe been born a generation later . . . he would have been spared a life of anguish, and his good name would have been unsmirched. Modern science understands the peculiar pathological case he presents and tempers its moral judgements with its alleviating remedies. . . .
>
> "In his poetic prayers and fantasies the neurologist can see suffering and recognize the feeling of hopelessness ever present in the victim of dipsomania . . . . The dipsomaniac attacks," we read, "are symptoms of disorganized brain cells."[21]

These were not isolated incidents. In "The Confessions of a Dipsomaniac," an alcoholic autobiography and recovery narrative published in 1904, a "well-known and gifted novelist and essayist" portrayed "the horrors of that impulsive form of insanity called dipsomania." Although the anonymous dipsomaniac portrayed his condition as alternately "a strange and weird insanity," "a symptom of disease, not the disease itself," and "alternating personalities," he was sure that he suffered from a disease and that his drinking binges were a result of his impaired nervous system.[22] Advocates of the disease concept were not above using popular literature such as this in their cause. Praising "The Confessions," the editor of *The St. Paul Medical Journal* in Minnesota wrote of its author:

> Had he been a medical man he could not have better portrayed the symptoms of the psychic epilepsy that was the basis of his affliction.... If we realized that he was as cruelly afflicted with a mental disease as is the cripple with a physical condition, we would be in a position properly to appreciate the actual facts and to apply the proper therapeutics.... It was only when he was considered to be the victim of a disease and treated as such, that his normal personality was able to gain the mastery.[23]

The message was clear: the inebriate's moral paralysis could only be addressed when he realized that he suffered from a disease. The inebriate hospital was the best place for this to happen, added the editor. It would take three more years of lobbying for Minnesota to pass legislation to create an inebriate hospital and farm colony, following in the footsteps of other states such as Massachusetts and Iowa. In each of these cases, the concept of dipsomania played a key role.

Why was the term *dipsomania* so important to the inebriate reform movement? In its most common usage, whether to refer broadly to the disease of inebriety or to refer to a special type of diseased drunkenness, *dipsomania* stood for a type of alcoholism that possessed three important characteristics: it was a form of insanity; it was often hereditary; and it affected the middle and upper ranks of society. Several scholars—William F. Bynum, Roy MacLeod, and Mariana Valverde—have traced the concept of dipsomania back to the French alienists Philippe Pinel, Jean Etienne Dominique Esquirol, and Jean Etienne Georget, who promoted the concept of monomania, *manie sans délire*, in French psychiatry in the first half of the nineteenth century. Called "moral insanity" in Great Britain and the United States, monomania was a form of partial insanity that required the physician's expertise for correct diagnosis. While a dipsomaniac's intellect could be fully functional, his emotions and willpower could still lead him to commit impulsive acts. For Esquirol, dipsomania was one form of monomania, distinct from other types such as kleptomania, pyromania, suicidal mania, and homicidal mania. If kleptomania was defined in the late nineteenth century as a disease of the middle-class or affluent woman, dipsomania was believed to affect mostly middle- and upper-class men *and* women. Professional men frequently depleted their small stores of nervous energy through overexertion in the workplace, while women's delicate and cyclical reproductive systems predisposed them to periodic cravings for alcohol.[24]

By stressing the condition's ties to insanity, albeit the hotly contested psychiatric concept of moral insanity, inebriety doctors hoped to advance their cause. Indeed, one of the founders of the AACI and its first president, Willard Parker, firmly believed that moral insanity was a legitimate form of mental illness.[25] Another AACI member, Edward Mann, medical superintendent of the New York State Emigrant Insane Asylum at Ward's Island, concurred: "The manifestation of moral insanity may be a simple perversion of some sentiment or propensity, under certain exciting causes, and I think this exactly comprehends cases of dipsomania with less of self-control and perversion of the moral senses."[26] By labeling dipsomania (and by extension, inebriety) a form of insanity, American physicians were placing it within the medical domain. If dipsomania was insanity by another name, then physicians were the natural caretakers of its sufferers and should be granted the right to confine inebriates within an asylum. It was, these medical reformers claimed, in both the drunkard's and the public's best interest.[27]

Heredity, too, was a key element in defining dipsomania. Although the disease was not always thought to originate in people's hereditary endowment, physicians claimed that this was frequently the case. Moreover, most disease concept advocates were in agreement about the dipsomaniac's ability to transmit his or her disability to subsequent generations in the form of a defective nervous constitution—a body predisposed to defective, degenerate, and delinquent behavior. Both of these issues impinged not only on the inebriate's responsibility for his condition but also on the physician's responsibility to his patient and to the nation's health. Francis Galton may have coined the word *eugenics* in 1883, but the preservation of the public's collective germ plasm was of concern to psychiatrists and neurologists well before the "science" of the well-born made its debut. Many physicians argued that dipsomaniacs and inebriates should not have children, while others believed that drunkards simply should wait to reproduce until they were cured of their condition.

Among those discussing dipsomania's hereditary nature and dangers, none was more prominent than French psychiatrist Benedict-Augustin Morel. If dipsomania's ties to insanity were born in the asylums of Great Britain and France, so too was Morel's influential theory of hereditary degeneration, which found an attentive American audience. Doctors in the United States cited the Frenchman's work routinely in their discussions of dipsomania. Physician John O'Dea, who spoke to New York's Medico-Legal Society on methomania (a less-popular name for dipsomania), paid homage to Morel in his speech, then added, "Few allied questions are more fully supported by evidence than the transmissibility, not merely of the drunkard's degenerate constitution, but even of his appetite for intoxicating drinks."[28] What passed from one generation to the next, according to O'Dea (and Morel), was a neuropathic constitution that presented in different forms of mental disease and aberrant behavior. Alcoholic excess in one generation could be followed by hysteria in the next, hypochondria and depression in the third, and idiocy and sterility in the last.[29] Thus, not only did the dipsomaniac's drinking have immediate consequences for those closest to him or her—penury, disgrace, troubled marriages—but it could also undermine the health and well-being of future generations. As we shall see, the focus on the hereditary transmission of dipsomania helped bolster the argument that inebriety was a significant public health problem.

Although Joseph Parrish had argued that knowledge of one's hereditary predisposition to habitual drunkenness made a person more, not less responsible for preventing it, most physicians who wrote about dipsomania presumed that few dipsomaniacs would pay heed to their heredity on the way to the saloon. Temperance advocates were correct in believing that most physicians tended to regard drinkers who had inherited their thirst for alcohol as *less*, not more responsible for their condition. The views of neurologist L. W. Baker, superintendent of the Private Medical Home for Nervous Invalids in Baldwinville, Massachusetts, were typical of those advocating medical care. In 1888 Baker acknowledged the importance of the dipsomaniac's responsibility for acts committed while under the influence. He noted, however, that excessive use of alcohol was "usually transmitted from a previous generation . . . [and was] often only a symptom of an inherited defective nervous system, the morbid actions of which are not always within the control of the individual."[30] Selling this liberal, secular message to a public seasoned by WCTU diatribes against the "drink evil" was not an inviting or easy task, but the proposition was easier to make, it turned out, when the patient was born into wealth. For late Gilded Age and Progressive Era physicians, class was an essential diagnostic marker for dipsomania.

As early as 1867, Charles Nichols, M.D., superintendent of St. Elizabeth's Hospital for the Insane in Washington, D.C., remarked, "It is because we so often see, in dipsomaniacs, the sacrifice of the extraordinary capacities and opportunities for usefulness to other fellow men, which are afforded by liberal education, wealth, and social influence, that these cases excite our deepest interest."[31] Nichols's views were typical of most turn-of-the-century psychiatrists concerned with habitual drunkenness. They revealed a sympathetic concern for the mental malaise of the wealthy, a segment of the population that psychiatrists and neurologists would increasingly court. Nearly a decade later, famed New York neurologist George Miller Beard (best known for defining neurasthenia, or nervous weakness), explained the apparent increase in habitually drunk Americans in terms that were similarly prejudiced toward the "higher classes." Beard referred to *inebriety* rather than *dipsomania*, but his words were reminiscent of Nichols's own. He believed that "drunkenness the vice" was on the rise in the biologically inferior lower classes, who had reached an evolutionary dead end in the previous century. These people still possessed the rugged physiques, primitive and sensual instincts, and tolerance for alcohol that had characterized previous generations, while among the upper classes, the situation was quite the reverse: "Drunkenness, as a vice, among the better classes of civilized lands, is then decreasing, while drunkenness, as a disease, inebriety, is increasing."[32] The more affluent and accomplished classes, according to Beard, had achieved their social status by dint of their exquisitely sensitive nervous systems. These systems were susceptible to alcohol, rendering most of their owners unable to drink without suffering debilitating effects. Thus, reasoned Beard, most of society's higher ranks chose a temperate life, but when they did drink, they were especially vulnerable to "drunkenness the disease."

As late as 1907, when the term *dipsomania* was little used, Yale-trained psychologist George Cutten observed that

> dipsomaniacs are not infrequently persons of extraordinary mental ability.... Whether he be physician, artist, musician, or literateur, he is living at a high nervous tension, his nervous energy is easily exhausted, and his reserve brain power is soon expended. No other class of partakers of alcohol is composed of such bright and intelligent men, or men who both by nature and education are better equipped morally.... It will be some advance when dipsomania is clearly diagnosed, not only by physicians but by the courts, and it is recognized that during the paroxysms the individual is really insane and irresponsible for his acts.[33]

The dipsomaniac was the sickest, most accomplished, and most innocent of the different identities proposed for the habitual drunkard. As such, dipsomania made the strongest case for medical treatment.

Over time, physicians concluded that the number of dipsomaniacs—hereditary, middle- or upper-class, and insane—was relatively small compared to the number of other inebriates. It is difficult to say why the term *dipsomania* faded from view. It may have had something to do with the increasing influence of the "new psychology" in the 1880s and 1890s and the rise of psychodynamic psychiatry in the early twentieth century in the United States. Changes in the intellectual foundation of psychiatry shifted physicians' attention away from the defective heredity that was essential to the dipsomaniac's makeup.[34] Increasingly, mental science and mental health stressed the importance of individual experience over heredity. If an inherited neuropathic constitution relieved the affluent dipsomaniac of responsibility, it ultimately suggested a more pessimistic prognosis than that of the simple inebriate who brought about his own decline. Physicians could do little to change a dipsomaniac's heredity, but psychodynamic psychiatry offered him a method to uncover the neurotic conflict that lay beneath his chronic heavy drinking.

## Inebriety

*Inebriety* was the most common term used to describe habitual drunkenness between 1870 and 1920, and it is used primarily that way in this book as well. However, like *dipsomania*, it had both specific and general meanings. Though it is unclear when the term *inebriety* was first used, it achieved wide circulation in the 1870s and 1880s and continued to be commonly used until Prohibition.[35] Initially, *inebriety* referred to the disease of habitual drunkenness caused either by defective heredity or repeated debauch. Physicians frequently referred to the former as hereditary inebriety or as periodic inebriety (and sometimes, of course, as dipsomania). Simple inebriates or chronic inebriates were usually those who compromised their health through routine indulgence. Discussions of inebriety, like discussions of dipsomania or intemperance, focused on a wide range of predisposing causes that ranged from excess brainwork to exhausting physical labor—each promoting the desire for an alcoholic respite that physicians frequently, if incorrectly, called alcoholic stimulation. The process by which drinking passed from vicious habit to debilitating disease was a staple of most discussions of inebriety. Neurologist L. W. Baker offered the typical trajectory of the diseased inebriate when he asserted in 1888 that "the habitual use of intoxicants is very likely, sooner or later, to become a fixed and uncontrollable desire, as the natural result of frequently repeated impressions upon the central nervous system. That which was at first but a mere habit, may, in this way, pass beyond the control of the individual and become a confirmed neurosis. A distinction must be made between the self-controlling vice of drunkenness . . . and the irresistible impulse of disease."[36] Baker's example, of course, did not include the hereditary inebriate, whose uncontrollable desire was an inborn pathology and who did not require prolonged and repeated indulgence to cultivate the disease.

Class, too, remained an essential feature of the discussions on inebriety. Even those who questioned the validity of the disease concept perceived that social class mattered, and the rich were once again deemed more diseased than the poor. Consider the comments of C. W. Earle, M.D., the president of the Illinois State Medical Society in 1889. Earle remained skeptical of the disease concept, believing that it exonerated habitual drunkards of responsibility for their condition. He also felt that the motives of disease-concept advocates were tainted with filthy lucre: "I am not aware that anybody ever thought drunkenness in a poor man a disease—it is always in a class that are able to pay large sums for specialists, and those . . . coddled with the idea that they have a peculiar pathology, that we find the largest number of advocates."[37] Though Earle misjudged inebriate-asylum advocates, who also considered the impoverished inebriate sick, he correctly discerned that most inebriate-asylum superintendents ran private institutions catering to the more affluent classes.

In his exhaustive *Inebriety or Narcomania—Its Etiology, Pathology, Treatment and Jurisprudence*, British inebriety expert Norman Kerr distinguished among myriad forms of the disease, including lunar inebriety, seafaring inebriety, constant inebriety, solitary inebriety, social inebriety, neurolytic inebriety, traumatic inebriety, and reproductive inebriety. For Kerr, there were as many types of the disease as there were reasons for drinking.[38] American physicians listened to Kerr, making him an honorary member of the AACI.[39] Between 1902 and 1920, Iowa's state hospital physicians and social reformers followed Kerr's practice and employed a bewildering array of terms to refer to their habitually drunk charges: periodical inebriates; environmental or associational inebriates; vicious and incorrigible inebriates; incurable inebriates; hopeful and respectable inebriates; weak and self-indulgent inebriates; nervous, impulsive, and easily led inebriates; chronic, selfish, ignorant, lazy, and criminal inebriates; gentleman tippler inebriates; and honest, hereditary victim inebriates.[40] These classifications provide a sense of the moral shroud that cloaked the condition. Even as it was discussed and defined by physicians, one's diseased status depended heavily on one's behavior and social class. Of course, the variety of terms also suggests the inconsistency of inebriety's clinical nomenclature.

So great was this terminological confusion that physician Lewis D. Mason made it the subject of his presidential oration, delivered at a special meeting of the American Association for the Cure of Inebriates in 1903. Written some thirty-three years after the AACI's founding, Mason's speech revealed that inconsistencies of language posed a significant problem for AACI members and medical reformers who promoted the disease concept. The public as well as a good portion of the medical profession defined *inebriety* as intoxication from any cause—not a disease. Consequently, they adhered to its literal meaning: "The term inebriety in its etymological sense or strict definition does not convey the idea that we desire it to convey.... We apply the term inebriety to a class of persons who are irresponsible [meaning not responsible for their condition], and who involuntarily, and not as a matter of choice, have become habituated to the use of alcohol, in excess, either periodically or more or less continuously."[41] Recognizing that the lay public and the general medical profession used *inebriety* in a variety of ways, he urged that fellow AACI members (who engaged in the same practice) "define what we mean by the term ... and determine whether or not such a use of it will be accepted by those who have a right to receive or reject it, or perchance who may offer us a better term or name to express the condition under consideration."[42] While it might be a stretch to say that Mason recognized the socially negotiated nature of medicalization, it is clear that he was sensitive to the give-and-take involved in promoting the disease concept. He appreciated the utility, if not the necessity, of promoting a consistent message.

The battle was not just for inebriety the disease versus inebriety as a synonym for intoxication, however. By the time Mason questioned the viability of inebriety as a diagnosis in 1903, the term had taken on a new meaning in some circles: it was used to denote a global concept of addiction. As early as 1871, George Miller Beard grouped those forms of addiction stemming from tea, coffee, cocoa, chicory, tobacco, opium, hemp, coca—and, quite remarkably, lettuce (reported to have soothing properties akin to opium)—along with alcoholic inebriety.[43] In 1893, *Quarterly Journal of Inebriety* editor Thomas Crothers advanced one of the broadest definitions of inebriety in his edited volume, *The Disease of Inebriety from Alcohol, Opium, and Other Narcotic Drugs.* Crothers chose to include ether, cocaine, chloroform, coffee, tea, nicotine, cologne, arsenic, and ginger (an intriguing, if unevenly appealing, assemblage).[44] Kerr's *Inebriety or Narcomania,* published one year later, was of a similar stripe—comparing alcohol addiction with dependencies to a wide range of psychotropics. Lest anyone question the basis for grouping these substances together, Catholic theologian, English professor, and American inebriety expert Austin O'Malley justified the inclusion of a special chapter on nonalcoholic addictive substances in his 1913 *The Cure of Alcoholism* by arguing that the "cure of morphinism, cocainism, and similar drug-addictions is, medically and morally, the same as that of alcoholism."[45]

O'Malley's remarks underscore the importance of alcoholic inebriety to the definition of addiction generally and to its treatment in Gilded Age and Progressive Era America. Mariana Valverde has claimed that drinking offered 1880s physicians and scientists "the paradigmatic case" of inebriety, that is, the one by which addiction to other substances was judged. There is an even stronger case to be made that alcoholic inebriety provided the archetype for addiction between 1870 and 1920. Consider historian Caroline Acker's comments on "competing" theories of opiate addiction at the turn of the century: "In the first decade of the twentieth century opiate addiction was variously considered an example of inebriety (a psychiatric condition that included alcoholism), a functional disorder of disturbed physiological processes, or a moral failing involving a collapse of will."[46] All three of the alternatives Acker presents here were encompassed *within* the inebriety paradigm. Alcoholic inebriety was a debilitating psychiatric condition treated at sanitaria and at general and state hospitals. The disturbed physiological processes of the habitual drunkard were a staple of the articles and monographs on intemperance, dipsomania, inebriety, and alcoholism written by physicians, physiologists, and temperance workers between 1870 and 1920.[47] Finally, from 1870 on, the moral paralysis of the inebriate's will, most prominent in the case of dipsomania, was an essential element of this hybrid medico-moral disease. The general term of *inebriety,* rooted in the model of alcoholic inebriety, served at least until the time of Prohibition as the template for addiction. It was a complex model encompassing body, mind, and morals. However, within the psychiatric, neurological, and general medical communities and within the public domain, *inebriety* faced a competitor as the nineteenth century came to an end: *alcoholism.*

## Alcoholism

Although Swedish physician Magnus Huss coined the term *alcoholism* in 1849, Americans had used it sparingly throughout the second half of the nineteenth century.[48] In the first decade of the twentieth century, however, the term began its ascent to the dominant linguistic place it holds today. In the years leading up to Prohibition, *alcoholism* fit neatly with the WCTU's and Scientific Temperance Instruction's depiction of alcohol as a universally corrupting poison. It was also easily aligned with the Anti-Saloon League's campaign against the sale of alcohol. For temperance reformers, the onus was right where it belonged: on King Alcohol's shoulders.[49]

Of course, *alcoholism* proved useful in light of *inebriety's* broader meaning, for it specified the substance responsible for intoxication. *Alcoholism* became a routine way to describe habitual drunkenness during the mid-1890s and the first decade of the twentieth century, at roughly the same time as Crothers and Kerr were promoting their inclusive vision of *inebriety*. The term *alcoholic inebriety* may have served as a segue to *alcoholism*. By 1914, Crothers acknowledged that "the terms alcoholism and inebriety are often used synonymously."[50] Crothers, however, preferred to employ *alcoholism* to designate the condition of those "poisoned by spirits, using it constantly in small quantities, never intoxicated or stupid, and often able to carry on [their] business, and appear reasonably sane." For Crothers, the "inebriate," had distinct drink "paroxysms" that were followed by periods of sane, sober, and rational thought and conduct.[51] According to him, this was the result of an underlying neuropathic constitution that the "alcoholic" lacked.

Few besides Crothers, however, drew this distinction. His minority voice nonetheless highlights an interesting example of terminological drift. As the term *dipsomania* faded from view, *inebriety* seemed to absorb its meaning; that is, the dipsomaniac became the periodic inebriate who drank in unexplained binges punctuated by extended periods of sobriety, as opposed to the chronic inebriate, who drank daily and regularly became drunk. Crothers's concerns suggest that with the arrival of the term *alcoholism*, the periodic inebriate (formerly the dipsomaniac) became "the inebriate," and the chronic or simple inebriate, who brought about his own disease through routine drinking, became "the alcoholic."

But why did *alcoholism* begin to eclipse *inebriety* in the first place, eventually replacing it? Several factors were responsible. First, the most notable change in the shift from *inebriety* to *alcoholism* was the switch from process to substance. *Intemperance, dipsomania,* and *inebriety* were dependent on a state of mind (intoxicated, insane) and were achieved through behavior (excessive drinking). By contrast, the key concept denoted by *alcoholism* was poisoning by alcohol. Comparable terms such as *morphinism* and *cocainism* denoted addiction to other psychoactive substances. This change from process to substance fit well with the growing anti-alcohol campaigns of the last decade of the nineteenth century and the drive for a dry America that gained momentum in the first two decades of the twentieth century, culminating in Prohibition. Groups such as the WCTU, the Prohibition Party, and the Anti-Saloon League did a great deal to demonize alcohol and to promote an understanding of it as a universally corrupting poison. To refer to habitual drunkenness as alcoholism was consistent with the emphasis on alcohol's corrupting properties that characterized the campaign for Prohibition.

While it is easy to see how the emphasis on alcohol rather than on the process of getting intoxicated might relieve the drinker of part of the responsibility for his condition, blame was not always lifted from the alcoholic's shoulders. The Pennsylvania State Sabbath School Association, for example, declared "Alcoholism a Disease," but added, "Alcoholism cannot be taken as smallpox or scarlet fever by contact with others. It must be taken through the mouth, consciously and deliberately, by each individual so that this disease is more easily preventable than many others."[52] Thus, the focus on alcohol could redirect attention from the underlying pathology of the drinker to the drinker's decision to imbibe a poison required to bring on the disease.

Another explanation of the increasing popularity of the term *alcoholism* is that it eliminated any confusion created by the narrow definition of *inebriety* (relating only to alcohol) versus the general definition of *inebriety*. If *inebriety* could encompass the diversity of substances described above, then the use of the separate terms *alcoholism, morphinism,* and *cocainism* helped eliminate some of the uncertainty. Furthermore, after the passage of the Pure Food and Drug Act in 1906 and the Harrison Narcotic Act in 1914, there was another distinction to be made: that between legal and illegal substances. The years between these two pieces of legislation witnessed the alignment of two concepts: narcotic addiction and social deviance.[53] Whereas a significant portion of the American population consumed alcohol, an increasingly marginalized minority population used other newly criminalized drugs.[54] Alcoholics were not breaking the law where prohibition was not in effect unless they appeared drunk in public or committed crimes while intoxicated. Thus, to refer to habitual drunkards as "alcoholics" may have proved less stigmatizing than to use the less-selective "inebriates."

The medical meaning of *alcoholism* remained relatively constant for the very last years of the nineteenth century and the first two decades of the twentieth century, but there was one change that took place at the dawn of Prohibition. Psychiatrists, neurologists, and psychologists began to apply psychoanalysis to the drunkard's case and to interpret men's and women's inebriety in terms of latent homosexuality. Freud's theories of sexual development were applied to the alcoholic. Consider this analysis by neurologist L. Pierce Clark:

> If we start from the fact that there is a bisexuality inherent in everyone, the homosexual component in emotional development must manifest itself even in adult life. If this can not be shown openly then masks and symbols must be used. ... Can it be merely chance that men so much enjoy being among themselves and drinking together, sometimes roughly, sometimes in more refined manner...? Gambrinus and Bacchus are the gods and guardians of alcoholic masculinity. It was no chance or fancy that made male deities the patrons of this particular custom. Gambrinus and Bacchus themselves are only symbols and objectivations of the homosexual.... Every drinking bout has a touch of homosexuality.[55]

Nor was the psychoanalytic view that alcoholism was rooted in latent homosexuality confined to men. Clark discussed women's alcoholism in terms of the "New Woman's" entry into areas formerly restricted to men.[56] He asserted that the crumbling of social barriers, the opening of new occupations, and the general desire to do what men do could explain part of women's growing interest in drinking. For most of the nineteenth century, social etiquette frowned on women's consumption of alcohol. In this earlier period, claimed Clark, "the formerly unjustly laughed-at social tea was the sublimated expression of feminine homosexuality." But by 1919, alcohol had proved more appealing than tea: "The virile component of women is stirred today and this helps to explain women's increased turning to alcohol. The more the virile works itself out, the more these expressions and symbols will be required.... Women who have a strong desire for liquor are likely to prove homosexual."[57]

Unconscious homosexuality was not the only psychological frame that psychiatry, neurology, and psychology placed on alcoholism. Sadomasochism, masochism, autoeroticism, and mother-fixation were likewise invoked by healers of the mind as the underlying causes of alcoholism. If neurologist and psychoanalyst William A. White observed that there were as many types of alcoholic as there were reasons for drinking, this did not prevent him and other physicians and psychologists at the turn of the century from viewing alcoholism as a form of flight—an effort to escape from circumstances that were troubling to the alcoholic and to which he could not adjust. These same physicians viewed some cases of alcoholism as a *symptom* of one or more underlying psychopathologies: "It must be borne in mind that indulgence in alcohol is oftentimes the expression of a neurosis or psychosis. For example, the recurrent attacks of manic-depressive psychosis may be ushered in by alcoholic indulgence, and if one is not keenly observant he may easily suspect that he is dealing with an alcoholic psychosis rather than with a manic-depressive."[58] Thus, while they acknowledged the devastation wrought by the habit of chronic and excessive alcohol consumption, and while they referred to alcoholism as a disease, psychoanalysts and psychologists often viewed alcoholism as a disease secondary to a primary neurosis or psychosis.

Finally, by no means should we assume that the influence of psychoanalysis and psychology on the definition and treatment of alcoholism was an exclusively secular and medical endeavor, devoid of moral or religious influence. The Emmanuel movement furnished some of the alcoholic's greatest friends, counselors, and healers. Begun in Boston, Massachusetts, in 1906 at Rev. Elwood Worcester's Emmanuel Episcopal Church, the Emmanuel movement brought together Protestant clergy, psychiatrists, neurologists, and psychologists to treat nonorganic or functional diseases of the body and mind. Although relatively short-lived, Emmanuelism spread across the country.[59] A blend of nondenominational Christian wisdom, pragmatism, and psychoanalysis, the movement was an experiment in public health that sparked controversy in the psychological, the medical, and the clerical professions. As fate would have it, Emmanuelism was also one of the greatest boons to psychotherapists nationwide, for it popularized psychotherapy as the medical profession had not. Indeed, it forced the medical profession to come to terms with psychoanalysis and its utility in treating both organic and functional disease. Up until 1908, American medicine, dominated by what William James called "the materialistic or mechanistic-deterministic enemy," had all but ignored psychoanalysis.[60]

The alcoholic was well served by this pastoral brand of psychoanalysis. At the Emmanuel "mother church" in Boston, one parishioner, a businessman named Ernest Jacoby, established a club for reforming and reformed alcoholics. "A club for men to help themselves by helping others," the Jacoby Club was an organization for alcoholic congregation members. According to Emmanuel historian John Greene, "it was believed that through the assistance rendered by such a club it would be easier for the leaders [of the Emmanuel psychotherapy and education classes] to prevent victims of alcohol from relapsing after the conclusion of the treatment and before they had acquired complete self reliance."[61] The Jacoby Club not only assisted the Emmanuel Church, but it collaborated with the Norfolk State Hospital (formerly the Massachusetts Hospital for Dipsomaniacs and Inebriates) by referring alcoholics to Norfolk and helping with the aftercare of those released from the institution. It specifically addressed "the most necessary steps and acts in effecting a recovery from moral illness or bad habits," and it maintained that "the first thing necessary was the formation of such a moral character as will resist disease."[62] Alcoholism was just such a medico-moral condition, and one whose definition and treatment reflected a variety of contemporaneous concerns about social norms and values.

## Female Inebriety

Early-twentieth-century psychoanalytic studies of alcoholism were not alone in their focus on sexuality. Nowhere were physicians' concerns with social and gender norms more evident than in their discussions of women who had become habitual drunkards. Throughout the Gilded Age and Progressive Era, the terminology used to describe the disease of habitual drunkenness was much the same for men and women. Physicians described women with drinking problems as "intemperate," "inebriate," "dipsomaniac," and "alcoholic." However, a woman who drank to excess defied social expectations, and physicians regarded her as significantly more deviant than her inebriated brethren.[63] Two images dominated the public's perception of the relationship between women and alcohol: women who were the victims of their husband's, father's, or brother's drinking; and the WCTU's women in white who campaigned against the liquor traffic under the "home protection" banner. Habitually drunk women, on the other hand, represented a third, less common and distinctly more stigmatized image—one that was synonymous with loose morals, sexual promiscuity, and ruined motherhood.

Estimates of the ratio of male to female drunkards in the United States during this period range from 3:1 to 9:1.[64] There seems little doubt that women drank less than men, although among certain ethnic groups—the Irish, in particular—the numbers for alcohol-related deaths were closer for men and women.[65] What female drunkards lacked in numbers, however, they possessed in ignominy. The words of Horatio Storer, a professor of obstetrics and medical jurisprudence at Berkshire Medical College, offer a sense of how disturbing habitual drunkenness was for Gilded Age physicians and for middle-class society generally when it affected the "fairer sex": "A debauched woman is always, everywhere, a more terrible object to behold than a brutish man. We look to see them a little nearer to the angels than ourselves, and so their fall seems greater."[66]

Storer's sentiments, written in 1867, were echoed thirty-two years later in a description of women's drinking problems appearing in *Catholic World*. There, M. E. J. Kelley argued that because American women possessed superior moral sensibilities, "drunkenness is so much worse for a woman than for a man."[67] As late as 1919, the municipal-court physicians of Boston argued that addiction to alcohol in women "means a further step downward than it does with men." For this reason, they continued, "we should expect to find among women taken up from the streets and from cafes because of drunkenness a larger percentage of mental defect and disorder."[68] Routinely, physicians described women with drinking problems as having grosser pathology than men who drank repeatedly to excess.

Physicians' assumption of more severe pathology in the female drunkard than her male counterpart was closely related to her breach of social role. Popular nineteenth-century definitions of womanhood that regarded females as keepers of the home, purveyors of morality, and pedestaled vessels of virtue helped shape the perception of the woman who drank to excess.[69] The growth of industrial capitalist society in the early decades of the nineteenth century was accompanied by a transformation of men's and women's roles, particularly within the middle class. Historians have called this division of labor and influence "separate spheres." According to this argument, men took possession of the public realm outside the home, for increasingly they left the home to earn a living in the marketplace. Women, in turn, gained authority over the domestic realm, and they made the home a sanctuary from the outside world and tended to the moral and spiritual growth of the family. Over the years, historians have revealed the many ways in which this neat division was transgressed by both men and women, but the fact remains that the first half of the nineteenth century witnessed significant changes in men's and women's social roles. Within this context, the "Cult of True Womanhood" proclaimed women's innate moral superiority.

The female inebriate violated her socially prescribed (and proscribed) role by leaving the domestic sphere to enter the manly, homosocial world of drinking—not to reform men, but to indulge herself. It was not just the violation of gender roles that made the female inebriate a disgraced and pitiable figure in the eyes of physicians. It was her adoption of a particular style of masculinity—dissolute rather than self-disciplined—that disturbed the doctors and the public. "Drunkenness, prostitution, and lawlessness of all forms in women are unmistakable signs of disease and early dissolution," claimed Thomas Crothers.[70] Faced with the aberrant spectacle of the inebriated woman, physicians concluded that her case was more severe than that of the male drunk. After all, they reasoned, no respectable middle-class woman would drink to excess; indeed, most would not drink at all. Thus, to the turn-of-the-century physician it appeared that only the most severe pathology could account for women's deviance.

What was the nature of this pathology? Throughout the late nineteenth century and into the first decades of the twentieth, explanations for heavy drinking in women often focused on their unique and cyclical physiology. At once the basis for woman's biological and social roles and the source of her most frequent medical complaints, the female reproductive system was said to be intimately linked to her nervous system. Turn-of-the-century physicians deemed the nervous systems of women finer, more irritable, and more prone to overstimulation and exhaustion than men's.[71] Likewise, the cyclical nature of women's reproductive systems conspired to make them more prone than men to dipsomania, or deranged periodic cravings for alcohol, claimed physicians, some of whom argued pessimistically that women were more likely than men to fall swiftly and irreparably down the slippery degenerative slope from inebriety to prostitution or insanity. In 1884 the *Quarterly Journal of Inebriety* reported: "In certain times of life women find themselves in physical and moral conditions, which seem to demand relief. . . . Dipsomania appears often at puberty and at the time of menstrual troubles, or at the menopause, or at the approach of old age at the decline of life."[72]

If women's physiology cut across class lines, class still remained a powerful diagnostic aid. During the 1870s, 1880s, and 1890s, middle- and upper-class women with drinking problems were often labeled dipsomaniacs, in contrast to working-class women "drunks," whom Thomas Crothers referred to as "the mere wreckage of worn-out foreign families far down the road to extinction."[73] Here the comparison with kleptomania is instructive. Dipsomania, like kleptomania, was a diagnosis of the middle and upper classes. Both supposed especially unstable nervous and reproductive systems in women of society's upper ranks, individuals for whom drunkenness or theft did not make intuitive sense to Victorian psychiatrists and neurologists. Why, after all, would women who lived in very comfortable circumstances wish to steal or drink? By defining such violations of respectable womanhood as disease, physicians both upheld turn-of-the-century notions of class and gender and protected women of their social strata from the legal repercussions faced by society's less privileged ranks.

Yet physicians and social workers did not invariably link a woman's class to her physiology. Other class-related explanations for women's drinking problems abounded in the Gilded Age and Progressive Era—ones that paid more attention to environmental circumstances. Among the middle and upper classes, noted one inebriate asylum matron, "the life of drift and lack of interest in anything practical" might easily lead a woman to the diversion of alcohol.[74] Her colleague, a physician, concurred. Heywood Smith, a gynecologist, proposed that the "new custom" of drinking champagne and wine with dinner represented an equally dangerous element in the daily lives of well-to-do women.[75] Eau de cologne drinking also figured among society's affluent ranks, according to a *Quarterly Journal of Inebriety* article: "The American inebriate, if a man, is not likely to use this perfume very long as a drink, but if a woman, it may be taken for years in secret. Obscure and complex nervous disorders in a woman that uses cologne externally should always suggest the possibility of its internal use."[76] Cologne offered one of several options for women wishing to keep their drinking hidden.

M. E. J. Kelley saw women at risk no matter where they stood on the social ladder. In 1899, Kelley perceived that habitual drunkenness was growing among "the very rich, who devote themselves almost entirely to the amusements of society, the theatre, the dinner party, the ball, the afternoon tea, the charity entertainment, and the host of other wearying activities which make up the daily routine of the society woman; the middle-class women, who live in comparative ease and comfort, but whose lives are monotonous; and the poor, who live in tenements in large cities."[77]

According to most physicians, working women stood in double jeopardy— at once threatening their health by taking on work outside the home and thus overburdening their more worn-out and pathology-prone constitutions, and by mixing with an often "unsavory" crowd at mills, factories, department stores, bars, and other public places. Indeed, more than one commentator singled out the mills for their toxic influence on the young women who flocked to them for employment. Lucy M. Hall, the physician-in-charge at the Reformatory Prison for Women in Sherborn, Massachusetts, lamented in 1883 that "the great factories of the State, are the *foci* in which drinking and dissolute habits are formed. . . . Nowhere is there an active work of reform more needed than among the mill population of our State."[78]

For Hall and for others, the migration of thousands of single young women (and men) from the country to the city for employment posed a threat to their morals and their health. Physicians and others often romanticized rural village life, where they believed strict standards of morality obtained, standards that fell by the wayside once the young women adapted to the looser urban lifestyle of their new companions. "A striking instance of this process of evolution was furnished by two sisters who came from a little place in Western Pennsylvania to do housework in New York," offered the ever-observant Kelley: "They brought with them some strong convictions that it was shocking for women to drink. In a year they had reached the stage where they kept a bottle of whisky and some quaint little glasses in their own room to treat their friends who came to see them. They regarded the country visitors who had known them in their old home, and who pointed out the dangers of the practice, as 'awfully countrified and bigoted.'"[79]

The transition from rural community to urban society figured regularly in discussions of young working women's drinking problems. Relatedly, physicians pointed to women's drinking as one of the disastrous consequences of their increasing independence in modern society. They reasoned that as women assumed the rights and responsibilities of citizenship formerly confined to men, they abused their newfound freedoms and frequently overtaxed their nervous systems.[80] Still, other important medical voices claimed that only through such independence could society evolve and become more sober. "The emancipation of women from the slavery of caste and ignorance, and the steady upward movement in mental and physical development, will prevent the general increase of alcoholism or inebriety," argued Thomas Crothers in 1892. Women, he thought, were "far more sensitive than men to the evolutions and revolutions of daily life. The constant educational forces of travel, of lectures, of the theatre, of literary societies, of churches and reform movements, of public schools, and the possibility of leadership and prominence in many directions, all lead away from alcoholism."[81] It was an interesting line of reasoning, hinging upon Crothers's belief that women's exquisitely sensitive nervous systems would promote their evolution away from inebriety. Although men had for some time enjoyed all of the privileges Crothers listed, their coarser nervous systems had not allowed them to evolve toward sobriety.

Regardless of their opinions on the pernicious influence of urban culture or positive influence of women's expanding role in public life, most medical experts believed that the female drunkard courted biological and social disaster. Turn-of-the-century physicians asserted that a woman's drinking made her "useless" toward fulfilling her domestic duties and likely to harm the health of any children she conceived and carried in her toxic womb or nursed from her besotted bosom.[82] Indeed, "race suicide" was a common theme in much of the literature on women's drinking between 1870 and 1910. Brooklyn physician Agnes Sparks, addressing the Medico-Legal Society of New York in 1894, considered alcoholic women the source of "a diseased, depraved progeny that tends to curse every community with a physical and moral blight, the extent of which is beyond compete, and for which no other agent for ill can compare."[83] The female inebriate jeopardized the nation's germ plasm, and thus the nation's future.

Though the American medical profession's concern for the female inebriate never reached the fevered pitch that it did in Britain, physicians addressed these troubling cases of deviant behavior and pathology in private homes for nervous invalids, in state hospitals for the insane, in state inebriate hospitals, and in private mental facilities.[84] In each of these venues, physicians also learned that they were in part responsible for the problem. Listing the chief causes for women's drinking problems in 1901, physician Heywood Smith lamented, "I am sorry that I cannot exclude the not unfrequent carelessness of our own profession, for medical men are too apt, should they deem it necessary to prescribe alcohol, to say, 'Oh, take a little wine,' leaving it to the patient to determine the dose."[85] In 1887, C. W. Earle offered additional revelations: "If I make the statement that two or three in each hundred in our profession use alcohol or opium to excess, and there are those who believe that an estimate of ten per cent would not be an exaggeration, I must also say . . . that a considerable number of the non-medical people who succumb to the seductive influence of alcohol and opium took their first dose from a physician.[86] Between 1850 and 1890, alcohol was an essential element of the stimulative therapy employed by physicians and at hospitals across the United States. "Whiskey" and "brandy" remained in the *Pharmacopoeia of the United States of America* until 1916, when they were dropped from the list of standard drugs.[87] Popular in Britain and France as well as the United States, alcohol was prescribed for much of the nineteenth century as a heart stimulant, as a nutritive tonic, as a nervous depressant, as an anti-pyretic, in cases of snake-bite, as a stimulant in breast-feeding, in combating the germs of infectious disease, and for the relief of pain, especially menstrual pain. Its widespread use provided ample opportunity for iatrogenic addiction, that is, addiction that had its origins in medical treatment or a doctor's prescription.

The censure that women risked should they reveal their dependence on alcohol deterred many—particularly among the middle and upper classes—from seeking help. These women often hid their addiction to drink, and it was relatively easy to do. A variety of household items and proprietary remedies available through the druggist or grocery store, or by mail order, contained large amounts of alcohol. Proprietary remedies and tonics such as Hood's Sarsaparilla Cures, Howe's Arabian Tonic, Scotch Oats Essence, and Hostetter's Stomach Bitters contained as much as 44 percent alcohol. Lydia Pinkham's Vegetable Compound, aimed specifically at women and their "female complaints," contained 20 percent alcohol.[88] Cologne and household cleaning solutions—other "discreet" ways to consume—often contained methyl alcohol. Their consumption, of course, placed women at increased risk of severe neurological problems. As a result, when women alcoholics finally did fall under the clinical gaze, their condition was frequently more advanced than it was for alcoholic men, whose drinking in public places was more likely to attract early judicial if not medical attention. So physicians who portrayed the female inebriate as more impaired than her besotted brothers could have been correct, if not necessarily for the reasons they proposed.

The irony was that many women of the middle and upper classes who could afford medical care felt they could not afford to breach the social mores of the day, and this prevented them from seeking help. Working-class women, for whom drinking appears to have been less stigmatizing, rarely had access to the few sanatoriums and private homes for nervous invalids that treated inebriety. Public inebriate facilities, few and far between, usually didn't admit women. So most working-class women were either jailed for their condition, if caught in public, or sent to state or municipal insane hospitals to dry out—

both remedies that physicians abhorred. In the end, Gilded Age and Progressive Era society had little to offer the turn-of-the-century female alcoholic besides disgust and denial. However, the public image of the female inebriate served many political purposes. Temperance leaders incorporated her into their morality tales to portray the ravages of physician-prescribed alcohol and men's drinking that led to their wives' and daughters' addiction.[89] The WCTU and muckrakers alike pushed forward the Pure Food and Drug Act of 1906 with reference to women who were the victims of patent medicines and their undisclosed contents.[90] Physicians advanced their eugenic concerns, reminding Americans of the dangers of race suicide through women's habitual drunkenness. Reform-minded physicians, social commentators, and charity workers alike decried the state of the workplace in mills and the alcohol-associated leisure activities of the young who flocked to the cities for employment.[91] In short, the deviant image of the female inebriate proved a powerful aid in launching some significant reforms—from workplace safety to the purity of drugs—even if alcoholic women themselves received little help.

## Professional Identity: The Scientific Clinician and Healer of the Body Politic

By redefining habitual drunkenness as a disease in both men and women, physicians staked a claim to social and cultural authority over a resilient social problem; they asserted their professional expertise on a matter that had long been the domain of the church, the court system, and the temperance movement. Doctors' claims revealed much about the medical profession's aspirations at the century's end, particularly with regard to its changing role in matters of public health. Who or what was to blame for the inebriate's diseased state also determined which party should take responsibility for his cure. This was made plain in a letter written in 1895 to the editor of the *Journal of the American Medical Association* (*JAMA*)—a letter in which W. P. Howle, a physician from Oran Scott, Missouri, sarcastically objected to *JAMA*'s endorsement of medical care for the inebriate: "I think the editor is striking out in the proper direction. If we can convince the people that the inebriate belongs exclusively to us we will soon have plenty of practice and money, too. The inebriate has paid the legal profession and courts good wages all the time, and if we can turn this great source of wealth into its proper channel it will be a blessing to the profession."[92] Though Howle may have thought the medical profession's interest in treating the inebriate unwise, he was very much aware of the professional tensions raised as doctors addressed this perennial problem. Indeed, the literature addressing intemperance, dipsomania, inebriety, and alcoholism as a disease also focused on a range of professional concerns. Among these were the responsibility of the physician and the scientist in healing and assisting the body politic, and relatedly, alcoholism's status as a public health problem and eugenics issue.

As they defined inebriety as a disease, physicians revealed a great deal about their general conceptions of health and illness at a transitional time in the history of American medicine. The Gilded Age and the Progressive Era witnessed tremendous change within the medical profession. Medicine was becoming "scientific." The discoveries taking place within bacteriology and physiology laboratories—including the germ theory of disease, an effective range of anesthetics, antisepsis and asepsis, the creation of antitoxins and other specific chemotherapeutic agents, and the discovery of X-rays—promoted an optimistic, reductionistic, and expansive vision of medicine's capabilities. The rediscovery of Mendelian genetics likewise fueled this materialist perspective and the burgeoning sense that scientific and medical knowledge might be capable of redirecting human progress. Mid-nineteenth-century sectarian fighting among allopaths, homeopaths, and assorted others gave way to the consolidation of regular medical authority at the century's end. Laboratory-based medicine and public health revealed their mighty powers through triumphal events such as the manufacture of diphtheria antitoxin and the epidemiological sleuthing that discovered "Typhoid Mary" and led to her long-term quarantine.[93] These discoveries and developments allowed physicians to achieve unprecedented progress in helping their patients and in improving their own professional status. Physicians were poised to expand their medical reach into the social arena and offer their expertise on a myriad of social problems and issues that seemed to demand the scientist's attention. The problem of inebriety was such an issue.

Their willingness to incorporate the moral with the medical did not prevent inebriety physicians from framing their work in the most scientific terms. Willard Parker's distinction between temperance societies and the AACI was an important one to the new association's membership. As the AACI's founding president, Willard argued that its scientific orientation set his organization apart from the fevered pitch of temperance agitators, whose impassioned pleas for sobriety drew upon drama, emotionalism, and hyperbole—the very opposite of what he and other physicians understood as "Science." Joseph Parrish, who delivered one of the first addresses at the group's inaugural meeting in 1870, saw the scientific approach as a third and more promising path than either the temperance movement's moral censure or the judicial system's punishment. Parrish believed that it was time to introduce an "objective," scientific perspective into an issue dominated by "moralists and jurists" who had treated the topic "as a social offence" but had not gone behind its "visible manifestations," to seek "absolute and primary causes."[94]

AACI physicians and other alienists and neurologists condemned the courts and legislators who supported punishment for habitual drunkenness. The law in most states held inebriates responsible for their condition, refusing to recognize the drunkard's inability to control his actions while under the influence. Fines and short jail sentences designed to stop the use of spirits only made the problem worse claimed Thomas Crothers in 1891, by "intensify[ing] the conditions that impel the drunkard to drink."[95] Fines levied against inebriates often affected individuals and families who were already in financial straits due to the expense of drinking. Jailing inebriates with petty criminals whose bad habits might rub off on them and keeping them locked up with little or no medical support—these measures were ineffectual at best, physicians maintained. At worst, they hastened the drunkard's degeneration and his family's descent into penury.

As they invoked science to support their claims, the physicians who wrote about inebriety and treated inebriates in their practices referred to the inebriate asylum as their laboratory. The "scientific method" employed in the "asylum-laboratory" was Baconian, a sort of clinical induction. Crothers described this process in 1891 in the *North American Review:*

> The scientific method to be pursued in this study is the same as in all other physical problems. First, gather and tabulate the histories of a large number of inebriates; then make comparative studies of these records, and ascertain what facts, if any, are common to all of them. . . . The more exhaustive these facts are, the more accurate the conclusions. From a grouping of a large number of such histories a startling uniformity in the causation, development, and termination appears.[96]

The asylum was not the only laboratory to which the inebriety doctors turned. In promoting their scientific approach to inebriety, physicians frequently recited the findings of European and American physiologists and alienists—works that focused on alcohol's nourishing, debilitating, stimulating, and narcotizing properties.[97] Understanding the nature of alcohol was an important preliminary step to discerning its effects on humankind. Broadcasting this knowledge of alcohol's physiological and psychological effects was at once an effort to educate the public, curtail the use of liquor, and confirm physicians' professional expertise.

Realizing that they faced a skeptical audience that was often quick to condemn the drunkard, the early physicians who promoted the disease concept portrayed themselves as the persecuted voice of truth. In not unflattering terms, they compared their research on alcoholism to the struggles of Galileo and Darwin, scientists who had brought forth heretical—not to mention revolutionary—findings in spite of the church's censure. Crothers pleaded with readers to "lay aside all theories of religious teachers and reformers, and examine inebriety from the side of an exact science. . . . Like a problem concerning the stars above us, our only approach to its solution is along the line of accurately observed facts." In the end, science, not religion, revealed the truth, argued Crothers, who also reminded readers that the widely held view that inebriety was a moral failing, a spiritual disorder cured by conversion, was "a striking repetition of history in the efforts to treat insanity as a moral depravity and possession of the devil."[98] The message was clear: doctors, as medical scientists, had a right to intervene in affairs of state; and inebriety was a major national concern, for doctors framed it not only as a clinical problem but also as an important matter of public health.

As they wrote about dipsomania, inebriety, and alcoholism, physicians also portrayed their work as a grand public-works project, one the state had a moral obligation to support because most states reaped the rewards of liquor taxes. The physicians who treated inebriates maintained that they were not just healers of individual drunks. In the late nineteenth century, they saw themselves as part of an ongoing effort to nurse the nation back from the ravages of the Civil War. Indeed, numerous physicians and popular authors drew the analogy between habitual drunkenness and slavery, both evils to be eradicated in modern society.[99] By promoting the idea of inebriety as a public health problem, physicians invited the state to participate in containing this medico-moral contagion. Indeed, the AACI's founding principles urged states and large cities to build inebriate asylums and hospitals where the sick might be "quarantined" until well and no longer likely to spread the vicious disease of alcoholism.

This was a clever, if not wholly successful, strategy; some, but not many, municipalities and states did respond to the repeated demands of reformers—moral, legal, and medical—to create specialized facilities for inebriates with physicians at the helm. More states and cities simply put inebriate laws on the books that permitted the "quarantining" of inebriates in existing mental health and reformatory institutions, pending the recommendation of the court and the certification of physicians. Still, by casting habitual drunkenness as both a disease and a public health problem, doctors scripted a state-sanctioned role for themselves within which they would have the power to diagnose, help confine, and cure habitual drunkards.

One of the most forceful images of inebriety as a public health problem came from Thomas Crothers, never at a loss for words when it came to his favorite cause. Speaking before the American Social Science Association in 1883, Crothers observed that "inebriety moves in waves and currents, or, like an epidemic, prevailing for a time with great activity, then dying away; both endemic and epidemic at times.... Inebriety can be diminished and controlled with the same certainty as smallpox, or any contagious fever. The principle is the same, viz., to remove the causes, and quarantine the victim, in the best possible conditions for returning health." Crothers recommended committing inebriates of all classes to hospitals in lieu of sentencing them to jail or the insane asylum. Lest concerns be raised about the civil rights of the habitual drunkard, Crothers insisted that it was essential "to recognize the fact, that the inebriate, whether continuous or periodic, has, to a greater or less degree, forfeited his personal liberty, and become a public nuisance, and a great obstacle to all social progress and civilization." After all, he remarked, the inebriate's condition affected not just that individual, but every member of the community, weighed down by the drinker's lack of productivity and inability to fulfill his social role. "It is simply carrying out the highest principle of self-preservation, to take care of this class, and thus protect them and the community in which they live."[100] And yet in most cases the inebriate was not insane, was not a criminal apart from his public drinking, and not infectious in a literal sense. To demand the confinement of inebriates for indefinite periods of time was to enforce a remarkably coercive policy. As we will see, few states were willing to support such broad powers for physicians treating inebriates.

The 1880s were pivotal years for public health and the control of infectious diseases. Between 1880 and 1890, laboratory scientists isolated the microorganisms responsible for typhoid fever, leprosy, malaria, tuberculosis, cholera, and diphtheria. With the discovery of the germs that caused the century's major life-threatening illnesses, the analogy between alcoholism and infectious disease became more symbolic than real, but the public health framework held fast. Physicians were not quick to abandon the metaphor of contagious disease. Instead, they expanded their repertoire and discussed new links between "gin and germs," urging that the giddiness associated with the discovery of germs not distract medicine from the importance of the "soil as well as the seed." As J. W. Grosvenor reminded members of the American Academy of Medicine in 1897: "The zeal and persistency of our boards of health in hunting for the ubiquitous microbe call forth our admiration. It is suggested that they spend a larger portion of their time than is used at present in searching for those conditions of the body in which micro-organisms flourish most luxuriantly.... Oftentimes it will be found that these conditions are caused by the alcoholic drinking of the individual or the ancestral alcoholism.... It is part of wisdom to look for the *cause* of a cause."[101]

From the late 1880s on, inebriety experts highlighted the specific damage alcohol caused to drinkers and those around them. "It perverts digestion. It depresses and weakens the heart action. It decreases the capacity to do muscular work," warned physician Frederick Peterson, president of New York State's Craig Colony for Epileptics, in 1909. "It brings about slow far-reaching anatomical changes, such as fatty degeneration of the heart, kidney disease, diseases of the blood vessels," he continued. "Its habitual use lessens the normal defenses of the organism against infectious diseases, especially tuberculosis."[102]

Physicians devoted special attention to the relationship between alcohol and tuberculosis.[103] Reminiscent of Grosvenor's remarks above, physician Henry Smith Williams warned *McClure's Magazine* readers that alcohol paved the way for tuberculosis by "impairing the tissues" so that they easily succumbed to the bacilli.[104] Philadelphia physician Thomas Mays wrote that his experience at the Clinic for the Home Treatment of Consumption had taught him that alcoholism and consumption were "allied to each other as cause and effect," and that the latter was frequently "the indirect product of the pernicious influence of alcohol on the nervous system."[105] A great debate ensued between the temperance movement and the medical profession over the exact nature of alcohol's physiological effects. Was liquor a stimulant or depressant? A food or a poison? Medical friend or foe?[106]

Usually medical journals, whether general, psychiatric, or concerned specifically with inebriety, portrayed the alcoholic as irresponsible and potentially abusive; inebriety was seen as a well-trod path to poverty and to the pathologies that accompanied lives without good habits, wholesome food, and a decent place to live. This gave public health claims a greater sense of urgency, for doctors argued that treating alcoholism kept drinkers from joining the ranks of the chronically unemployed and insane, and it preserved drinkers' families from requiring poor relief. Nor was it just physicians who made this argument. It found great support within the rising profession of social work. Preventive medicine and psychoanalysis gave this helping profession specialized bodies of knowledge that furthered its own claims to professional status, and alcoholism fit within both of these sociomedical contexts.

No one felt the necessity of addressing alcoholism more than Homer Folks, secretary of the New York State Charities Aid Association, who wrote: "If we mean what we say when we talk about controlling the preventable causes of poverty; if we are really in earnest in our campaigns for the prevention of tuberculosis and insanity; if we have not lost all vital faith in remedial measures; if we are, socially, progressives, not to say, insurgents; we must look squarely in the face the relation of alcoholism to our various problems."[107] Folks's 1910 rallying cry drew the causal arrows from inebriety to a multitude of community health problems, but others felt that the question of which came first, alcoholism or poverty, was moot.

Certainly, poverty and the "discouragement and despondency" that resulted from it led many to the saloon to forget their woes, acknowledged preventive medicine and public health advocate J. W. Grosvenor.[108] He and other public health reformers, social workers, and inebriety experts argued that for this reason there was still a place for the sanitary reforms that so characterized the early public health movement. With the discovery of the germs that caused infectious disease, dirt and squalor might no longer be the scourge of civilization; the miasmatic theory of disease had been laid to rest. Yet a psychological miasma still permeated the tenement housing of the cities, plaguing the spirits of men and women until they sought solace in alcohol.

In 1917, social work pioneer Mary Richmond urged that with each inebriate case, one must always ask: "Are the home conditions such as to incline him to seek the saloon as more cheerful? Is the home situated in vicinity of saloons? Is it squalid and in disorder? Does he take his meals at home? If so, are they well cooked?"[109] Richmond's questions revealed as much about the gendered assumptions of social workers caring for inebriate cases as they did about the living circumstances that were thought to contribute to drinking. For both physicians and social workers, the archetypal inebriate was male. Doctors and social workers frequently assigned blame for men's drinking to their wives, who failed to keep house, nagged too much, or didn't show their husbands proper deference. If the WCTU urged Prohibition for the sake of "home pro-tection," social workers and physicians believed in rescuing the distressed male drinker from his wife's potentially corrosive personality and poor house-keeping habits.

Relations between men and women were indeed problematic in turn-of-the-century discussions of inebriety. This was especially true when it came to the issue of sex and procreation. The habitual drunkard was not just a victim of his or her own drinking; the alcoholic's children suffered too, for they were believed to inherit a neuropathic constitution. Recall the degeneration theory of Benedict-Augustin Morel. In the words of British psychiatrist Henry Maud-sley, the alcoholic's descendents were the "step-children of nature . . . [who] groan under the worst of tyrannies—the tyranny of a bad organization."[110] The physicians who promoted the disease concept of inebriety expressed their hereditarian concerns as a way of galvanizing support for their cause and driv-ing home the potency of this threat to the public's health.

Consider the words of New York City physician Stephen Rogers, read before the Medico-Legal Society of New York in 1869: "It is a grave and most inter-esting fact, not in this instance only, unfortunately, that the parent who destroys the organization of his nervous system by alcoholic excesses, is ex-ceedingly liable to transmit to his offspring disordered nervous systems, which become manifest in almost any form diseases take."[111] Rogers joined others, such as the superintendent of St. Elizabeth's Hospital for the Insane, W. W. Godding, who believed that society demanded protection "against that race deterioration—those inherited neuroses, chorea, epilepsy, idiocy and insan-ity—of offspring begotten in a debauch."[112] To Godding's mind, the social compact between the individual and the state was in the nature of a "public trust." The inebriate broke this trust by losing his self-control and discarding the responsibilities of citizenship. He became, as an AMA subcommittee on inebriety put it, "a public nuisance to be abated."[113]

As early as 1887, doctors advocated a type of "hereditary containment" pol-icy for habitual drunkards. Writing in the *Alienist and Neurologist*, L. W. Baker reasoned that by caring for inebriates in state-run inebriate hospitals, the state would ensure that the drinker would no longer be able to injure himself, his family, or society through alcohol. In the inebriate hospital, Baker hoped, the alcoholic would have a chance to recover, and if he were deemed incurable, he would be isolated in his "permanent asylum home," thus "prevent[ing] the transmission of the insane or inebriate diathesis to a succeeding genera-tion."[114] President of the Indiana Medical Society Gonzalva Smythe echoed Baker's eugenic concerns, observing in 1891: "Most rapid advances have been made recently in the treatment of the insane, the epileptic and the inebriate .... They are correctly regarded as sick with a physical disease and treated ac-cordingly—kindly and humanely. This course should be continued ... but no license to marry should ever be issued, and every means necessary to prevent reproduction should be vigorously enforced."[115] Indiana would become one of the greatest champions of eugenics, or the science of good breeding, in the early twentieth century, enacting some of the nation's most rigorous steriliza-tion laws as the germ theory gave way to the germ plasm and eugenics became part of the public health movement.[116]

Perhaps the most widely publicized eugenics study of alcoholism was *The Nam Family.*[117] The turn of the century marked the heyday of family studies in eugenics: The Jukes (1888), the Kallikaks (1908), and the Ishmaelites of Indi-ana (1888) became household words by the 1900s, synonymous with heredi-tary and social degeneration. The Nam Family, although never as popular as the Jukes or the Kallikaks, was one of several clans that eugenicists identified as afflicted with bad heredity, or as neuropsychiatrist E. E. Southard put it, cursed by "cacogenics." The fundamental traits of the Nam Family, as reported in 1912 by Arthur Estabrook and Charles Davenport, were their "alcoholism and lack of ambition." The Nams, it seemed, had descended from "a roving Dutchman, who had wandered" into western Massachusetts in the early eigh-teenth century, only to settle down with "an Indian princess" and produce several children, known in their environs as "vagabonds, half farmers, half fishermen and hunters" and individuals who on occasional visits to nearby settlements "were apt to fall into temptation and rum."[118] Branches of the Nam family—the Naps, the Nars, and the Nats—settled in New York state, but the greatest concentration of Nams made New Hampshire their home, clus-tering in a village known as "Nam Hollow."

Estabrook and Davenport traced the histories of 784 descendents—Nams and Naps, Nars, and Nats—from the original Nam union, eclipsing Reverend Dugdale's record of 540 Jukes by about 45 percent. Although the Nams' social traits stacked up well in comparison to the Jukes'—less poverty, prostitution, and venereal disease—the prevalence of alcoholism and indolence was "extraordinarily high." Eugenics experts, Estabrook and Davenport acknowledged that they could not tease out the relative effects of environment and heredity in producing these two traits, but they maintained, naturally, that heredity played a heavy hand. Even the few Nams who escaped Nam Hollow to start new lives in Minnesota faired only marginally better, claimed the scientists, due to their "more ambitious make-up." Estabrook and Davenport were forced to conclude that alcoholism and indolence were "traits present in the germ-plasm, which plays a dominant part in the behavior and reactions of the individual."[119]

By way of "social prophylaxis," the Cold Spring Harbor eugenicists maintained that letting the Nam family take care of itself—that is, allowing the group to live their lives to extinction in Nam Hollow—was insufficient. The more ambitious, after all, would carry the Nams's alcoholic germ plasm to other parts of the country. Improving the environment in Nam Hollow through "trained nurses who should teach housekeeping and elements of hygiene," would have a positive, if insufficiently permanent effect: "Most of the work would be supplying a veneer of good manners to a punky social body." Likewise, breaking up Nam Hollow, and scattering its inhabitants in the hope that they might marry outside the clan, might multiply "the centers of indolence and alcoholism." The best solution, according to Estabrook and Davenport, was to detain in "State Villages for the non-social" those children whose family histories offered them little hope of becoming "parents of socially desirable stock." The scientists believed that detaining the young Nams through their reproductive years would meet with more public approval than sterilization.

Only the second memoir of the Eugenics Laboratory at Cold Spring Harbor, Estabrook and Davenport's study testified to the centrality of alcoholism within the hereditarian discourse of eugenics (as well as the weakness of the science behind the eugenics movement and the extreme measures advocated). Also striking is neuropsychiatrist E. E. Southard's association with the study, even if only to donate the term "cacogenics." Southard, who had charted his own geographical map of mental deficiency in Massachusetts between 1901 and 1910, was living testimony to the ability of psychiatrists to embrace both eugenics and psychodynamic psychiatry—heredity and experience. He and many of his colleagues diagnosed and treated the behavioral problems of everyday life and expanded their normative professional and social roles.[120]

Increasingly, psychiatry and psychology came to play important roles in public health. Indeed, the "New Public Health" of the early twentieth century emphasized preventive medicine, sanitary hygiene, and the modification of individual behavior through education and sometimes through more coercive measures such as the sterilization of the feeble-minded. The framing of inebriety followed suit, as did treatment strategies (discussed in chapters 3, 4, and 5).

Thus, at a time when the American medical profession was engaged in a campaign for professional legitimacy and authority, the physicians who promoted the disease concept of inebriety were charting a new course for doctors of the mind—that of social engineer. The inebriate needed medical treatment, they argued, and by providing it, physicians might stem the rising tide of insanity, poverty, and crime in America. Disease-concept advocates portrayed their work as a progressive, preventive, public health measure that broadened the scope of state medicine. An inebriate's physician needed to be both healer and educator, for both the patient and the public required re-education. As Edward Mann reminded readers of the *Journal of the American Medical Association* in 1894, the founding members of the AACI had envisioned the inebriate asylum as the school "in which knowledge is to be acquired for the enlightenment of the medical profession, the information of the public mind, and for the guidance to a better system of legal enactments upon the subject."[121] Irwin Neff, the superintendent of the Foxborough State Hospital (formerly the Massachusetts Hospital for Dipsomaniacs and Inebriates), highlighted the same duality of role when he told the Boston Society of Psychiatry and Neurology in 1910: "The medical profession is expected to take a leading part in any plan which is inaugurated for the prevention or amelioration of inebriety. Physicians should accept this obligation and discharge it to the best of their ability; it rightly belongs to medicine, as it is first of all a remedial measure, and secondly, educational along the lines of preventive medicine."[122] For Neff and for others, as we shall see, the inebriety physician's role did not end with healer and educator. It also included public administrator and system builder.

CHAPTER TWO

# Cultural Framing of Inebriety

The physicians who described habitual drunkenness in the Gilded Age and Progressive Era wrote about it in cultural as well as medical terms. They saw inebriety as a disease of modern civilization, much as George Miller Beard had described neurasthenia or modern nervousness. Modern industrial society's frenzied pace, urbanization, immigration, and changing gender roles all posed obstacles to sobriety, observed many physicians. Joseph Parrish condemned the frenzied pace at which Americans led their lives in 1870. This was no short-lived hustle and bustle, according to Parrish, but a "form of method, of business system" that had "infused an impetuous inspiration into the whole texture of society."[1] The struggle for wealth, power, and social position; unhealthy rivalries for the sake of display; and all the commotion of city life depleted people's nervous energy and prompted them to search for stimulants, Parrish claimed, and routinely they turned to alcohol.

Albert Day, superintendent of the Washingtonian Home in Boston, suggested in 1876 that the Civil War had exerted a similar effect. To Day's mind, the "constant excitement" to which the nation had been subjected between 1861 and 1865—"the recruiting of men for the navy and army, their thrilling deeds upon land and sea, their fatigues and exposures to the summer's heat and the winter's cold, the breaking up of families, the scattering of households, disruption of business relations, and the numerous other exciting influences that attend the pomp and circumstances of war"—had created an appetite for action and a demand for stimulants.[2]

Beard concurred, writing in the *Quarterly Journal of Inebriety* that same year that there were powerful reasons for the progress of civilization and the increase in nervous disease to run hand in hand. In words that echoed Parrish, Beard maintained that overexertion of the nerves in the quest for "wealth, fame and social standing" prompted society's better ranks to seek solace in alcoholic stimulants. Beard added the element of technology to his explanation for the apparent increase in inebriety and other nervous disability. Three inventions—the printing press, the steam engine, and the telegraph—had made it possible "to concentrate an enormous quantity and intense quality of work in a short time," claimed the New York neurologist.[3] Late-nineteenth-century Americans might reap the rewards of these technological marvels, Beard allowed, but they also bore the burdens of a society striving intently toward advancement in the arts, religion, politics, and science: "Diffusion of knowledge and of freedom are followed by diffusion of care and responsibility; in America everyman is king and bears the burden of the republic."[4]

If this fast-paced American society took a toll on its citizens, none of the medical commentators suggested that the pace would or should change, for in their eyes it was an inevitable price of progress. Indeed, they argued, the progress of technological civilization might contribute to inebriety, but only an advanced society recognized inebriety as a disease. "Medicine has given to the world the substantial basis of a new reformation. It has sounded the alarm; it offers the remedy, and, on this score, we think humanity is on the road to safety. This is one of the jewels we place in the crown of Nineteenth Century progress," offered St. Louis neurologist C. H. Hughes in 1894.[5] For physicians such as Hughes, it was up to the medical profession, public health experts, and social workers to engineer ways to educate the population about the nature of alcohol, to suggest alternatives to alcohol in reviving depleted constitutions, and to rehabilitate inebriates through medical and moral care. These physicians held that professional and scientific expertise might help society adjust to the waves of social, economic, technological, and political change that swept over the nation at the turn of the century.

In addition, there were a number of social Darwinists who suggested that alcohol would act as "a potent agent of elimination" and that in time weak-willed drunkards would be neatly eliminated from the nation's hereditary stock. These laissez-faire hereditarians, however, could not claim the majority voice within the medical discourse on inebriety and heredity. More frequent were the calls for negative eugenics measures such as vasectomy and the confinement of incorrigible alcoholics to prevent them from procreating. By the end of the nineteenth century, however, it was even more common to find voices such as that of New York City social worker Lillian Brandt, who believed that it was naive and simplistic to consider alcohol abuse a means to weed the unfit from the human race. After all, she told readers of *The Survey,* such an argument disregarded the fact that alcohol's destructive influences harm the strength of everyone, directly or indirectly. With "rational reform" as her watchword, Brandt observed that "society is not made up of two rigidly separable groups: 'the fit,' to be conserved, and 'the unfit,' to be eliminated." Wise social measures, she suggested, would not interfere with evolution. Instead, expert interventions might "determine the place above which evolution shall be allowed to go on by its own slow and costly methods, just as they may determine in business relations the plane above which competition may go on unrestricted."[6] For sociomedical reformers such as Brandt, these wise measures involved both the rehabilitation and the detention of inebriates to restore them to society's productive and employable ranks.

The relationship between industry and inebriety attracted a great deal of attention throughout the late nineteenth and early twentieth centuries.[7] Modern civilization's fast pace was in tempo with the advance of technology and the mechanization and routinization of the industrial workplace. Employers, in the name of "industrial efficiency," were more loath than ever to keep alcoholic workers on their payrolls in the closing decades of the nineteenth century. As many reformers were quick to note, the nature of industrial work had changed. According to Brandt in 1910, "Factory work, work on railways, office work, intellectual work," were all incompatible with drinking.[8] The expansion of lower white-collar employment in conformity-oriented corporations and the increasingly mechanized shop floor each problematized alcoholism within American business. Concerns about human and industrial efficiency peaked in the Progressive Era, as systemic management in industry gave way to the scientific managerial practices of Frederick Taylor.[9] "Now the quickness of the time-reaction is important to every mechanic," observed Elizabeth Tilton, social worker and prohibitionist in 1914.[10] P. A. Lovering, a medical inspector for the United States Navy, drew similar conclusions for the armed services. He observed that in the course of his three-decade-long career at sea, "the necessity for temperance has increased very greatly." The "huge floating machines of war" have grown in size and force and "require more intelligent and skillful handling" to remain effective.[11] Thus, the alcoholic's dulled thought processes and reflexes posed a threat, not just to the sailor, but to his ship's crew and the nation's security. Even if the flow of alcohol remained constant in society, civilization itself was changing, and drunkenness posed both an obstacle to industrial efficiency and to worker safety. By the end of the nineteenth century, as Roy Rosenzweig observed, "most employers tended to view workplace drinking as part of a bygone era."[12] A government survey of 30,000 employers in 1897 indicated that three-quarters took prospective employees' drinking habits into account when hiring new workers.[13]

Also embodied in the 1870s and 1880s image of the inebriate was an anti-industrial or anti-modern vision of the habitual drunkard as automaton. Albert Day described the inebriate in 1876 as an "automatic subject; a mere puppet to be pulled by suggesting wires, capable of being played upon by everyone who shall make himself master of the springs of action."[14] Day's description of the habitual drunkard echoed concerns of workers and artisans who had lost autonomy and been "de-skilled" within the expansive industrial workplace. On the shop floor, drinking might compromise a worker's safety and efficiency, but Day was concerned that alcohol made the drinker more, not less machinelike. Mechanization, whether literal or symbolic, posed a threat to individual autonomy in the late nineteenth century.

Day's use of the image of "man, the machine" to describe the inebriate's general loss of willpower was not idiosyncratic. Subsequent discussions of inebriety employed the automaton metaphor to address the inebriate's legal status, particularly in cases of "criminal inebriety," where individuals violated the law while intoxicated. Physicians in the last two decades of the nineteenth century and the first decade of the twentieth described an alcohol-induced state of mind that was divorced from consciousness as "alcoholic trance," "alcoholic somnambulism," "alcoholic hypnotism," "alcoholic cerebral automatism," and "double personality of alcoholism." Robert Louis Stevenson's *Dr. Jekyll and Mr. Hyde,* published in 1886, presented a popular treatment of the alcoholic alter ego at its most nefarious, with the evil Mr. Hyde appearing only after Dr. Jekyll has ingested a special potion. The two most prominent inebriety specialists in America and Britain, Thomas Crothers and Norman Kerr, maintained that American nervousness and the extremes of climate in the United States predisposed its citizens to alcoholic automatism more than drinkers in other nations.[15]

Regardless of whether the drinkers toiled in the mechanized workplace or resembled automatons themselves, work and the self-discipline necessary to retain employment and remain a productive member of society were central to the definition of problem drinking and its treatment throughout the late nineteenth and early twentieth centuries. As a pathology of consumption (the act of drinking), inebriety comprised a potent symbol of modern consumer society run amok. Although the origins of consumer society may be traced to the early nineteenth century, it was not until the Progressive Era that the national marketplace was truly consolidated, with the great department stores of cities and retail mail-order houses making a wide array of goods available to both urban and rural customers. As mass consumption eclipsed property ownership and economic autonomy as the foundation for Americans' definition of freedom, the vision of the alcoholic's excessive, self-destructive consumption proved especially troublesome.[16] The drunkard's drinking habits offered a stark contrast to the "healthy" consumption of goods that was to sustain the industrial capitalist economy.

Excess consumption of alcohol consumed drinkers themselves and often transformed productive workers into "burdens on the community every year, and on the tax payer," noted Thomas Crothers in 1883. Crothers was hardly alone in his observation. Inebriety specialists highlighted the financial drain of alcoholism in their strategies to gain the public's support. Lewis D. Mason submitted in 1904 that when facing the average taxpayer or a committee of ways and means for a legislative body, arguments grounded in morality, spirituality, or human compassion would usually fail. The only successful argument that one could bring forward under these circumstances was "one based on *financial considerations and on dire and urgent public* necessity. You must appeal as a political economist and show by fact and figure, that which is good business policy, and that money is saved to the state or municipality, and consequently to the taxpayer."[17] If enlightened public sentiment couldn't carry the day, the economic argument might prevail. Here medicine adopted the tactics of scientific charity and rational Progressive reform. It was not enough for physicians to be healers and educators; they were to be political economists and lobbyists, too!

As they adopted the tactics of the scientific charity organizer and political economist, however, physicians began to make distinctions between those habitual drunkards who were worthy of treatment, the "hopeful" group that might be cured, and those who were not likely to benefit from treatment, the group they called "incorrigible" inebriates. This distinction mirrored charity workers' own discrimination between the "worthy" and "unworthy" poor.[18] It was only natural for some cases of inebriety to prove more or less challenging than others, but making the distinction between worthy and unworthy patients was open acknowledgment of the disease's moral shroud, and it opened the project of medicalizing habitual drunkenness to opposition. How could one distinguish between good inebriates and bad inebriates if they suffered from the same disease? If inebriety was a treatable disorder, as physicians claimed, was it worth treating at the state level when such a large percentage might relapse and return to their "bad habit"? Would offering medical care to inebriates offer savings to the state if they could not be cured permanently? Although many inebriety physicians viewed medical treatment for habitual drunkards as an important step toward preventing insanity and poverty, physicians were forced to acknowledge that, like insanity, inebriety was a chronic disease.

For as much as physicians and reformers attributed pauperism to alcoholism, they also recognized that the rampant poverty in cities was an important factor in *producing* inebriety. Lack of fresh air, cramped housing, and bad nutrition all figured in alcoholism's etiology.[19] Albert Day implicated workmen's poorly ventilated and crowded shop floors in 1877, conditions that left laborers feeling "fretful and uneasy," and likely to resort to "ardent spirits" for temporary relief from the "monotony and listlessness of their pent up lives."[20] Others maintained that deficient home conditions "produce inefficient people, and the saloons exist in proportion to the inefficiency of the community."[21] This focus on urban poverty highlighted another way in which modern civilization, with its burgeoning cities and mill towns, was to blame for alcoholism. As Lucy Hall and M. E. J. Kelley had observed in the case of inebriate women, so psychiatrist Stewart Paton concluded in his 1905 textbook, *Psychiatry:* not only did the poverty of the city threaten one's mental well-being, but the very transition from rural to urban living could also be problematic. In discussing the origins of alcoholism, Paton noted that "removal from country to city life—[was] a factor of great importance and should receive most careful consideration."[22] Indeed, the city came to embody the essence of modern civilization's best and worst attributes. Nowhere else were progress and poverty so evident.

Poverty was as old as the cities in which it thrived, but the Gilded Age and Progressive Era ushered poverty of mythic scale into the nation's urban centers. Several factors contributed to the high levels of impoverishment and its perception as a formidable social problem allied with alcoholism. First, the expansion of American industry brought huge numbers of rural inhabitants to the cities to find their fortunes. In 1860, only 13 percent of Americans lived in cities; by 1900, 40 percent were urban dwellers. More than one hundred cities doubled in size in the 1880s when manufacturing overtook agricultural production as the greatest power in the economy. Amidst this change, social philosopher Henry George maintained that the laboring classes reaped a disproportionately small advantage compared to the growth of great industries and their increased levels of production.[23] Harsh industrial cycles of boom and bust exacted a heavy toll on precariously employed workers. Economic depressions besieged the nation in 1873, 1883, and 1893, leaving more of the urban masses "huddled" and destitute than ever before.

A large proportion of the laboring classes were new immigrants who arrived in America penniless and ready to work, no matter what the conditions. And as muckraking journalists and novelists revealed in exposés of the Standard Oil Company, the patent medicine business, the meat-packing industry, and countless other industries, the conditions were often dreadful and the gulf between management and labor great. At the high tide of immigration, from 1900 to 1910, the 8.8 million immigrants received into the United States comprised more than 40 percent of urban newcomers. Many of them hailed from southeastern European cultures where drinking (although generally moderate drinking with meals) was the order of the day—a fact that did not escape the inebriety physicians' attention.[24] The immigrants from Ireland, England, Germany, and Scandinavia who had preceded the Italians, Slavs, and Poles were certainly not teetotalers. Surrounded by saloons, which afforded tired workers recreation and an opportunity for political organization, working-class neighborhoods were places where alcohol flowed freely. In 1886 Henry William Blair illustrated this point in his history of the temperance movement by drawing a map of saloon locations in New York City. Blair revealed that in one working-class neighborhood, the first assembly district, where there was but one public school for every 14,566 residents, the ratio of saloons to residents was one to 41.[25]

Middle-class reformers and medical professionals were concerned with the plight of their less affluent neighbors, even if only out of fear that violence might arise from an oppressed and discontented working class—a realistic concern, given the Haymarket Affair of 1886, the Homestead Lockout in 1892, and the Pullman Strike of 1894. Beginning in the 1880s, the middle class, especially middle-class women, became better acquainted with the poverty of their less-fortunate fellow citizens. Between 1880 and 1910, middle-class urbanites entered the social reform "business" as never before. The professionalization of social work, medical social work, and nursing (especially visiting nursing) brought thousands of trained "career" women and professional men into the homes of the immigrant and working classes, where alcohol was often present, even if it was not problematic. Visiting nurses and settlement workers witnessed poverty and alcoholism firsthand as they attempted to teach immigrant newcomers and the persistently poor how to live according to the "healthier" ways of the American middle class.[26] In many respects, these new explorers of the city wilderness followed where reform organizations such as the Woman's Christian Temperance Union had blazed the trail, lending professional status to work the WCTU had already commenced in its departments of "health and heredity," "missionary work to Ellis Island," "Christian citizenship," and "temperance and labor." Uniting social betterment (as they defined it) with social control, private agencies such as settlement houses, associated charities, and the WCTU promoted temperance teachings alongside education in nutrition, home and personal hygiene, and "Americanization" throughout the early twentieth century. As champions of scientific and rational charity, these emerging professional groups took an active interest in the medical side of alcoholism, even if their interests did not always promote medical care for inebriety.

If private philanthropy directed its attentions toward "Americanizing" the nation's newest and poorest citizens, the physicians who were treating alcoholics of all classes concerned themselves with the nature of citizenship itself.[27] Medical discussions of inebriety routinely focused on the compact between the individual and the state. To do so was a necessity, for physicians portrayed alcoholism as a public health problem, and they enjoined state legislatures across the country to pass laws permitting them to commit inebriates to medical facilities. Neurologists, psychiatrists, and general clinicians alike wished to control the inebriate as they had managed the insane and quarantined infectious smallpox patients. To treat the inebriate as they desired meant obtaining the power to institutionalize the drunkard for long periods of time. Thus, physicians mobilized to convince municipal and state officials that it was crucial to alter the laws governing habitual drunkenness—laws that usually levied a short jail sentence or fine on the public inebriate but rarely facilitated his medical care.

When the American Association for the Cure of Inebriates formed in 1870, an interventionist state was hardly the order of the day, particularly with regard to the public's health. The Metropolitan Board of Health, the nation's first, came into being in New York City in 1866. Massachusetts established the first state board of health three years later. On the heels of these developments, disease-concept advocates such as Joseph Parrish and Stephen Rogers urged the state to play a greater role in its citizens' health and to enlist physicians in this cause. As Parrish made plain in his first address to the AACI, it was appropriate for the medical profession to judge inebriates in their relationship to society because physicians were "part of the social organism, and are related to them, and they to us in the compact." No matter what the cause—sin and/or disease—if inebriety affected "the peace and safety of the community," it was a social offense and subject to law, Parrish concluded. The law provided only imperfect penal solutions.[28] Alienist Orpheus Everts concurred, adding in 1883 that the inebriate's "inhibition" (what we would call "detention") was more viable than the "prohibition" of alcoholic beverages for whole states or the nation: "It is to the interest of the individual who is not capable of such self-control as enables him to conform to the social laws and customs regarded by a large majority of his fellow-men as essential to the peace and happiness of society, that he be so far constrained by force as to prevent constant collisions between himself and society, by which both suffer, and neither are benefitted."[29] Everts recommended that the inebriate be classified as a "protected citizen" whose rights would be temporarily cut short until he became a sober and productive member of society. State inebriate hospitals would prove most effective in rehabilitating men and women to this point, he maintained, for they would provide their "inmates" with more opportunities for employment. The inebriate, then, was to become a new ward of the state when the court determined that he could no longer care for himself or his family and was in need of protection from himself.[30]

It was not simply out of concern for the inebriate's and the community's health that the state was to play a larger role in combating inebriety, argued disease-concept advocates. The state had a moral *obligation* to address the problems of the inebriate because its treasury reaped great financial gain from taxes on the sale of alcoholic beverages. The state and the liquor trade had formed a profitable alliance at the inebriate's expense. This partnership, argued Gonzalva Smythe in 1891, made the state *particeps criminis,* or an accessory to the inebriate's decline and potential crimes while under alcohol's influence. Smythe's comments brought home concerns commonly voiced in late Gilded Age America about the mounting power of the corporation, corporate monopolies, and the state's collusive role in furthering both. Certainly, the WCTU and the Anti-Saloon League addressed the "big business" side of the alcohol problem more directly, routinely lambasting the alcohol "traffic." Yet it was not until the Gilded Age gave way to the Progressive Era that politicians, the judiciary, and social workers joined physicians en masse in their efforts to check habitual drunkenness. In the 1890s and 1900s, each of these parties began to see an essential role for the state in solving the sociomedical problem of inebriety, an important part of the nation's larger "alcohol problem." This enthusiasm, of course, was related to the mounting distrust of local authorities to successfully administer to the growing population of diseased and dependent people who sought or required medical and welfare assistance. Indeed, as early as 1864, Sylvester Willard, the secretary of the New York State Medical Society, observed that local institutions were marred by the "gross want of provision for the common necessities of physical health and comfort."[31] Willard was a major proponent of New York's State Care Act, passed in 1890 and aimed at providing all of the state's insane with an adequate and uniform standard of care.

Social problem? Medical problem? Moral problem? Penal problem? No matter how reformers framed the "liquor problem" in the Gilded Age and Progressive Era, it was closely related to the definition of citizenship, democracy, and the relationship between the individual and the state. The liberalism of the Gilded Age favored a laissez-faire government that let the market determine a living wage, but growing unrest within the labor movement and the expanding economic and political divide between wage earners and those who employed them urgently demanded intervention as the nineteenth century came to a close. Seen in the context of Gilded Age immigration restrictions, court rulings such as *Plessy versus Ferguson,* and the rise of American imperialism abroad, Everts's, Parrish's, and Smythe's calls for the inebriate's forfeiture of civil rights and prolonged institutional care may appear yet another measure designed to curtail the rights of citizenship. Seen in the context of the Progressive Era's state and federal initiatives to better the public's health, regulate the conditions of labor, and break up the trusts, the same arguments about the necessity of state care for impoverished inebriates may appear prescient and empowering (even if Progressive reforms were largely unsuccessful). Regardless, to propose that inebriate reform demanded the state's attention was not such a stretch for physicians and reformers in the early years of the twentieth century. After all, these were the years in which the WCTU and Anti-Saloon League gained the public's support for Prohibition—an unprecedented step in the state policing of the public's health and morality. During the Progressive Era, it became much easier to see inebriety as a disease of modern society and the state as an appropriate actor in redressing this national threat.

Greater familiarity with the alcohol problem, a more activist state, and alcohol's ties to the labor problem and poverty were not the only factors influencing the public's and politicians' views of inebriety treatment. American psychiatry was coming of age in the early years of the twentieth century and expanding its therapeutic domain beyond the asylum walls and neurologists' urban offices. When the AACI began its campaign to medicalize habitual drunkenness in 1870, American mental medicine had been in a shambles. Asylum alienists acted more like accountants than physicians, balancing the budgets of their bloated, largely custodial institutions. For their part, neurologists, buoyed by their work in the Civil War, searched (mostly in vain) for the specific nervous lesions underpinning contested psychiatric diagnoses. By the 1890s, however, matters were beginning to change, making the intellectual climate of mental medicine far more hospitable to the medicalization of inebriety. The "new psychiatry" got its start in the 1890s as individuals such as Adolf Meyer, S. Weir Mitchell, William Alanson White, Smith Ely Jelliffe, and Elmer E. Southard brought together these two specialties—asylum medicine and neurology—with general medicine.

Drawing on the work of the Chicago school of sociology and the functionalist "new psychology" of William James, G. Stanley Hall, and John Dewey, the new psychiatry addressed the whole personality—psyche and soma—of the individual rather than his or her specific mental faculties. This new approach to understanding mental health and disease viewed the individual as part of the social organism. Mental illness "was not a defect either of mind or body, but a lowering of an individual's ability to function in the struggle for existence—a struggle that for modern man was intimately bound up with his success in social relations."[32] Psychiatry's new mission, according to one of the new mental medicine's greatest exponents, Adolf Meyer, was to assist persons of all types adjust to their positions within society. Maladjustment and maladaptation required the "re-education" of the individual.

Meyer's maladjustment model was nothing short of brilliant in its ability to unite asylum medicine, neurology, and general medicine, for it positioned psychoses and neuroses on the same continuum. This allowed psychiatry to move into new domains, addressing domestic problems, workplace dynamics, educational concerns, and public health (especially mental hygiene). The new psychiatrist attended to the individual's maladies, and in so doing, promoted social and societal health.

"Conduct is the measure of the functional capacity of the central nervous system," declared Baltimore psychiatric consultant Stewart Paton in 1905, adding, "The addiction to alcohol is a symptom of a functionally unstable nervous system." According to Paton, the stress and strain of "modern civilization" affected "many individuals in the social organism," threw them out of harmony with their surroundings, and led them to drink. Alcoholism, he claimed, was simply a phase of "general mental and physical instability."[33] Within the new functional psychiatry of the Progressive Era, alcoholism was both a personal and social problem that demanded the physician's attention.

## Alcoholism and the Crisis of Masculinity

By the end of the nineteenth century, the consumption of alcohol had come to signify the end of the workday and "the passage to play." This too was an important element in the advance of modern industrial society, for it was only with the rise of the routinized, disciplined factory workplace that work and leisure were clearly delineated.[34] As the laboring classes stepped out of the factory and into the saloon, they proceeded from work time to leisure time. Inside the saloon, the act of drinking could turn a roomful of working men into bonded brothers and offer men a chance to demonstrate their masculinity. As novelist Jack London observed in his autobiographical *John Barleycorn*, published in 1913, the saloon served as a sort of working man's club for urban dwellers, where drinking was a sign of social solidarity: "When I thought of alcohol, the connotation was fellowship. When I thought of fellowship, the connotation was alcohol. Fellowship and alcohol were Siamese twins. They always occurred linked together."[35]

Through "treating" or "clubbing"—the two most popular practices of buying people drinks—saloon goers strengthened their ties to one another. The former practice involved one drinker buying a round for his imbibing companions at the bar; he would receive subsequent "free" rounds through the treating of other drinkers. With clubbing, groups usually prepaid their drinking tab and might retire anywhere within the saloon to drink as a collective unit, a "club," or they might take their alcohol outside the barroom to celebrate special occasions en masse. Both of these practices inspired solidarity through the exchange of imbibed gifts.[36]

This is not to say that the young working-class women who flocked to the mill towns and factories did not enjoy the same division of work and leisure as men, or even similar saloon-based camaraderie with their working sisters. Yet, as we have seen, when women wandered into the public realm of drinking in saloons, and still worse, found themselves dependent upon alcohol, physicians regarded them as more deviant by far than their male counterparts. Public drinking in the Gilded Age and Progressive Era may not have been perceived by social reformers and physicians as desirable, but both of these groups—along with the public—grudgingly regarded it as the province and the right of men.

The inebriate's gendered identity makes it important to view physicians' discourse on inebriety, and the plentiful references to the inebriate's loss of self-control, in the context of a larger discussion of the "crisis of masculinity" that characterized the Gilded Age and Progressive Era. By crisis of masculinity, I refer to publicly voiced concerns over the fate of American manhood's potential for self-actualization between 1870 and 1920. Fears about the feminization of American culture and the workplace were occasioned by a range of societal and social developments. These included the decline of opportunities for self-employment in the growing industrial economy; the rise of the corporation, with its enforcement of workplace conformity and discipline; the entrance of thousands of immigrants, African-Americans, and women into the urban labor force; and the general perception that the outlets for nurturing the heroic virility of self-made men—deemed a vital source of national strength in previous generations—were dwindling.[37]

On the one hand, physicians throughout the Gilded Age and Progressive

Era maintained that drunkenness stripped men of their reason, enfeebled their wills, and rendered them dependent not only on the bottle but also on the state's coffers. Basically, it was felt, alcoholism stripped men of their masculinity. On the other hand, to drink in the company of men was to promote "a glorious manhood unfettered by the nagging demands of women, who would, had they their way, ensconce men at home, squander their wages, forbid them to drink—in short, emasculate them."[38] The saloon, while offensive to the middle-class reformers who championed a "respectable manliness" centered around familial and social obligation, nevertheless offered working men a forum for egalitarian social exchange—a haven from the ever more mechanistic, routinized, and disciplined corporate or industrial workplace.[39] At stake in drinking, and thus in inebriety, was the meaning of manhood itself.

As we have seen, many physicians were aware of (if not always sympathetic with) the special hardships endured by the working classes, much as they were sensitive to the demands placed on the more delicate and refined nervous systems of the middle and upper classes. In the end, however, doctors interpreted the inebriate's nervous weakness as the basis for his inability to adapt to modern society, and they advanced a definition and ideal of sober masculinity, or respectable manliness, that revealed their own professional, middle-class values. The secret to success at the state hospital for inebriates in Massachusetts, maintained settlement house pioneer Robert Archey Woods, was that patients were treated "like men and expected to live up to the part."[40] Woods and hospital superintendent Irwin Neff found inebriety's cure in patient "re-education." Part paternalism, part social control, part psychotherapy, this reeducation program urged men to reclaim (or claim) a respectable masculinity, in which they learned to prioritize the middle-class virtues of sobriety, personal responsibility, fiscal independence, and civic duty. Physicians deemed the dissolute masculinity of the working-class saloon less than useful in modern society's "struggle for life."[41]

Competing cultural definitions of masculinity and femininity exerted increasing influence on the late-nineteenth and early-twentieth-century world of psychoanalysis. This was especially true as psychiatrists, neurologists, and psychologists in the United States and abroad began to turn to psychoanalysis for insights into the everyday lives of men, women—and, in turn, of alcoholics. The rise of the New Woman fanned the flames of masculinity's turn-of-the-century cultural crisis, and in this context, psychoanalysts often regarded alcohol as a marker of sexual deviance for both their male and female patients.[42] Psychiatrists such as Freudian Karl Abraham believed "the homosexual components which have been repressed and sublimated by the influence of education become unmistakably evident under the influence of alcohol."[43] Thus, alcoholism often indicated homosexuality in patients. On the other hand, Abraham observed, "Men turn to alcohol because it gives them an increased feeling of manliness and flatters their complex of masculinity." A careful psychiatric examination of the individual was crucial to diagnosis and treatment, although for Abraham at least, the situation was less complex for women. Those women "who show a strong inclination for alcohol always have a marked homosexual component in them," he urged.[44] Others might associate women's heavy drinking with prostitution, but in either case, alcohol-related deviance from society's sexual norms demanded psychiatric attention.

The sexual deviance believed to underlie alcoholism was not just the subject of psychoanalytic, psychiatric, and psychology journals. In the first two decades of the twentieth century, *Everybody's Magazine, The Literary Digest, The Arena, The American Magazine, McClure's Magazine,* and a host of other popular journals ran regular articles on inebriety, many of which dealt, some more openly than others, with inebriety's relationship to sexuality and masculinity. Through these articles, and especially through the first-person narratives of recovering and recovered inebriates, the general public was exposed to the early-twentieth-century medical and psychiatric theories concerning inebriety.

## Domesticity and Inebriety: The WCTU and Public Discourse on Drunkenness

For the general public at the turn of the century, the bulk of their information regarding alcohol came from organizations such as the Woman's Christian Temperance Union, not from the medical establishment. The relations between men and women were at the heart of the WCTU's critique of alcohol, but it was women's victimization by male drinkers that received the lion's share of the organization's attentions when it came to reforming the drunkard. Although the physicians who promoted the disease concept were aware of the domestic problems caused by alcoholism, they found the WCTU's broad focus on prohibiting the manufacture and sale of alcohol, its public instruction in temperance physiology, and its appeal to address habitual drunkenness through mostly moral and penal treatment especially annoying. The WCTU's rise to power in the public discourse on alcohol coincided with the campaign to provide medical care for inebriates, but the group's size, organizational strategies, and loud public voice frequently overshadowed the efforts of inebriate reformers.

In the 1870s, the drive for a dry nation regained some of the momentum it had lost in the years preceding the Civil War. During the 1850s and 1860s, the wet-dry divide was superceded by the growing schism between northern and southern states. Temperance support was a significant casualty of the Civil War. The 1860s, however, were not adverse times for the manufacturers of alcohol. The United States Brewers' Association was born one year into the war, and by the war's end drink was on the rise again. Industrial breweries employed the nation's growing immigrant population and used expanding rail systems to distribute their wares widely, making beer twice as popular in 1865 as it had been in 1850. Per capita consumption of the beverage increased by 300 percent between 1850 and 1870, a trend that did not go unnoticed. Temperance observer Daniel Dorchester labeled this development "one of the saddest phases of American life during the last thirty years."[45] Others shared his view.

If the Midwest became one of the nation's brewing capitals, it was also a cauldron for grassroots temperance advocacy in the last thirty years of the nineteenth century, with women leading the way. In Ohio, under the state's new Adair law, women were able to sue saloon keepers for the damage caused by the publicans' sale of liquor to their besotted husbands. Other postbellum women, galvanized by the temperance cause, began to collectively confront saloon keepers, holding "pray-ins" outside their shops to plead for the saloon's closing. The "Women's Crusade" of 1873–74 culminated in Cleveland, Ohio, with the formation of the Woman's Christian Temperance Union. The WCTU expanded its mission in the 1880s under the direction of Frances E. Willard. Willard promoted a wide range of reforms for drunkards, most notably "gospel temperance," a variety of moral suasion that reclaimed drinkers through their public confession, pledging of sobriety, and pursuit of other drinkers in need of reclamation. The WCTU promoted gospel temperance through many avenues; in particular, it worked closely with the mostly male Red Ribbon and Blue Ribbon reform clubs, which offered a community of support for anyone willing to take the pledge of total abstinence. For men such as Joseph Parrish, the pledge and the pleadings of temperance organizations were at best naive, and at worst the source of much misinformation. Referring to gospel temperance advocates of all persuasions, the *Quarterly Journal of Inebriety's* editor, Thomas Crothers, opined, "The literature of these movements is the strangest compound of errors and misconceptions that are repeated without a question or doubt of their reality."[46]

The relationship between advocates for the disease concept and the WCTU was checkered at best. A familiar refrain in the literature on inebriety derided the temperance movement generally, and the WCTU in particular, for their distortion of scientific truth and their heavy emphasis on the moral dimensions of drunkenness. In fact, the WCTU's presentations to the public did rely on actual science, but they were slanted to serve the WCTU's prohibitionist purposes.[47] WCTU president Frances Willard's belief in "the Gospel cure as the only true deliverance" from inebriety did not win her or her organization many friends among the physicians who promoted the disease concept of inebriety.[48] Eventually, even the general medical profession took umbrage at being "instructed" on the nature of alcohol by "temperance ladies."[49]

Unfortunately for the physicians who promoted the disease concept, the WCTU had a remarkably powerful propaganda machine. The "women in white" were a far more visible presence than the asylum alienists and urban neurologists who worked daily to bring sobriety, health, and order to the lives of their inebriate patients. Frances Willard possessed the entrepreneurial skills of an Andrew Carnegie or a Henry Ford. Her "do everything" strategy placed "white ribboners," as the WCTU members were called, in a remarkable variety of public and private reform contexts. By 1906 the WCTU could boast thirty-eight separate "departments" devoted to concerns as varied as "Work among Colored People," "Legislation," "Health and Heredity," "Work among Railroad Employees," "Medical Temperance," "Work among Miners," "Fermented Wine at Sacrament," "Rescue Work," "Purity in Art," and "Press." Indeed, almost two decades before the ratification of the Eighteenth Amendment, every state in the Union had passed legislation requiring public-school temperance physiology lessons, thanks to the lobbying efforts of the WCTU's department of scientific temperance instruction. Here, as Philip Pauly and Jonathan Zimmerman have argued, the women in white engaged in a fierce battle with physiologists and medical scientists over both the nature of alcohol—did it have medicinal properties or was it simply an evil substance?—and the public meaning of physiology.[50] Two of the strongest campaigns in the WCTU's war against "demon rum" were scientific temperance instruction and medical temperance, or the practice of medicine without recourse to alcoholic therapeutics.

Had the WCTU confined itself to these two issues, the organization would still have received more than a few disparaging remarks from disease-concept advocates. Consider the words of Irwin Neff, one of the most conciliatory of inebriety experts—a physician who believed that the inebriate required both medical and moral attention and that physicians needed to cooperate with temperance organizers to solve America's alcohol problem: "Sentimental antagonism and generalized and exaggerated statements have formerly operated against any decided advancement in the way of popular education on the dangers and abuse of alcohol."[51] Neff's critique was typical. Medical authors used the views of the WCTU as a foil for their expert opinion, contrasting physicians' "rational," "practical," and "scientifically accurate" positions on inebriety to the "emotional," "sentimental," "exaggerated," "distorted," and "dogmatic" views they believed the WCTU pressed upon the public. No doubt there was a significant gender element to this debate between professional men of science and lay women reformers. The accusations of temperance "emotionalism" certainly suggest this, since Victorian culture held that women were the keepers of sentiment and emotion. But the medical critique of the WCTU's scientific temperance instruction and medical temperance campaigns focused on claims of accurate versus distorted facts, and scientific versus moral reform.[52] The more salient tension appears to have been between lay and professional authority, with the WCTU using science to support its religious and moral platform for the reform of society and the enactment of legislation to eliminate alcohol generally, and inebriety physicians employing science to heal the inebriate, educate the public, and reform society.

Of course, the WCTU did not confine its efforts to either scientific temperance instruction or medical temperance—to do so would not have been to "do everything." Frances Willard and her temperance legions did have views about the reform of the habitual drunkard, and they did their best to promote them, which is where they came into the most direct conflict with physicians who wished to treat alcoholics. Speaking in 1897 to the Fourth World's WCTU Convention, the last such convention held before her death, Frances Willard allowed that the physicians and "famous experts" had "proved that alcoholism is a disease, while the studies of religious and ethical experts have proved it a crime against both natural and spiritual laws." To Willard's mind, the most interesting change in public sentiment was that politicians and statesmen had begun to regard the drunkard as "an unmitigated nuisance and a danger in the home" and proposed to deal with the habitual drunkard "as one who commits a crime against society" and establish industrial homes for their detention.[53]

Two things are striking about Willard's position, and they afford some insight into the struggle to medicalize habitual drunkenness. First, Willard regarded the studies of religious and ethical experts as no less valid than those of physicians and scientific experts. This was a decided obstacle to the physicians who were promoting the necessity of science-based medical care to address the problem of habitual drunkenness. For while she might champion medical care, moral solutions were equally appropriate in her mind. For Willard, habitual drunkenness remained a moral condition first, and a medical condition second. Moreover, she was hardly alone in her views, which at the very least exercised a tremendous influence over tens of thousands of women across the country.

Second, Willard did advocate "treatment" in the form of detention for the male drunkard who "commits a crime against society." The similarity between "inebriety: the public health problem" and "inebriety: crime against the state" are obvious, but in the end, Willard clung to an older vice-crime perspective. According to Willard's biographer, Anna Gordon, this was a perspective influenced by her great friend, Lady Henry Somerset, who developed one such industrial detention home for women inebriates in Great Britain. In these institutions, medical care was one element, but work and moral reform were the order of the day.[54]

Nowhere was Willard's broad but traditional approach to the drunkard's reform more evident than in her comment that the WCTU favored a "do everything policy" for the drunkard: "We favor the Keeley Cure, the Christian Home for inebriate men, the Gospel Meeting, the temperance pledge.... We also believe that to arrest the drunkard, no matter what his social position, and to place him in custody, would greatly deter the ignorant and thoughtless from looking lightly upon such a brutalized condition, and would be of incalculable service in stamping the drunkard with the displeasure of the community."[55]

Here we get a sense of the WCTU's reform priorities when it came to the drunkard. He—and the "habitual drunkard" was a "he" for Willard, who referred to women alcoholics, not as "drunkards," but as "inebriate women"—

was best served through some medical, some penal, but mostly moral measures. The same moral emphasis, however, was also true for the few "homes for inebriate women" established in the 1880s and 1890s. According to historian Catherine Murdock, the WCTU founded at least five homes for women alcoholics: two in Chicago and the others in New York, Philadelphia, and Manchester, New Hampshire. In these homes, "women who are victims of strong drink" could have "the chemical cure applied to their diseased bodies and the gospel cure to their diseased souls."[56] The story was no different in Boston, where concerned citizens in 1879 established the New England Home for Intemperate Women, an institution that "relied upon two great instrumentalities—labor, and the moral and religious influence of a Christian home."[57]

The WCTU's position on physician Leslie E. Keeley's popular proprietary cure for alcoholism, which is discussed in detail below, further illuminates the relationship between the WCTU and the medical profession, for the WCTU's position was not completely dissimilar to that of the AMA and the AACI. At the WCTU's annual meeting in 1893, Willard denounced the journalists who had portrayed the WCTU as the enemy of Keeley. Like the regular medical profession, Willard doubted Keeley's exaggerated claims of success (95 percent cured). She said that bichloride of gold—advertised as the Keeley Cure's key ingredient—was assuredly not "a medical finality for the cure of inebriety" and that Keeley should not keep his medical compound a secret and profit from it if it were successful. Believing that many had been helped through Keeley's treatment institutes, however, Willard issued a challenge to the doctor: "If Dr. Keeley would communicate his prescription to an accredited physician in every locality, our Society would gladly pay for as much of the medicine as might be needed to help all men not able themselves to pay the price of the remedy."[58] To no one's surprise, Keeley declined the offer, but it was an interesting one that revealed the unique position of the WCTU. Willard attempted to support a remedy that received great press and public support, while maintaining a stand that was consistent with organized medicine's condemnation of the false claims and profiteering of medical hucksters.

A resolution adopted in 1894, however, put the WCTU's qualified endorsement of Keeley in perspective. The resolution stated: "While friendly to all institutions having for their object the restoration of the drunkard, we do not recognize in them a cure for the saloon evil. Neither do we encourage local unions to adopt this work as any solution of the temperance problem, so long as the licensed saloon exists."[59] In the end, shutting off the supply of alcohol at its source, the saloon, was what mattered most to the white ribboners. The reform of inebriates was secondary, and not worthy of local chapters' time and effort.

It is easy to see the sources of antagonism between the WCTU and the disease-concept advocates. Willard's qualified endorsement of Keeley, her advocacy of moral and penal measures for the habitual drunkard, and the low priority given to the inebriate's reform—each of these issues placed the two groups at odds. Moreover, because inebriety specialists were mostly physicians and almost uniformly members of the AMA, they were obliged to take a stand against the WCTU's lay efforts to end the profession's use of alcohol as a medicine—the "medical temperance" campaign—even if many members regarded alcohol as an unnecessary remedy. No professional group wished to be hamstrung by lay agitators, let alone "lady reformers."

Instead, Nathan Smith Davis, former president of the AMA and one of the founders of the AACI, organized the American Medical Temperance Association in 1891 to create a professional group devoted to establishing the true nature of alcohol and its appropriate use or disuse in medicine. With this, the profession could police itself with regard to the prescription of alcohol. In 1904, this group merged with the AACI, by then called the American Association for the Study and Cure of Inebriety. The name of the new organization was the American Medical Society for the Study of Alcohol and Other Narcotics. In the end, Thomas Crothers, the editor of the AACI's *Quarterly Journal of Inebriety,* and Lewis Mason, one of the founding members of the AACI, became confirmed scientific temperance supporters and even served on the WCTU's scientific temperance advisory board.

Historians have mistaken Crothers's temperance enthusiasm and the merger of the two societies as an indication that advocates for the disease concept had "switched course and embraced Scientific Temperance," a decision that supposedly "stemmed from the conviction that 'success would be in prevention rather than cure.'"[60] While it may be true that the merger of the two societies resulted in the perception that the AACI had switched allegiances from cure to prevention, this was not the case. In 1911 Crothers published his magnum opus, *Inebriety: A Clinical Treatise on the Etiology, Symptomatology, Neurosis, Psychosis, and Treatment and the Medico-Legal Relations*—hardly the work of a man who had given up on curing alcoholics. Both he and Mason continued to write extensively on the etiology and treatment of inebriety throughout the first decades of the twentieth century.

Although the *Journal of the American Medical Association* posted notices from the Society for the Study of Alcohol and Other Narcotics under such dismissive headlines as "Foes of Alcohol Meet" and "Progress in Temperance," this did not mean that the drive to provide medical care for inebriety suffered. Instead, it is likely that these headlines indicated a changing of the guard. The AACI had started the campaign to view the inebriate as a diseased individual, but after nearly two generations, its members—save Crothers—were no longer leading the way. Instead neurologists, psychiatrists, and public health physicians were at the forefront as they began to focus their attentions on alcoholism. Indeed, the AMA devoted increasing attention to the problem of inebriety throughout the early twentieth century. In 1906, *JAMA* published a long series of papers on alcoholism read at the annual meeting of the AMA's Section on Hygiene and Sanitary Science, devoting some fifty pages to the subject. *JAMA* editorials likewise discussed the need to bring the public and the general medical profession into line on the inebriety problem and the necessity of state medical care for inebriates.[61]

If the association with the temperance movement cast doubt on the credibility of the Society for the Study of Alcohol and Other Narcotics, it did not prevent many of the society's members or other public and mental health advocates from addressing the disease of alcoholism within the general medical press. This new wave of alcoholism treatment advocates, like Irwin Neff, attempted to draw a boundary between themselves and the temperance cause, and they urged that physicians, not temperance reformers, take the lead in educating the public about alcohol and alcoholism.

## Inebriety, Commercial Success, and Professional Quackery: The Keeley Cure

At the turn of the century, cures for alcoholism grew at a lightning pace. In addition to private inebriate homes and public inebriate hospitals, proprietary cures and their affiliated mail-order remedies did a tremendous business, and these widely available cures were often the first contact the general public had with treating alcoholism as a disease. Some of the proprietary cures—the Oppenheimer Institutes, the Neal Institutes, the Gatlin Institutes, and others—opened "franchises" across the country, expanding their services to a national clientele. None met with more spectacular success than Leslie E. Keeley's Bi-Chloride of Gold Cure for the treatment of alcoholism, morphinism, and tobacco addiction. An estimated half-million alcoholics and drug addicts "took the Keeley Cure" between 1880 and 1920.[62] Like the WCTU, Keeley commanded more public attention in his campaign to treat inebriety as a disease than did the disease-concept advocates, who regarded Keeley as a proprietary-cure huckster. Widely publicized, Keeley's labors unquestionably promoted the disease concept within the public arena, but Keeley advanced a vision of the condition as a specific disease cured through specific remedies, an image that contradicted the less-popular approach of regular physicians, who considered alcoholism a treatable but chronic disease.

Born in Ireland, Keeley immigrated to the United States and studied medicine at Rush Medical College in Chicago; he served in the Union Army as a surgeon during the Civil War. In collaboration with a well-known temperance speaker and former minister, Keeley developed a specific chemical compound for the treatment of alcoholics and found that it worked on the local vagrants in Dwight, Illinois, where he lived. Boldly declaring in 1879, "Drunkenness is a disease and I can cure it," Keeley opened the first Keeley Institute, where he began to treat those who came for help with their alcohol and drug habits. Keeley believed that drunkenness was a disease; he implicated heredity as the cause early in his career, only to abandon this explanation for that of gradual poisoning. He thought that the body's cells were poisoned by toxic substances found in alcohol, opiates, cocaine, and tobacco. Keeley's treatment was meant to rid the body of the toxins and restore health.[63]

Not until the early 1890s did Keeley become a household name in the United States. With the Dwight facility flooded with inebriates desiring his help, Keeley decided to franchise the business and open institutes across the country. In 1891, Keeley became a national celebrity when he issued a challenge to the editor of the *Chicago Tribune,* Joseph Medill. He told Medill to send him the worst drunks the Windy City had to offer, and he promised to cure them. Medill sent drunks to Keeley one after another and reported to the *Tribune's* readers that "they went away sots and returned gentlemen."[64] By 1893, there were 118 institutes across the nation. The 1893 World's Fair in Chicago even hosted a "Keeley Day."

The essence of Keeley's treatment was four daily injections of his specific (and secret) bichloride of gold formula and a relaxed and informal alcohol-free, tobacco-free, gambling-free "home away from home." Patients paid between twenty-five and fifty dollars a week during the traditional month-long treatment. Colorado, Louisiana, Maryland, Minnesota, North Dakota, and the

Oklahoma Territory passed legislation to provide assistance to their poorer citizens who desired treatment. Reformed patients formed "Keeley Leagues," fraternities and sororities designed to keep members sober, to spread the message of hope through Keeley treatment to drunkards everywhere, and to educate America's youth about the pernicious effects of alcohol and drugs. Keeley claimed that the cure rate for his patients was at least 95 percent. Even though most of the patients who entered Keeley Institutes did so voluntarily, this rate of success was unprecedented and was strongly challenged by the medical profession. In part because of the medical profession's criticisms, in part because of various scandals surrounding Keeley, and in part because of the public's realization that many of the estimated 500,000 Keeley "graduates" relapsed, the Keeley Cure declined in popularity. By 1900, the year of Keeley's death, the 118 institutions that bore his name had dwindled to 50.

Much as psychiatrists and the WCTU typically framed inebriety as a condition affecting mostly men and closely related to men's social interactions, Keeley's regimen at the Dwight institute and at other franchises such as that in Frederickton, New Brunswick, Canada, also relied heavily on replacing the manly camaraderie of the barroom with that of treatment.[65] Indeed, several elements of the Keeley treatment were designed to restore manhood as well as sobriety. Three or four times a day, men lined up in the main building (called the "shot tower") for their hypodermic injections of strychnine (supposedly mixed with incidental amounts of gold and sodium chloride). Every two hours, patients took a dram of tonic (called "the dope"), which was said to contain atropine, strychnine, cinchona, glycerin, and gold and sodium chloride.[66] One had to possess a strong constitution to withstand the treatment. According to Chauncy Chapman, a former physician at Dwight, uncooperative patients were subjected to a primitive form of what today's psychologists would call "aversion therapy." Individuals resisting the shots and dope were encouraged to drink as they pleased, but they were unknowingly given a vigorous emetic, apomorphine, at the same time. The nausea and vomiting would then be associated with the alcohol, eventually producing an aversive reaction to drink.[67] As Cheryl Warsh has argued, "photographic images of scores of tough-looking, mustachioed men lined up for their shots manifest part of the secret of the Keeley system's success: its appeal to manly dignity."[68] Keeley himself portrayed his cure in similar terms, comparing it favorably to the asylum treatment of addiction, which he believed stripped men of their manhood as they rested under the vigilant nurse and "guard." Lambasting traditional asylum care in *Opium—Its Use, Abuse, and Cure* (a text in which he also discussed treatment for the abuse of alcohol, chloral, and cocaine), Keeley observed of the asylum patient: "He is made to feel from the outset that he is a lower animal, incapable of thinking, acting or doing rationally for himself. This blow, struck at his manhood, degrades his self-respect."[69]

As suggested earlier, for the medical profession and the physicians who wished to promote the disease concept of inebriety, Leslie Keeley was both friend and foe. J. M. French, the medical director of the Elmwood Sanitarium for Drug Habitues observed in 1896, "The subject of the disease of alcoholism and its medical treatment has been brought prominently to the notice of both the general public and the medical profession, by the rise, progress and decline of the so-called "gold cure" of Dr. Leslie E. Keeley." Keeley's well-publicized campaign to cure inebriety across the land garnered national attention and support for the treatment project. However, continued French, who as the director of an asylum for drug habitués was clearly *not* an uninterested party, Keeley's "absurd claims of 100 percent of cures and not over 5 percent of relapses, his use of secret formulae, and his offensive and unprofessional methods of advertising, all stamp him as a charlatan and pretender of the first degree, rather than a scientific physician and lover of truth." Keeley's greatest contribution, French maintained, was the change he had affected in the public's mind. The great masses now believed that inebriety could be cured by medical treatment, and Keeley's work had likewise stimulated the regular medical profession to pursue a better understanding of the pathology and treatment of alcoholism. Indeed, for this reason, others within the medical profession recognized that Keeley and his proprietary colleagues were not an "unmixed evil."[70]

The American medical profession was initially loath to take the Keeley Cure very seriously. The glacial pace with which general medicine had approached inebriety was indicative of their level of interest, and it rankled the president of the AACI in 1903. President Lewis Mason maintained that for most of the nineteenth century treating the inebriate had been a matter of passing the buck: "The chief reason why the medical profession have, until within quite a recent period, ignored the inebriate is that medical science has from time immemorial refused to admit that the inebriate is a diseased person, and that inebriety is a disease to be studied, but has turned him over to the law and the moralist; and latterly the moralist, and, in a measure, the law, has retired in favor of the charlatan."[71]

H. R. Chamberlain offered a slightly different explanation of the profession's lack of interest in inebriety, noting that the AACI had tested some forty widely advertised nostrums and quack remedies only to discover that they contained inert and "useless" ingredients or alcohol itself. Chamberlain supposed that such disclosures prevented the medical profession from paying Keeley much heed, though the gold cure had captured the imaginations of many in the west.[72] With Keeley's increasing popularity, however, the medical profession was forced to take a position on the doctor and his 95 percent cure rate. After all, Keeley breached two principles that the AMA held dear in its quest for professional power and cultural authority: he maintained secrecy in his treatment, and he advertised his medical services. More importantly, the medical profession could not address Keeley without discussing the "true" nature of inebriety and its treatment. Keeley, after all, was placing inebriety alongside microbial diseases, much as physicians had argued in the pre–germ theory days. Keeley, however, was suggesting that alcoholism was similar to germ-based diseases because each was a type of poisoning—"ptomaine poisoning" in the case of microbial disease, alcohol poisoning in the case of inebriety.

Indeed, part of Keeley's popular appeal lay in his ability to capitalize on scientific medicine's great triumphs. Readers of his optimistically entitled *The Non-Heredity of Inebriety* learned that Keeley did in fact acknowledge hereditary influence. He argued that only certain drinkers became inebriates "simply because so many of the people have inherited a greater or less protection against the poison of alcohol that drinking does not make them drunk nor make them drunkards."[73] However, Keeley focused on alcohol's poisonous effects for the unfortunate one out of twenty-five drinkers (his suggested susceptibility ratio). Observing that "the discovery of the microbe as a cause of disease and its final acceptance by the medical profession, and the crowning of the mythical deities of medicine and hygiene with the laurel of science" were all within recent memory, Keeley continued: "There is but one cause, respectively, for consumption, typhoid fever, small-pox, or other microbe diseases. The remedies, if any are known, are also differentiated and single cures. The same rule holds good in the disease of inebriety. The single cause of inebriety is alcohol. There can be but a single remedy, if the remedy is a scientific cure."[74]

The scientific remedy, of course, was Keeley's bichloride of gold cure, designed to rid the body of the alcohol poison. *The Non-Heredity of Inebriety* was riddled with self-contradiction and the expropriation and distortion of contemporary scientific ideas such as atavism, the law of entropy, and the survival of the fittest. Moreover, in touting his own work as part of the great advance of science, Keeley chastised the medical profession for applying "symptomatic" remedies to alcoholism and insisting that the condition had multiple origins. Such thinking, he claimed reflected old-fashioned, pre–germ theory reasoning. Keeley's chastising of regular medicine for holding a multicausal explanation of alcoholism indicates the degree to which specific causality had become the norm within medical explanations. Ever sensitive to the best marketing strategy, he adopted a framework that was the most "scientific" possible, even if it was not the most appropriate one for viewing alcoholism.

The response to Keeley and his proprietary cure colleagues was typically vitriolic. "It is difficult to see how he can claim permanency of cure for so large a percentage unless the patients are all dead," quipped Elon Carpenter in 1891.[75] New York neurologist William Hammond was also dismissive: "As to the specific influence of the nitrate of strychnia, which has had its day, and the double chloride of gold and sodium, which is now being palmed off on the public as a certain cure for drunkenness, I have only to say that their use in such a connection is most irrational, unscientific, and delusive."[76]

Yet derision was not the only reaction Keeley elicited from the medical profession. Recognizing that "there is no other class of cases that comes under the care of the physician that presents greater difficulties in the way of treatment than that of sufferers of alcoholic inebriety," neurologist Frederick Peterson offered *JAMA* readers an explanation of why Keeley's cure might succeed in some cases. It was not that the specific compound gold bichloride was effective, Peterson suggested, but that it accomplished some success "by virtue of the support by faith or suggestion given the weak will of the patient." What's more, Peterson confessed, though the remedies of "the advertising quacks" were no more effective than regular medical care, and really no different from remedies the regular medical profession had used for years—with the exception of their hyped-up advertising and claims of success—the quacks took "more pains with each individual case than we do." Inebriates were suggestible patients and benefited greatly, Peterson believed, by receiving continuing individual attention from the physician.[77] Others concurred. Years later Irwin Neff, who ran the state hospital for inebriates in Massachusetts, pronounced any remedy that failed to consider the individuality of inebriates doomed to failure.[78] Likewise Neff established a program for the inebriates' aftercare that continued the attention and support begun during patients' stays at his hospital.

Most proprietary "cures" such as Keeley's consisted of the administration of some nauseant—often strychnine, atropine, or apomorphine—some substance that would make the inebriate ill if he were to drink. With these compounds were combined tonics, laxatives, wholesome meals, and the psychical influences exercised on the patient by the physician and these interventions. As we have seen, the Keeley Cure also relied on the camaraderie among patients. These elements of the treatment were psychologically important, for they won the patient's confidence in a way that "the most honest physician" could not, claimed the eminent New York neurologist Charles Dana. What Dana also remarked upon in 1901, a few years after the heyday of the Keeley Cure, was the climbing relapse rates of inebriates taking proprietary cures once their vogue had passed and the prestige associated with them had faded. Under such circumstances, the all-important element of suggestion, what we might call "the placebo effect," was not as powerful.[79] Of course, rising relapse rates may have contributed greatly to the initial decline of the cures.

The relationship between Keeley and the regular medical profession was a complicated one. Without question Keeley was a thorn in the side of a medical profession whose leadership was avowedly scientific and determined to win popular approval and cultural authority. Keeley reminded the public and the medical profession that the proprietary, commercial side of medicine was still alive and well and had not yet succumbed to the new scientific professionalism. Indeed, Keeley claimed that he was more scientific than disease-concept advocates who failed to accept his specific remedy. For all their efforts to discredit Keeley and his fellow proprietary physicians, the AACI and the AMA members who treated inebriates were often assumed to be the same breed of commercial healers as Keeley. In the last quarter of the nineteenth century, the public had little way of distinguishing between the two as each maligned the other. Was Keeley an iconoclastic, self-made medical genius fighting the establishment to save inebriates from themselves everywhere? Or was he a crank, a quack physician out to make a dollar by preying upon the inebriate's (and his family's) desperate hope of finding an easy and sure cure? Whether they liked it or not—and they surely did not—this tainted image of medical hucksterism and quackery hung over the AACI's membership because of Keeley's popularity, even though it distorted the nature of their work.

Yet without Keeley's national fame and extensively franchised cure, the public would have been less likely to embrace the disease concept to the extent that it did in the late nineteenth century. In the 1890s, the AACI appeared to have enough momentum to continue in its own efforts to lobby for state care for the inebriate and to develop new modes of treatment. The general medical profession was prompted to take action against Keeley and an interest in inebriety, if only to rid itself of the physician from Dwight and his proprietary chain. As the AMA editorialized in 1895, at the height of the Keeley campaign, "Within a comparatively recent period, it has become a well recognized fact that the inebriate is diseased and curable by medical means and measures.... The wild wave of gold cure specifics was only possible through the failure of physicians to recognize the physical nature of inebriety, and failure to give these cases the care and attention they deserve."[80]

In 1911, Irwin Neff observed the same "general awakening of the medical profession" to the realization that the inebriate required medical care. He optimistically maintained that the quacks would leave when medicine developed a successful regimen for treating the inebriate. In spite of the strides Neff made at the state hospital for inebriates in Massachusetts, he would have a long wait. Prohibition spelled the end to most state and private facilities for inebriates. Leslie E. Keeley and his successors, however, continued to operate the Dwight, Illinois, institute until 1966.

By and large, then, the general public at the turn of the century had a perspective on alcoholism that owed as much, if not more, to the WCTU's activities and the popular proprietary cures of the day than to disease-concept advocates within the medical profession. The public was further influenced by the many magazine stories about recovered alcoholics, which echoed physicians by advocating psychiatric treatment for alcoholism and by often speaking to the relations among alcohol and masculinity, femininity, homosexuality, and heterosexuality. Reformers and others added to the framing of alcoholism by illuminating the ways in which the stress of modern life and the structure of modern work contributed to both a rise in alcoholism and to the more detrimental effects that alcoholism had in a regimented society.

It was within this intellectual and cultural context that a wide range of institutions for the treatment of inebriates proliferated. That the Progressive Era state was engaged in its own campaign to consolidate and expand its authority within the realms of mental health, public health, penal reform, and social welfare, and was creating new hospitals and reformatories at an unprecedented rate, also helped justify the "institutional solution" to inebriety.

# CHAPTER THREE

# Institutional Solutions for Inebriety

In mid-September 1913, reformed inebriate Thomas Rand returned to New York City from a three-day excursion to Massachusetts. On his trip, Rand had stopped in to see his remaining friends and staff members at the state inebriate hospital in Foxborough. He had also visited his old hometown of Randolph. Upon his return to Manhattan, Rand penned a letter to hospital superintendent Irwin Neff. "My Dear Doctor," he began, "I am writing you these few lines as I promised, to let you know that I arrived back all safe." Reflecting on the previous week's visit, Rand reminisced that long ago in Randolph he "had been made an outcast all from my besetting sin of drink." The response Rand received this time was different, he confided to Neff: "I tell you I felt mighty proud as I walked down the Main Street and was met with the glad hand on all sides, my wretched past forgotten and a warm welcome extended to me wherever I went." For Rand, the trip had been a success, as had the social visit to the inebriate hospital. There he had been able to trade baseball trivia with the staff and see his old friends still in residence. He also had a long chat with the superintendent: "I was deeply touched by your kind words and the confidence you have in me," he told Neff, adding, "I will never go back to the old life and I mean never to betray that confidence and prove to you that you made no mistake when you said I was a reconstructed man and would never drink again." Rand allowed that his "old desire for the stuff" had been extinguished, and "with His hand to guide me," he testified, "I have no fear of the future."[1]

Irwin Neff replied promptly, as was his custom with almost all patients. "My dear Thomas," he began,

> I assure you that I am as much pleased as yourself to know that you have regained your self-respect and that you are able to compete with others in the stress of living. As I told you, after all the real purpose of living is to secure as much happiness as possible. This is accomplished by congenial work and an ethical method of living, by the latter I mean the retention of one's self-respect and conformity to the rules of society. I feel quite sure that your re-education is final and that you will guard against over-confidence. I want you to write to me occasionally and I assure you that anything that we can do for you will be cheerfully done.

Neff followed these words with some local news and then urged his former charge to write to another patient, James Rowan, for "it offers him considerable encouragement and will be a help to him in his fight which he is now making."[2]

This exchange took place in the context of what historians of the early treatment movement have called an "industrial inebriate hospital." As initially envisioned by *Quarterly Journal of Inebriety* editor Thomas Crothers and later characterized by historians, such facilities "blurred any practical distinction between medical and correctional institutions."[3] Yet there was far more going on at Foxborough than prison discipline. The exchange between Rand and Neff illustrates the range of treatment traditions aimed at the inebriate between 1870 and 1920—mutual aid, moral reform, moral or psychiatric therapy, religious salvation, and aftercare.[4] Although all were present in Foxborough's Progressive Era inebriate program, each of these therapeutic traditions characterized different forms of institutional care for inebriates between 1870 and 1920, a period in which inebriate treatment echoed the variety of perspectives reformers entertained with regard to habitual drunkenness itself.

As intemperance gave way to dipsomania and inebriety, and finally to alcoholism, the array of treatment options for inebriates expanded. Homes for intemperate men, with either loose or tight ties to Christian fellowship, were joined by private sanatoriums and asylums for the medical treatment of dipsomania, inebriety, and other nervous diseases. In the 1890s these institutions faced stiff competition from proprietary cures designed to quickly and painlessly slake the inebriate's thirst for alcohol. Finally, in the last decade of the nineteenth century and the first two decades of the twentieth, cities and states began to construct the facilities that they had contemplated for years.

Like the terminology used to describe habitual drunkenness, many of these institutional forms coexisted, and they shared key features. In the early 1870s, for example, Philadelphia's Franklin Reformatory Home, a Christian fellowship–based lodging home for inebriates, offered its patients supportive care while they pursued jobs within the city. At the same time, Joseph Parrish's Pennsylvania Sanitarium just outside Philadelphia offered patients sequestered residential medical care for more protracted periods of time. Although they differed in their emphasis on the inebriate's interaction with the outside workaday world, these facilities both prioritized the role of religion in reforming their patients. Likewise, institutions that began under the auspices of mutual aid societies or Christian philanthropy sometimes attracted state monies, gaining semi-public status, at least temporarily. Boston's Washingtonian Home, founded as the Home for the Fallen in 1857, is a case in point. The home lost its state subsidies when Massachusetts began to contemplate building its own hospital for inebriates, but the Boston Home's methods of care continued to evolve with time. By the early 1900s, it had come to resemble more an acute medical facility for alcoholics than a sober, Christian-influenced boardinghouse for intemperate men who worked in Boston. It even began to coordinate its efforts with those of the State Hospital for Inebriates in the second decade of the new century.[5]

All of this is to say that the varieties of inebriety discussed in chapter 1 were well matched by the varieties of inebriety treatment, and yet these treatments shared common ground. During the Gilded Age and Progressive Era, there was a general movement from morally grounded to medically governed therapeutics—as seen in the conceptual evolution from intemperance to alcoholism. Often, however, the changes taking place in treatment were more a matter of putting old wine in new bottles, of repackaging or emphasizing different elements of a fairly standard, holistic approach to reforming the inebriate. In fact, with few exceptions, habitual drunkenness was recognized and treated as both a vice and a disease. Just as John Shaw Billings had contended in 1876, "The vice is the disease. They are the same thing," so the editor of *The Outlook*,

Lyman Abbott, advocated medical care for inebriates in 1911, while maintaining: "Inebriety, though a disease, has been produced by vice and leads to crime."[6]

Of course, during the Progressive Era individually tailored rehabilitation and reform was the ethic that informed both medical and penal institutions. Indeed, the distinctions historians have drawn between the Christian or morality-based home, the medical somaticism of the asylum, and the penal discipline of the industrial hospital have obscured the degree to which these institutions shared therapeutic orientations and emphasized moral agency as well as biological determinism. As with the disease concepts of inebriety, treatment regimens were heavily influenced by the gender and class expectations of Gilded Age and Progressive Era America. This chapter examines the variety of private facilities created for the treatment of inebriety between 1857 and 1920.

## Early Calls for Inebriate Institutions

Members of the American Association for the Cure of Inebriates were not the first to suggest that habitual drunkards would be served best by institutional treatment. At the dawn of the nineteenth century, physician and temperance agitator Benjamin Rush recommended the construction of "sober homes" where inebriates might receive medical care.[7] Alienist Samuel B. Woodward recommended the same in a series of essays initially published in 1833 in the *Boston Daily Mercantile Journal*. Woodward, the first superintendent of the state insane asylum at Worcester, Massachusetts, had collaborated in 1830 with Eli Todd and other members of the Connecticut Medical Society to investigate alternatives to incarcerating habitual drunkards or sending them to the almshouse or insane asylum. The Connecticut Medical Society's recommendations of 1830 revealed the influence of French alienist Phillipe Pinel and British physician William Tuke. These men had led movements in the 1790s to free the mentally ill from their chains inside the insane asylums of Bicêtre and Saltpetriere and the Retreat at York. Pinel and Tuke advocated the use of "moral therapy" for the insane—a gentle treatment program of healthful food, fresh air, light occupations, and kind attention designed to restore mental and physical well-being—in lieu of keeping the mentally ill in heavy restraint. Similarly, the Connecticut Medical Society's report urged that the inebriate no longer be subjected to penalties such as fines and jail sentences that had "been proven positively injurious to its subjects and ultimately detrimental to society."[8] For Todd, Woodward, and their Connecticut Medical Society colleagues, to provide medical care for inebriates was not only an enlightened response to a tremendous medical and social problem, but it was a means of preventing insanity.

Thus, inebriate homes and asylums rose in the shadow of two other institutions for social deviants: the jail and the insane hospital. Indeed, the first advocates for "sober homes" or "inebriate asylums" were managers of madness, the superintendents of the young nation's first insane asylums. By 1833 Samuel Woodward had worked with the mentally ill long enough to argue that "a large proportion of the intemperate in a well-conducted institution would be radically cured, and would again go into society with health reestablished, diseased appetites removed," and that they would be "afterwards safe and sober men."[9] Inebriates appeared easier to cure than the severely mentally ill, and assisting them might prevent alcohol-related insanity.

Similar experiences also led men such as Rush, Todd, and Woodward to appreciate the advantages of treating inebriates in institutions distinct from the insane asylum. As Nancy Tomes argued in her biography of Thomas Kirkbride, the superintendent of the Institute at Pennsylvania Hospital, inebriates were among the most recalcitrant patients when committed to an asylum for the insane.[10] While intoxicated, they displayed any number of symptoms consistent with a *non compos mentis* diagnosis, but after a few days of confinement in the temperate environs of the insane asylum, no judge would sanction their commitment. The "dried-out" and protesting inebriate became a vociferous obstacle to institutional order.

Decades later, in 1875, when the Association of Medical Superintendents of American Institutions for the Insane passed a resolution sanctioning the creation of inebriate asylums, the topic was raised within the context of the threats to asylum discipline posed by unruly patients such as Elizabeth Packard. In 1860 Packard's husband had her committed to the Illinois State Hospital for the Insane for three years. At the time, the law permitted the confinement of married women and invalids without any of the legal safeguards in place for adult men and single women. Packard and the confinement laws governing the insane in most states received notoriety in the 1870s, when she led a public campaign to reform commitment laws nationwide, following her own allegedly wrongful institutionalization.[11]

In this context, St. Louis physician Charles Hughes, a former asylum superintendent and editor of the *Alienist and Neurologist,* believed that the practice of accepting inebriates and discharging them as "not insane" when they had sobered up led "some suspicious disturbers of public confidence in our asylums" to declare that insane hospitals were in the practice of accepting people who were not mentally ill. "A man insane from the poison of alcohol is as really deranged in mind as from a blow, or the poison of syphilis," argued Hughes, but the insane asylum was not the appropriate place for him.[12] Habitual drunkards were troublesome and unwelcome asylum cases. Alienists paid a price for declaring that inebriety was a form of insanity.

Still, most inebriates were not sent to the asylum but to the local jail—a place equally inappropriate in the eyes of social and medical reformers who were advocating state-based medical care. Often arrested by the police for the misdemeanor of public drunkenness, inebriates were usually charged a fine and released when sober, or they served a short sentence in lieu of a fine. Within the municipal lock-up, lamented Todd and Woodward, inebriates were subjected to the detrimental influences of "real criminals" or "companions who have lost all self-respect and respect and regard for others."[13] What's more, incarcerating or fining inebriates imposed an unfair tax upon their relatives. When the household breadwinner or a wife and mother was jailed or asked to pay a fine, their families suffered too—not just from the moral stigma associated with imprisonment but also from the family member's absence and the loss of emotional and financial support.

The pernicious consequences of incarcerating the inebriate with the criminal were brought to the fore by inebriates themselves at the founding of the American Association for the Cure of Inebriates in 1870. According to patients at the Pennsylvania Sanitarium: "In common life, so intimately mingled is the vice of intemperance with some of the offences of the professional criminal, that to most persons they are but synonyms. The one is but too often added to the crimes of the other, and appearing as they do, thus yoked, in our courts and penitentiaries, it is hardly strange that even the good and virtuous should esteem them identical."[14]

Not only might the association with criminals corrupt the inebriate whose drinking was often the lone offense, but also, claimed the patients, the career criminal's tendency to drink helped shape the public's impression that inebriates were by definition vicious cases. Indeed, as we will see, the first efforts to provide institutional care for inebriates began in the 1840s and 1850s and saw volunteers combing Boston's police courts for public drunkards who might be reformed rather than "recidivized." The tensions between the penal system and medical institutions for inebriates would remain a constant and significant factor in the institutions' evolution.

## Mutual Aid Societies and Liberal Christianity: Homes for Inebriates

The earliest attempts to provide institutional care for inebriates may have taken a page from the moral therapy of the asylum, but they took equal inspiration from the fraternity and confessionals of the Washingtonians, a grassroots temperance movement of the 1840s.[15] An estimated 600,000 individuals joined the Washingtonian movement between 1840 and 1845, taking the group's pledge: "We, whose names are annexed, desirous of forming a society for mutual benefit, and to guard against a pernicious practice which is injurious to our health, standing and families, do pledge ourselves as gentlemen that we will not drink any spiritous or malt liquor, wine, or cider."[16]

Largely working-class and artisanal in socioeconomic composition, this secular temperance drive was a significant departure from its faith-based and elite-led temperance predecessors. There was another significant difference: although many members had not been habitual drunkards themselves, the focus of the group was on reclaiming the fallen. William White has identified seven important elements of the Washingtonian movement's "program of recovery": (1) public confession, (2) public commitment, (3) visits from older members, (4) economic assistance, (5) continued participation in experience sharing, (6) acts of service toward other alcoholics, and (7) sober entertainment.[17]

In practice, this meant that Washingtonians would come to one another's rescue in times of distress (especially temperance backsliding) by furnishing members with food, clothing, shelter, encouragement, financial assistance—and sometimes even employment—until they had recovered sufficiently to sustain themselves. Believing that successful reform came only with the adoption of a new and sober lifestyle, the Washingtonians also organized a range of "substitutes for the saloon," including temperance reading rooms, temperance fairs, temperance marches, temperance concerts, and, of course, temperance lectures.

The Washingtonians kindled public interest in the reform of inebriate men and women to an unprecedented degree during the 1840s. Nowhere was this concern felt more acutely than in Boston. There, three separate efforts in the 1840s and 1850s culminated in the rise of the nation's first home for inebriates. First, the good works of one John Augustus, "a humble mechanic," rescued over 220 inebriates from the clutches of the Boston Police Court between 1837 and 1841.[18] Second, the Washingtonian Hall, established in 1840, contained bunking quarters for new society members who were starting their journey to recovery. Between 1841 and 1845, the Boston Washingtonians were said to have "re-claimed" more than a thousand of the state's intemperate men, and they established three short-lived "lodging houses" for inebriates during these years. Both Augustus and the Washingtonians approached inebriates sympathetically, offering the fellowship, daily staples, temporary lodging, and in many cases, employment, that inebriates required to "resume their position[s] as useful citizens."[19] Finally, in 1857, Mrs. Charles Spear and Rev. Phineas Stone opened the Home for the Fallen in Boston. Though it bore no formal ties to the Washingtonian movement (all it shared with the earlier movement were the name and sympathy for the reformed inebriate), the Home for the Fallen was incorporated as the Washingtonian Home by the Commonwealth of Massachusetts in 1859.[20] From that point on, the institution cared exclusively for inebriate men. It received a state subsidy from the Massachusetts legislature from 1859 to 1871.

Care at the Washingtonian Home in Boston was based on a liberal Christian view of the inebriate as a fallen individual who could, through the exercise of his free will, return to sobriety and grace. This Christian orientation represented a strong departure from the original Washingtonians, who eschewed church involvement in the inebriate's recovery. According to the Washingtonian Home's first superintendent, Albert Day, "The first great principle of the Home was that the drunkard may be saved. . . . This was to be accomplished, so they professed, by the miracle of love which was begun by Christ and has been repeated daily ever since wherever Divine example has been followed. No restraint has been put upon the patient."[21]

Restraint was kept at a minimum, and patients were allowed to leave, go about their business in the city, and return to their "dry" lodgings on the honor system. Days began and ended with prayer, and group meetings occupied evenings as well; here new and old patients described their experiences to one another. Although the core of the therapeutic program remained more moral than medical, the balance would change under the direction of Albert Day, a businessman and legislator who became the institution's superintendent in 1857. Realizing that the medical needs of inebriates required special care, Day matriculated at Harvard University's medical department in the early 1860s and took his medical degree; he also wrote *Methomania* (1867), a guide for laymen to the treatment of inebriety.[22] If Day had attributed his earliest success in reforming inebriates to Christian philanthropy and "the law of human kindness," he had altered course by 1866, when he declared that "it is our rule that the foundation of all our efforts should be such medical treatment as will promote in the patient a healthy and vigorous action of all the organs of the body, thus indirectly by the close relationship between mind and body creating a moral tone and strengthening and invigorating the will."[23]

The neo-Washingtonian regimen of the Boston Home included group confessional, prayer, and pledging, but it was effective only if "the physical system is put in thorough repair and reinvigorated," added Day.[24] Moreover, the strengthening of one's moral tone relied in part on building up one's physique. In this regard, the Boston Washingtonian Home more resembled the private inebriate asylums, which espoused a view of the will that was rooted in biology. Yet, as Day and other members of the medical profession argued with increasing regularity in the 1870s, patients needed to cooperate in their treatment; they needed to earnestly desire reform. When he retired from the Washingtonian Home in 1893, Albert Day estimated he had assisted some 30,000 drinkers.[25] The Washingtonian Home became the Washingtonian Hospital in 1939 and continued to serve alcoholics until the 1980s.

The second neo-Washingtonian Home (also without connection to the Washingtonian temperance movement) was founded in Chicago in 1863 by "a few philanthropic men," including the first president of the AMA, Nathan Smith Davis; Robert A. Law, a Chicago publisher; and the Honorable Charles Hull, whose own home later became Jane Addams' famous settlement community, Hull House.[26] The Chicago founders used their Massachusetts counterparts as a guide in establishing their own home; they even adopted the Washingtonian name. Initially funded through the voluntary contributions of the Grand Lodge of Good Templars (a temperance fraternity that succeeded the Washingtonians) and the support of other friends, the Chicago home drew financial help between 1867 and 1875 through liquor license revenues collected by the City of Chicago. As was the case in Boston, the Chicago Washingtonian Home endeavored to make the inebriate "feel that he is regarded as 'a man and a brother.'"[27]

A social means of cure, "pre-eminent above all others," claimed Superintendent P. J. Wardner in 1870, "is a proper recognition of manhood in men. When we treat men as though they were men, they will feel and act as such. If they are cursed and trodden down by the better classes, as if to crush out all that remains in them of a manly nature, they will not come to a knowledge of truth."[28] Thus, for Wardner—as for many of his peers—recovery from alcoholism meant the recovery of one's manhood, and it was usually directed at men of "lesser" rather than "better" classes. It meant replacing a dissolute brand of masculinity with respectable manhood: a "moral manliness," in the words of Gail Bederman, that prioritized "strength, altruism, self-restraint, and chastity," as well as sobriety.[29] Fraternal temperance orders aided the Chicago Home's efforts to regain inebriates' lost manhood. A Lodge of Good Templars and a Division of the Sons of Temperance held weekly meetings in the Chicago Home, and the patients held their own gatherings for discussion, conversation, and social intercourse on Friday evenings, with religious meetings on Sundays. Sympathy and "the law of kindness," claimed Thomas Vancourt, the Chicago Home's physician-in-charge in 1871, were the most essential guiding principles for inebriate treatment. Without them, Vancourt maintained, all was lost. Medical treatment was useful, but it was secondary to "the moral and religious means to strengthen the purposes and resolutions of reformation."[30] Although the majority of patients treated at the Chicago Home were men, a branch of the home was established in 1869 as the Martha Washington Home for female inebriates. The project was abandoned after a few years but was retried in 1882 with success. Both the men's and the women's facilities were closed following Prohibition, but in their institutional lifetimes, they treated some 300,000 alcoholic men and 4,251 inebriate women.[31]

The Franklin Reformatory Home for Inebriates offers another picture of the mutual aid–based, Christian lodging home for urban inebriates—one that also brings to light the difficulty of balancing moral and medical therapeutic frameworks. Organized in 1872, the institution was named after Benjamin Franklin. The founding committee used the neo-Washingtonian Boston Home as its model, but where the Boston Home was growing more medical in its orientation during the 1860s and 1870s, the Franklin Home moved in the reverse direction, de-emphasizing the importance of medical care. Indeed, so far did the Franklin Home stray from the AACI's mission statement that the facility's annual report for the 1873–74 fiscal year, read at the AACI's yearly meeting, was returned to its author for revision before inclusion in the meeting proceedings. Moreover, it occasioned a heated debate within the AACI that continued into the following year.

In his annual report, Robert Harris, president of the Franklin Reformatory Home, stated that his institution did not "either in our name or management, recognize drunkenness as the effect of a *diseased* impulse; but regard it as a habit, sin and crime, we do not speak of cases as being cured, as in a hospital, but 'reformed.'" Medicine, in other words, might be necessary to address the drinker's infirmities, but morally, even after medical treatment, a man was "the same drunkard that he was before, unless you have in some way touched the root of evil in his moral nature," concluded Harris. For the Franklin Home medical director, it was always crucial to address the root cause of drunkenness: the train of thought that led the drunkard to drink.[32] At the Franklin Home, this meant the "temporary seclusion" of inebriates, followed by a vigilant effort—usually a few weeks in duration—to break all habits (especially tobacco use) associated with drinking.

Although Harris contrasted his institution's remedy with the AACI's approach (which was based on the disease concept of inebriety), the protocol at the Franklin Home, as Harris described it, seemed remarkably similar to the approach of the two other Washingtonian Homes. The primary elements of care were "to settle the stomach; check diarrhea if it exist; arouse the appetite; cause sleep, and overcome the depressing effects of alcoholic poisoning, and abstinence from food, by an abundant and well chosen diet frequently repeated during waking hours."[33] This treatment was applied during the patient's temporary isolation, usually just a few days. Thereafter, he was free to participate in the reformatory home's literary society, its temperance meetings, its religious services, and other social activities. As was the case in Chicago and Boston, patients pursued their work outside the inebriate home during the day. Given these similarities, why the controversy?

In the discussion that followed Harris's paper, Joseph Parrish suggested an answer. Those who were reluctant to accept "the disease theory . . . fear to acknowledge the fact of disease, lest it shall in some way compromise their views of the responsibility of the drinker. Their allegiance to the moral aspect of the subject is so strong that the real facts in the case are obscured." The truth of the matter, claimed Parrish, was that all of "our institutions, whether avowedly acting as reformatories, or as hospitals and asylums," treated the inebriate as a diseased person. The disordered appetite that Harris highlighted in his report was "the alcoholic diathesis" in most cases, observed the AACI president, and whether acquired or inherited, it was necessary "to cultivate a sensitive moral perception" in these patients and surround them "with the most ennobling social influences, that they may rise above the moral disorder which is their inheritance."[34]

Parrish's words were conciliatory. As president of a young organization promoting the disease concept of inebriety and a new specialty devoted to its cure, he stressed the common elements of care and the moral issues addressed within all forms of treatment. Other urban, Christian fellowship–based inebriate homes of the 1870s and 1880s also discussed their methods of care in moral terms. Consider the New England Home for Intemperate Women, founded in 1879. This institution relied "under the blessing of God, upon two great instrumentalities—labor and the moral and religious influence of a Christian Home." Like the other homes for inebriates, the New England Home for Intemperate Women applied a short course of medical treatment to restore the inebriate's health: "A bath, a strong dose of medicine, and a good sleep are the first things they need." As soon as their health had begun to return, however, the women were put to work in the laundry, kitchen, and sewing room, learning skills that they could apply inside their own homes or in those of potential employers. Cultivating habits of industry, claimed the home's matron, Mary Charpiot, was an essential part of reform, for "they begin to feel respectable, and to regain their self-esteem; and, when the proper time comes, they are prepared to go to such places as we may secure for them in the city, and more especially to the country."[35]

As with the men in the neo-Washingtonian and Franklin homes, the aim was to keep the inebriate women employed and sober. More than half of the 249 women treated at the New England Home between 1880 and 1881 were single, and some 80 percent were under forty. No doubt many had come to the city for employment. Where men customarily worked outside the inebriate home, women occupied their days engaged in domestic labor in various business settings—laundries, tailor shops, and hotels. This work was designed to retrain them for either fiscal independence or satisfactory homemaking. Charpiot's remarks above echoed the concerns of inebriety physicians such as Lucy Hall about the restorative properties of country life. A country "situation" was always a more desirable placement for these women, according to the New England Home's annual reports, not only because Hall and her medical contemporaries held modern, urban life responsible for neurological and psychological problems of all varieties, but because rural life was free of the seductions of the city—dance halls, pubs, saloons, and less than savory boarding houses.[36]

The promise that rural life held for restoring women to sobriety was matched only by the atmosphere of a Christian home, claimed Matron Charpiot. Indeed, "home" was more than a label for the neo-Washingtonian inebriate institutions. What Charles Rosenberg has observed for the early-nineteenth-century general hospital—that "the web of relationships within the hospital was always seen in terms of a family writ large"—also obtained for the inebriate home.[37] P. J. Wardner, superintendent of the Chicago Washingtonian Home, praised "the hallowed influence of a family circle" and endeavored to create an atmosphere in which men were made to feel comfortable, at peace, and supportive of one another. In its own efforts to recreate a homelike atmosphere, the Franklin Home hired a matron—"a zealous, sincere, Christian lady—seated with Bible in hand" who was reputedly beloved by all the "inmates," for she had "not only their temporal but their spiritual interests at heart, and could do no more for her own children than she does for them."[38] Likewise, D. Banks McKenzie, manager of the Appleton Temporary Home for Inebriates in Boston, concluded from experience that "an asylum in name, but a *Home* in reality, is necessary to the treatment of the inebriate."[39]

Several concerns made the image of the home a vital one to the rehabilitation of inebriates. Rosenberg has argued that for the first two-thirds of the nineteenth century, hospitals were organized along domestic lines, usually headed by a married male superintendent, who might or might not have been a physician; the superintendent's wife routinely served as matron, supervising the laundry, kitchens, and certain staff. Most people in the first half of the nineteenth century, of course, did not go to the hospital when they were sick; instead, they were treated at home by their families. If they could afford it, families sometimes employed a physician who came to the bedside. Hospitals were for the poor and those without friends or relatives to nurse and support them in their time of need. Large immigrant populations, who often lacked financial resources, and the mobile masses who left their countryside homesteads to pursue employment in the city created a growing demand for the "care of strangers." Thus, until late in the nineteenth century, when issues of medical authority, access to patients, and new technologies altered the nature of the hospital, it was organized around a domestic model.

While inebriate homes drew from the domestic traditions of the hospital, there was another compelling reason for providing a home treatment setting for inebriates. If the late-nineteenth-century hospital was transformed from an institution *in loco domūs* into a hall of science where patients received care unavailable to them in their homes—e.g., X-rays, antiseptic and aseptic surgeries, special hydrotherapeutic measures—this transformation had little application to the case of habitual drunkards. Reformers might have wished for an X-ray of the inebriate's soul to guide their efforts, but the treatment of this hybrid medico-moral condition was more a psychosocial rehabilitation exercise than an application of technology. Indeed, as early as 1866, Boston Washingtonian Home superintendent Albert Day observed that "medical science and pharmacy have never yet discovered a specific for drunkenness and probably never will."[40] As we have seen, regular physicians who treated the inebriate in the 1890s and the early twentieth century denounced the specific or patent cures widely promoted in local newspapers throughout the United States because they smacked of quackery. The same was true for the earlier Christian-influenced inebriate homes, which emphasized moral reformation in cases where "the sad experience of most of them, up to the time they come to us for shelter, has been such that they have forgotten all the good influences of their own homes, if ever they had any."[41] If moral education was at the heart of the inebriate's recovery, as the superintendents maintained, it was only logical that it would take place in a home, for home was where moral instruction and medical attention began for most people. Highlighting the importance of class in treatment, Joseph Parrish also emphasized in 1867 that "with inebriates, it is especially necessary to disturb as little as possible the domestic idea of life, in any transfer of residence that may be made."[42] Parrish regarded "home" as the most comfortable spot for recuperation, and he advocated the creation of separate cottages to keep men of the same class background together, for he assumed that they would feel more "at home" in their customary social milieu.

The "home" context of treatment made sense in yet another regard: intemperance destroyed homes. In the late 1870s, the WCTU would erect their anti-alcohol campaign on a "home protection" platform, highlighting the damage done to families by drink. Beneath the crime of public drunkenness that so concerned reformers lay private or domestic failure. Treatment at the inebriate home—with its patriarchal or matriarchal organization and its advocacy of fraternal and sororal relations—was meant to reawaken an appreciation for home life. It promoted intimate bonds and supportive sociability among men and among women, each group joined by their shared desire for sobriety.

For women inebriates, the home was, of course, of special significance. The gendered world of nineteenth-century America regarded women as the keepers of hearth and home as well as the protectors of morality. The habitually drunk woman had failed in both regards. To reformers, her recovery was commensurate with her ability to embrace and fulfill her domestic roles. Thus, remedial "home training" was an essential part of inebriate reform for women. While inebriate men worked outside their sober lodgings, the inebriate home offered women their room, their board, *and* their employment until they were deemed ready for placement in the alcohol-ravaged world outside. Rehabilitation through work and prayer, if successful, would leave the reformed women "filling useful situations in families, or enjoying the happiness of their own homes."[43]

Concerned with "home protection," WCTU leader Frances Willard urged her American temperance legions to consider modeling treatment for inebriate women after the Duxhurst Farm Home and Industrial Colony in England. There, the home matrons not only encouraged women to cultivate their domestic skills, but they also brought orphans from the Society for the Prevention of Cruelty to Children to Duxhurst to nurture the women's maternal instincts and serve as a reminder of "the awfulness of the sin that destroys the love of the mother for her own children."[44] At Duxhurst, there was no better way to prepare the women for return to their own homes or to those they would tend after treatment than to encourage their maternal instincts. Thus, if men were to engage in a heroic struggle to regain their manhood through treatment, women were supposed to return to their natural state of purity, virtue, and selfless concern for others through the cultivation of habits of industry and religious devotion.

With their emphasis on the moral and social dimensions of care, these early inebriate homes reflected the transition from "intemperance" to "inebriety." Though the AACI considered both constructions "diseases," intemperance prioritized the volitional and moral aspects of habitual drunkenness; the eradication of "bad habits" and "evil appetites" was the focal point of inebriate home care. Yet, if these institutions emphasized that it was never too late for the inebriate to reform, they also stressed the need for the patient to truly desire sobriety if treatment were to be successful. Medical care was an essential element of treatment, as the alcoholic's system was cleansed of liquor. Urban-based, Christian-influenced, voluntary, and rooted in the tradition of mutual aid, the inebriate homes put into practice elements of care that persisted into the twentieth century, even if the private asylums and state and municipal facilities that succeeded them emphasized the value of legally enforced long-term care to address the hereditary and physiological aspects of habitual drunkenness.

## Private Inebriate Asylums and the Somatic Approach

Founded upon the disease concepts of dipsomania, inebriety, and alcoholism, the private inebriate "asylum" was advanced by physicians as a truly medical institution, in contrast to the inebriate "home." Historians of the "inebriate asylum movement" have routinely drawn a line between the promoters of the urban homes just discussed and those who campaigned for asylum treatment: "Unlike the neo-Washingtonian physicians, they belonged to a younger generation less impressed with the spirituality of religion than with the materialism of natural science. They construed human behavior as an epiphenomenon of neurological activity that had no relation to an immaterial soul. . . . The somaticists dismissed metaphysical questions of residual freedom and human purpose. . . . [They] discounted and even pilloried anything that smacked of spirituality, moral heroism, or voluntarism."[45]

This distinction has some merit, for the early advocates of "home" treatment rarely if ever suggested the coercive confinement of inebriates that the asylum "somaticists" championed. Moreover, the managers of the neo-Washingtonian homes rarely regarded medical intervention to rebuild diseased wills as the key to therapeutic success. However, a strong case may be made for seeing this therapeutic distinction as one of degree rather than type. For example, far from pillorying "spirituality, moral heroism, and voluntarism," private and even public inebriate asylum superintendents and attending physicians often embraced religion as a valuable element in the habitual drunkard's reform. Even J. Edward Turner, the founder and first superintendent of the New York State Inebriate Asylum, solicited the support of the clergy for his new institution. A famous Unitarian pastor and president of the United States Sanitary Commission, Reverend Henry W. Bellows delivered several sermons on the necessity of a new asylum for inebriates, as did the president of New York City's Union Theological Seminary, Reverend Roswell D. Hitchcock.

Henry Ward Beecher, famed liberal Congregational minister and orator, like-wise contributed to the construction fund of the new asylum.[46] In fact, asylum physicians frequently construed the inebriate's recovery in terms of moral struggles, battles, and fights, and they much preferred voluntary patients. The first two inebriate asylums in the United States, the New York State Inebriate Asylum and the Pennsylvania Sanitarium, offer a glimpse into the difficulty of adhering to any sort of rigid interpretation of these institutions. Their thera-peutic strategies were more elastic and eclectic than previously portrayed.

The New York State Inebriate Asylum was the brainchild of one man, physi-cian J. Edward Turner, who pursued his goal of establishing the first asylum for the medical treatment of inebriates with missionary-like zeal. Turner, who had taken an interest in the morbid anatomy of inebriety while studying medicine, invested some eighteen years in the establishment of his asylum, which was finally situated in Binghamton, New York. Along the way, he enlisted the help of a variety of prominent individuals, including the acclaimed New York City surgeon Valentine Mott, who became the first president of the asylum; con-gressman, Civil War general, and Radical Republican Benjamin F. Butler; *Knickerbocker Tales* author and diplomat Washington Irving; and presidents Martin Van Buren, Millard Fillmore, and James Buchanan. A charter for the private asylum was given by the New York State Legislature in 1854, and Turner laid the cornerstone for the new institution in 1858. When the New York State Inebriate Asylum opened to patients in 1864, 2,800 voluntary ap-plications for treatment had been submitted.

Like the Chicago Washingtonian Home, the asylum at Binghamton derived much of its operating budget from the liquor trade. The New York legislature passed a law in 1859 that allocated 10 percent of the state's alcohol-related rev-enues to the asylum. But there were some significant differences, at least ini-tially, between the management of the neo-Washingtonian Homes and the direction of the asylum at Binghamton. As the hospital's first superintendent, J. Edward Turner successfully lobbied the state legislature to pass three laws that revealed the coercive nature of his treatment program. First, a bill was passed in 1864 that prohibited the sale of alcohol within one-half mile of the asylum; second, a provision of this law prohibited citizens from passing through the grounds of the asylum without written authority from one of its officers; and third, another provision gave the supreme court and the county courts the power to commit inebriates to the asylum for three months. The term of confinement was lengthened to one year in 1865.[47] Valentine Mott, the asylum's first president, recommended that the asylum be given permission to establish its own police force, "to guard the Asylum from without, and to protect its patients within," and the state legislature granted the asylum this power in 1866.[48] All of these measures are consistent with the image of the medically oriented inebriate asylum as a place of coercive medical authority.

Mott died in 1865 and was succeeded as president by Willard Parker, one of the original supporters of the inebriate asylum. Parker's and Superintendent Turner's visions of the ideal inebriate asylum differed, however, when it came to the levels of coercion and restraint required to successfully treat inebriates. Turner championed prolonged treatment (preferably a year) within a tightly governed asylum (viz. the asylum's police force). His methods were revealed and reviled by popular political journalist James Parton in his 1868 *Smoking and Drinking*: "His [Turner's] control and management of everything connected with the institution has been as absolute in fact, if not in form, as if he were its sole proprietor. . . . If he had a theory, it was that an inebriate is something between a criminal and a lunatic who is to be punished like the one and restrained like the other."[49]

Parker objected to this style of care, and many of the patients and their families found it excessive as well—particularly those inebriates who came from middle- and upper-middle-class families and were accustomed to governing their own lives.[50] Relations between Turner and Parker, each of whom was backed by factions of hospital trustees, deteriorated so badly that in 1867, when the main building at Binghamton burned to the ground, the board forced Turner out of his position and accused him of arson.[51]

Immediately following Turner's dismissal, Albert Day, formerly the superintendent of the Washingtonian Home in Boston, took the reins at Binghamton, and under his leadership the asylum replaced a coercive regimen with a style of therapy that more resembled the moral suasionist techniques of the neo-Washingtonians. Visiting the asylum in June 1868, James Parton observed that the patient population was almost exclusively voluntary, that "a large majority of the present inmates are persons of education and respectable position, who pay for their residence here," and that patient accommodations resembled "a spacious, handsome, well-arranged, and well-furnished hotel." Regarding inebriate "homes" and "asylums," Parton also commented that the institution was "in fact, as in appearance, a rationally conducted hotel or Temporary Home and resting-place for men diseased by the excessive use of alcoholic drinks."[52] In other words, Day was running the asylum much as he had governed the Boston Washingtonian Home, which Parton also praised. The superintendent attempted to treat each inebriate individually, and he confined men to their rooms only if they yielded to their craving and escaped to the city to drink. One in ten patients was subsidized by the state; the others paid according to their ability, and they paid three months in advance. This initial expenditure offers partial confirmation of both Parton's claim and that of an anonymous inmate at Binghamton who published an account of life at the asylum one year later: the gentlemanly, professional, and middling classes were well represented at the New York State Inebriate Asylum.[53]

In a sense, observed Parton, no "treatment" was offered patients, for there was no specific or patent medicine given to patients to "cure" them of drinking—a comment that revealed as much about popular definitions of medical therapeutics as it did about asylum life. Instead, the institution "simply gives its inmates rest, regimen, amusement, society, information. It tries to restore the health and renew the will, and both by rational means."[54] An anonymous patient, who identified himself only as "an inebriate at this Asylum—Congenital, Periodical, anxiously Hopeful," reported that the daily routine consisted of breakfast at 8 a.m., followed by prayers in the chapel, morning physical exercise and reading, an 11 a.m. mail call, and dinner at 1 p.m.. More exercise, light work or amusements in the afternoon, and a 6 p.m. mail call and supper preceded the evening's entertainments. Groups of inebriates met in the chapel to read poetry and discuss novels on Monday evenings. On Tuesdays and Saturdays, the literary society, the Ollapod Club, met to listen to patients give papers about a wide range of topics—from "The Hindoo Girl of the Period" to "Arctic Exploration and Adventure" to "George Frederick Cooke, Actor and Drunkard." Wednesday evening meetings were led by Day, who discussed different aspects of temperance "with all the plainness and good-humor, and much of the drollness, of the familiar 'Dutch Uncle'"[55] Thursday evenings were devoted to patient-led dramas and musical performances to which the townspeople of Binghamton received formal invitation. "If I were asked wherein lies the peculiar healing of this place," commented the anonymous patient in 1869, "I should answer in the profound impressions of its sympathetic intercourse; for here my trembling trouble is met with unstudied appeals transcending the eloquence of Gough [John Gough, the temperance orator] and confronted with pictures of pain beyond the eager, tearful utterance of Vine Hall. This anxious little world of ours is moved by the moral power of its own public opinion."[56]

By "public opinion," the anonymous author meant the peer pressure patients exerted on one another and the assistance each lent to the next man in his struggle for sobriety. This was no mere instance of goodwill. In their actions toward one another, the anonymous patient opined, the men at Binghamton presented "an example of pure democracy—quintessential Americanism, asserting itself in that freedom of opinion to which there is no limit but generosity." Here was "a free school of manners, equal rights, and common sense, where are taught the fair play of the Golden Rule."[57] Fraught with idealism, the patient's interpretation nonetheless suggests that inebriety treatment not only afforded men a microcosm of domestic relations but also offered them a model in republican virtue. As Jonathan Zimmerman has suggested, the science of inebriety may have been "peculiarly American" in its origins, but it was a moral and political science, as well as a medical one.[58] Under Albert Day, treatment at the New York State Inebriate Asylum was holistic and humanistic, revealing the power of one superintendent to radically alter the nature of care and the nature of the institution. The similarities between the neo-Washingtonian home ethos and the asylum under Day's management are evident.

The nation's second private inebriate asylum, the Pennsylvania Sanitarium, was established by the Citizens Association of Pennsylvania under the direction of Joseph Parrish in 1867. The group had received a charter to "purchase lands and erect buildings for the cure of the intemperate, to examine into the causes and statistics of pauperism, vagrancy, and crime, and do whatsoever may seem to be practicable for the prevention of these evils."[59] Fearing that the terms *intemperance,* and *inebriety* would prove too stigmatizing, the Citizens Association named their new institution without reference to either. In their view, alcohol was a product of nature; science imitated nature by developing alcohol in a variety of forms, some more useful and others more dangerous. To prohibit the consumption of alcohol was foolish and denied the natural status of both the substance and the human appetite for it; a much better solution was the creation of inebriate asylums to help establish limits for those who could not control their appetite.

Parrish and the Citizens Association planned to construct the sanitarium according to the "cottage system," which allowed for small residential buildings rather than "large and lofty institutional buildings."[60] Parrish's choice of the cottage plan was novel, and it may well have been the first time that it was employed in North America. New York State's Willard Asylum for the chronic insane, opened in 1869, was organized similarly, and the trustees of the state hospital for inebriates in Foxborough chose the same design.[61] The plan made sense from two perspectives, claimed Parrish. First, the cottage plan made it possible to segregate patients according to their type, "whether it be based on the form of alcoholism, or on social position, or on the term of residence."[62] Second, the smaller living quarters of each cottage resembled the domestic model of living that was essential to the restoration of patients: "each family will have a head, to whom should be entrusted the conduct of the household, and with whom, each member should be in daily intercourse." At all costs, the sanitarium was to avoid an atmosphere where patients felt "institutionalized," under the gaze of "a corps of attendants, with jingling keys, always in waiting to discover, and report improprieties."[63] The inebriate men had to be trusted and given responsibility if they were to become strong, trustworthy heads of households.

Indeed, "medical treatment" was more ancillary than primary, as Joseph Parrish pointed out in 1871, after discussing "the moral agencies which assist in the process of recovery." For Parrish, the mutual aid inebriates gave one another, the distance from their old surroundings, and the rest, healthful amusements, and regularity of schedule were all "moral agencies." In addition, added the Quaker physician, "Hygienic and medical appliances, such as are adapted to the individual necessities of inebriates, good diet, exercise, Russian baths, and suitable intellectual pursuits, are among the additional means employed, to restore the balance between the mental and visceral functions, and allow the patient to recover himself and prove his manhood."[64]

Again, as with the inebriate homes, the recovery process for male patients was framed in terms of proving one's manhood: that is, achieving self-sufficiency, adopting a stable and respectable lifestyle, and supporting one's family. Having outlined the course of therapy pursued at the sanitarium, Parrish directed his remarks to the inebriate himself: "You are not insane, and do not require the restraints of insanity. You are morally infirm, and need self discipline and culture. You are physically diseased, and need medical treatment. You cannot afford any longer to submit to the foe that has so long bound you. Your own secret yearnings for a better life tell you this."[65] Just as Parrish had addressed the opening meeting of the American Association for the Cure of Inebriates with a paper entitled "The Philosophy of Intemperance," so he continued to embrace the moral dimensions of "the disease."

It appears, then, that some of the most prominent private asylum superintendents were committed to a medico-moral model of inebriety that emphasized the moral struggle to regain one's lost manhood in a domestically oriented environment. Admittedly, patients may have paid their room and board up front and been expected to stay at the hospital for three months or longer, but most submitted to a new routine that was in the main not coercive. My intent here is not to argue that the asylums were identical to the homes, for they were not. It bears mentioning, however, that while the AACI's leadership advanced views that favored prolonged medical sequestration for the treatment of inebriety, *in practice*, their institutions combined medical and moral reform, harked back to the heroic struggle for manhood, and prioritized an atmosphere of routine domesticity.

An excellent example of the contrast between treatment rhetoric and actual treatment can be found in Edward Mann, who ran the Sunnyside Medical Retreat for inebriates in Fort Washington, New York, a private asylum. In *The Nature and Treatment of Inebriety also the Opium Habit and Its Treatment,* Mann insisted that "an inebriate asylum, where the patients cannot be detained by law until cured, loses much of its efficacy. It must be understood that the inebriate is suffering from a physical disease, and that he is committed to an inebriate hospital to stay until cured, whether that cure takes for its completion six weeks or six years; and that when in such an institution he is to submit to the discipline of the house."[66] Mann viewed discipline and confinement as essential to treatment. Yet, when he described his own asylum, Mann portrayed Sunnyside as a place "unsurpassed for its beauty of location and surroundings, and with handsomely furnished apartments" offering patients "all the attentions, comforts and attractiveness of a home, together with medical treatment." Billiards, rowing, driving, and horseback riding were all part of the therapeutic regimen at Sunnyside, for "physical and mental labor is absolutely necessary for restoration to health."[67] Were billiards and horseback riding part of "the discipline of the house"? Was Mann deliberately misleading patients—luring them to the retreat with promises of luxurious accommodations and pastimes, only to subject them to discipline and confinement?

The director of the private Walnut Lodge Hospital, Thomas Crothers, offered insight into this apparent conflict. In "Private Asylums and their Difficulties," Crothers observed that in order to succeed, private asylums "must have alternate restraint and liberty."[68] In a public asylum, noted Crothers, "the patient expects to be considered one of many and follow some general rules, but in a private asylum he assumes the air of a guest, and because he pays demands obsequious care and attention, and assumes that he can dictate in many things as a mere business." Yet patients were patients, not "customers," Crothers insisted, and when they found that they were feeling better after a few weeks but were unable to come and go from the inebriate asylum "at will ... as if in perfect health," they objected. Restraint, according to Crothers, was best applied at intervals to meet the special demands of each inebriate, and the appropriate application of restraint came only with understanding the idiosyncrasies of individual patients. Each case was unique and required different treatment to address "conditions of neurasthenia, hysteria, nervous instability and irritation, with all the complex emotional, nutrient, and mental disturbances."[69]

Thus, the managers of private asylums found themselves in the position of making their facilities as attractive as possible to their paying patients while trying to keep tight reins on the same individuals, avoiding their premature departure from the asylum—and as Crothers acknowledged, institutions often proceeded without sufficient legal authority to hold their patients. As we shall see, the problems that private asylums experienced in restraining their clientele paled in comparison to those faced by public institutions for inebriates.

There were, however, private asylums where treatment was relatively brief and patient retention was usually not a contentious issue. These were the proprietary clinics.

## The Business of Inebriety Treatment: Leslie E. Keeley and Charles B. Towns

Some 400,000 people received care at Keeley Institutes across the United States prior to Keeley's death in 1900; another hundred thousand had taken the cure by 1920.[70] Yet this popularity won few friends within the regular medical profession. Members of the AACI and their colleagues in general medicine approached proprietary cures such as the Keeley Cure with skepticism and ire. In a promotional brochure for the Walnut Lodge Hospital, Crothers summarized the orthodox position on proprietary cures clearly: "Walnut Lodge Hospital has no specific Gold cures or new mysterious drugs, to produce permanent restoration in a few weeks. Inebriety is a disease of the brain and nervous system, and there are no short cuts to health. Restoration and cure is only possible by the use of exact physical and psychical remedies, with the aid of nature in the best surroundings and conditions. This requires time and the best possible scientific judgement and experience."[71] Equally clear, however, was the fact that the public did not share these sentiments. Many believed in mysterious new drugs, and Leslie E. Keeley's bichloride of gold cure, relying on a relatively short, month-long stay, was immensely popular.

We know few details about what went on at the 125 different Keeley franchises operating across the United States and in Canada between 1880 and 1920, but according to William White, they used the "parent institute" at Dwight, Illinois, as their model, and treatment there was meticulously monitored. Patients arriving at the facility in Dwight were welcomed into an informal and hospitable atmosphere and were provided with as much alcohol as they cared for until their appetite for it had subsided. Patients boarded at a nearby hotel or in the private homes of townspeople, but all were required to appear four times per day for their bichloride of gold shots, lining up at 8 a.m., 12 p.m., 5 p.m., and 7:30 p.m. The injections were taken from bottles of *red, white,* and *blue* liquids and administered in varying quantities—therapy at the Keeley Institute, if not a microcosm of republican government, was patriotic. In addition, at two-hour intervals throughout the day, patients received individually tailored tonics.[72]

Patients were free to walk the grounds at the institute and to stroll through Dwight as they desired. Indeed, White has argued that the town of Dwight was an early "therapeutic community." The largest employer in the small town of Dwight, the Keeley Institute helped keep the local economy afloat with its patients and staff patronizing local restaurants, businesses, and lodgings. In turn, the townspeople kept an eye out for the patients. Addiction treatment became Dwight's "mission," and many patients chose to relocate to this "company town" following their four-week session at the Keeley Institute.[73]

Many parties, particularly within the regular medical profession, attempted to analyze Keeley's secret cure; reports of the ingredients varied widely and they included: "alcohol, strychnine, apomorphine, aloin from the aloe plant, willow bark, ginger, ammonia, belladonna, atropine, hyoscine, scopolamine, coca, opium, and morphine."[74] It is difficult to say how accurate any of the varying analyses were, since they usually stemmed from Keeley's critics—people who lumped his work together with the worst of the medical quacks. Historians have observed, as some of Keeley's contemporaries concluded, that there was much more to Keeley's gold cure than the medication.[75] Indeed, patients were required to attend institute lectures; they were prohibited from smoking cigarettes, drinking sodas, placing bets, driving automobiles, and fraternizing with patients of the opposite sex. Keeley also urged patients to rest, eat well, share their experiences with one another, and develop healthy pastimes. No doubt these measures conduced to recovery for some. Perhaps even more significantly, the treatment regimen at the Keeley Institute offered patients a psychological advantage. Remarking on this, famed New York City neurologist William Hammond concluded that "Strychnia or gold or almost anything else will, when used in the case of sensible persons who sincerely desire to be cured, and who are imbued with confidence in the physician, prove efficacious. I have cured patients with a daily hypodermic injection of a few drops of water combined with the influence or mental predominance which I happened to have over them."[76] Whether through simple suggestibility or holistic regimen, thousands of Keeley's patients found a path to sobriety. Treatment at the institutes took only about a month.

Keeping patients on this path was the greater challenge. By all accounts, it was the widely recognized relapse rate that hurt Keeley's credibility and initiated the decline of the Keeley dynasty by the nineteenth century's end. Yet the Keeley Leagues—local assemblies of former patients (and in the case of the women's leagues, reformed female inebriates and the wives, daughters, and sisters of reformed men)—offered patients a community and journal, *Banner of Gold*, to keep them thinking sober thoughts. Though the leagues were composed of a small fraction of the number who took the cure—about 10 percent—they offered patients an early form of aftercare and recognized the chronic nature of habitual drunkenness.

Other gold cures proliferated at the turn of the century, capitalizing on the metal's general currency within American culture in the late 1800s and on Keeley's success and popularity, for his was the first remedy of its type.[77] Even more numerous than the gold cures, however, were the hundreds of other proprietary cures—some institutes, some mail-order remedies that advertised daily in newspapers across the country.[78] The Gatlin Institute, the Neal Institutes, the Oppenheimer Institute, Alcola, Dr. Haines Golden Remedy, White Cross Anti-Liquor Treatment, the White Star Secret Liquor Cure, and Cravex are some of the better-known firms and products. None of them came close to achieving Keeley's fame or fortune.

One proprietary cure, however, did win a good measure of respect from within the medical profession: the Towns-Lambert Cure, offered first at the Charles B. Towns Hospital in New York City. In 1901, Charles B. Towns, by all accounts a charismatic man with a larger-than-life presence, left his native

Georgia to find his future in New York City. That same year, he opened the Charles B. Towns Hospital for Drug and Alcoholic Addictions. Much like Leslie E. Keeley's ascent to fame and fortune, Towns's own tale involved a serendipitous meeting with a stranger who claimed to possess a cure for the drug habit, which upon further investigation by Towns proved effective. Starting small, Towns's business grew by word of mouth, but his success was so pronounced that he soon purchased a permanent facility along Central Park West. Success with some morphine addicts who were also dependent on alcohol led him to offer his cure to alcoholics as well. In turn, the new hospital began to cater to the rich, famous, and addicted. Crossing the threshold of the Towns Hospital "gave one the feeling of entering, not a hospital, but a private and quite exclusive hotel," notes William White.[79] Eventually, Towns opened a hospital annex for less affluent alcohol and drug-dependent clients.

Towns viewed alcoholism as a kind of poisoning for which he could provide the best antidote. Writing on the eve of national prohibition in 1917, Towns rejected out of hand the possibility of inheriting a craving for alcohol or drugs: "One of the most difficult ideas that I have to remove from the minds of alcoholics and their friends is this idea of inheritance. A man who has a poisoned father, a father whose system is thoroughly saturated with 'booze' and tobacco, could and probably would inherit a defective nervous system. But he could not inherit a craving for drugs or drink." Instead, Towns maintained, alcoholism was the result of alcohol poisoning: "the unpoisoned alcoholic" did not experience "any physical or mental craving for alcohol."[80] Even an individual with the most neurotic of constitutions was not foredoomed to drink to excess. It was simply a habit that took root in a compromised constitution, often assisted by bad diet, poor digestion, overwork, or anxiety. Alcoholics could never be "cured" of alcoholism, but they could be freed of their craving for alcohol swiftly and effectively, Towns claimed, should they seek his "definite medical treatment."

By "definite medical treatment," Towns meant a formula of belladonna, xanthoxylum, and hyoscyamus administered hourly by hospital staff able to distinguish between belladonna-induced delirium and any caused by the alcohol. This routine was continued for one to two days or until the patient had eliminated everything from his digestive tract. Next came a large dose of castor oil, and finally a temporary stimulant—often strychnine—and something to stimulate digestion—often a mixture of rhubarb, sodium, and capsicum. In severe cases of chronic alcoholism, Towns guarded against delirium tremens with an alcohol-based formula of chloral hydrate, hyoscyamus, ginger, capsicum, and water administered alternately with alcohol alone. As with the less severe case of alcoholism, a cathartic was administered along with the hourly belladonna mixture. Patients underwent a few days to a week of aftercare following the detoxification, although Towns's definition of aftercare differed from that of most hospitals and sanatoriums. For Towns, aftercare comprised a wholesome diet, hydrotherapeutic and electrotherapeutic interventions, and massage—all of which were regarded as *part* of therapy by most asylum managers. As Towns put it, "Get him 'into shape' physically, and you will then be able to get a definite line on both the physical and mental man."[81] By 1917, Towns had established a "physiological department" to this end.

Several features of Towns's treatment distinguished it from other proprietary cures. First, Towns actively and successfully courted the support of the American medical profession. In 1904, he introduced Cornell University Professor of Clinical Medicine Alexander Lambert to the work done at Towns Hospital. Lambert, who was also a staff physician at New York City's public Bellevue Hospital, had a longstanding interest in the treatment of alcoholics.[82] Lambert could not have been better positioned to assist Towns, for he was also Theodore Roosevelt's personal physician, and through his intervention, Towns was able to meet then Secretary of War William H. Taft. Concerned about the opium trade in China, Taft arranged for Towns to open branches of his hospital in Beijing and a few other Chinese cities.

Meanwhile, Lambert convinced Towns that he should take his treatment public and introduce it to the larger medical community, avoiding the secrecy that had imperiled Keeley's stature with the medical profession. This Towns did, disclosing the composition of his "definite medical treatment" to the world in 1909 at the International Opium Conference in Shanghai, an engagement arranged through Taft. By the time Lambert became the president of the AMA a decade later, the Towns Cure had achieved some prominence as the Towns-Lambert treatment. Towns and Lambert disagreed on one aspect of the treatment, however: its success rate. While Towns maintained that his hospital brought about success in 75 to 95 percent of its patients, Lambert was less sanguine, placing that figure closer to 25 percent.[83]

Perhaps the difference lay in their definitions of "success," for Towns's definition was a limited one. He asserted in 1917 that there was "no such thing as 'curing' a case of alcoholism. There is nothing on earth you can do to prevent any human being from taking up the use of alcohol again, if he wants to do so and will 'go to it.'"[84] Towns claimed only that he could obliterate the craving for alcohol through his detoxification program; for Towns, this was the necessary first step in the alcoholic's reformation. His "definite medical treatment" might successfully detoxify a much greater percentage of patients than would pursue continued sobriety.[85]

Towns's view on the possibility of cure was one shared by many who treated alcoholics. Pronouncing alcoholics "cured" proved problematic for both private and public institutions for inebriates. As we have seen in the case of the Keeley Cure, hyperbolic therapeutic claims drew hundreds of thousands to treatment but created a backlash when "graduates" failed. Public institutions also found themselves in a difficult situation, for their superintendents needed to demonstrate their hospitals' utility to legislators and taxpayers, not just to patients and their families. Claiming effectiveness was essential in this regard, but rather than making false assertions of efficacy, these facilities defined "success" in ways that deviated from simple abstinence. In Massachusetts, for example, holding a job and being self-supporting, even if one drank occasionally, were considered within the realm of success and "material improvement." Hospital physicians likened their inebriate cases to their insane ones. If someone suffered a relapse of mental illness after leaving the hospital and functioning normally for a prolonged period, was their first treatment a failure? they asked. Whether working in private or public contexts, physicians determined that alcoholism was a chronic condition that would need to be addressed for the remainder of their patients' years.

In retrospect, Charles B. Towns's therapy was unquestionably "somatic," but as we have seen for the private inebriate asylums, the element of agency, of the patient actively choosing to get well, was also at play. After all, Towns insisted that nothing could keep a man from drink if that was what he willed. Towns believed that the inebriate had to be an active participant in his cure. The patient himself had to decide to enter treatment and to cooperate with physicians, according to Towns, and the patient also needed to commit himself to work. Towns firmly believed that idleness was a great threat to continued sobriety, and he insisted that employers should assist the state in creating jobs and training programs for inebriates that might bolster their self-esteem and improve their financial position. In the end, Towns's approach—like the approach of the homes and private asylums—embraced the whole person. In his own words, "a complete physical as well as mental change must be brought about in order to help [the alcoholic]."[86]

Finally, Towns's popularity and success—he was one of the few individuals able to sustain a specialized institution for alcohol and drug inebriates following the passage of the Harrison Narcotics Act in 1914 and Prohibition in 1919—and his alliance with AMA president Alexander Lambert promoted a medical framework for addressing alcohol problems, even if it did not further the disease concept itself. Although his hospital served private, paying patients, Towns was a tireless advocate for municipal and state facilities for inebriates of all economic means, arguing that every large city needed a clearinghouse for briefly treating and classifying alcoholics and that every state had best create an institution where alcoholics in need of medical care might be sent for longer stays. "I am going to keep hammering at state and municipal authorities all over this country until I put them all to thinking on this subject," he wrote. "To set state and municipal authorities right as to the best ways and means of dealing with the matter is, I feel, one of the most important undertakings my entire work has developed."[87] As important as Towns deemed the creation of public institutions for inebriates, only a handful of states and municipalities developed hospitals or farm colonies for inebriates between 1870 and 1920. The number of these institutions paled by comparison with the private asylums.

This brief "tour" through the early history of private and proprietary institutions for inebriates reveals that in spite of their differences—length of stay, level of coercion, location, segregation of patients—they shared many characteristics. Homes, asylums, and proprietary hospitals all focused on patients' bodies, minds, and spirits. Inebriates' moral stature and their physical status were both of concern. Domesticity also figured very importantly in care— whether in the construction of inebriate asylums or in the relations among patients and between patients and staff. Indeed, it would seem that the directors of the neo-Washingtonian homes and private asylums anticipated the rallying cry of the WCTU: "home protection." The histories of these institutions also suggest that institutional care could vary greatly from administration to administration, as was the case with the New York State Inebriate Asylum under the direction of J. Edward Turner and Albert Day.

Within the context of private institutional treatment, we also see that the process of medicalization was by no means advanced by the medical profession alone, nor was it uncontested. The neo-Washingtonian homes were established in most cases by people outside medicine; the first superintendents of the Chicago and Franklin homes were laymen. Treatment at these institutions and at some of the more prominent private asylums relied on prayer and moral uplift, just as it relied on medical interventions. Even proprietary physicians such as Charles B. Towns insisted on the therapeutic value of work in rekindling the inebriate's self-esteem and productive citizenship. For all their criticism of penal and moral approaches to the problem, inebriety specialists such as Joseph Parrish took a very conciliatory line on the moral and medical aspects of the disease. Clearly, this made inebriety's medicalization more difficult.

Was the situation any different in public asylums for inebriates? How did medical care for inebriates play out in some of the states and cities that attempted to build systems to manage their problem drinkers? The next three chapters will examine the efforts of four states (Connecticut, Minnesota, Massachusetts, and Iowa) and one large municipality (New York City) to construct public inebriate facilities and provide medical care for their charges.

CHAPTER FOUR

# Public Inebriate Hospitals and Farm Colonies

New York City, California, Connecticut, Iowa, Massachusetts, and Minnesota all chartered hospitals and farm colonies for inebriates. In California, however, a rapidly growing population with no shortage of mentally ill citizens led the board of trustees of the new Southern California State Asylum for Inebriates and Insane to change the institution's charter to accommodate only the insane even before the facility had opened.[1] Only Massachusetts and Iowa had longstanding inebriate hospitals (discussed in detail in chapters 5 and 6); however, Connecticut, Minnesota, and New York City were all able to develop and operate inebriate hospital farms. Their stories give us a sense of the factors—medical, moral, political, and financial—that shaped both the public's perception of inebriety and state and municipal treatment of the condition. The histories of these institutions also give us a sense of the similar styles of care afforded in private and public facilities.

The concept of a public institution for inebriates antedated the creation of any private asylum, but legislators were slow to embrace the cause. The memorials and petitions for a state inebriate asylum in Massachusetts—drawn up by groups of concerned citizens in the 1830s, 1840s, and 1850s—received no

enthusiastic embrace. Nor did similar requests to consider creating a state facility in Iowa in the following three decades. The battles waged on behalf of
public care for inebriates in Connecticut, Minnesota, and New York City were
similarly protracted. The construction of public facilities for inebriates appears
to have gained support where there were activist medical and judicial communities, drives to rationalize and reorganize state and municipal health and
social welfare institutions, and well-organized and active temperance and
charity communities.

## Connecticut

In Connecticut, the campaign for a public inebriate asylum began with Eli
Todd and Samuel Woodward's *Report of a Committee of the Connecticut Medical
Society Respecting an Asylum for Inebriates* in 1830. In spite of the favorable recommendations of the committee, the legislators and public took little interest
in the creation of a state inebriate hospital; even the Connecticut Medical Society voted in 1845 to recommend postponement of the project. Not much
changed until after the Civil War, when the state legislature began to consider
new proposals for an inebriate hospital, granting a charter to the Connecticut
Reformatory Home for Inebriates in 1875, renamed the Asylum at Walnut Hill
later that year.

The chartering of the Connecticut Reformatory Home came on the heels of
a series of legislatively mandated reports. The first, from the Committee on the
Necessity and Expediency of an Inebriate Asylum, appeared in 1870 and laid
out a strong economic argument for the construction of separate facilities for
habitual drunkards. The report highlighted the mental and physical illness,
crime, and poverty due to drunkenness. Placing dollar figures on the state and
industrial revenues lost because of intemperance demonstrated that inebriety
was a burden on the taxpayer. The committee paired this data with their own
observations of the Boston Washingtonian Home, Joseph Parrish's Pennsylvania Sanitarium, and the New York State Inebriate Asylum under Albert Day—
facilities that set the standard for care in the 1860s and 1870s. The success of
these facilities further convinced representatives that inebriates could be reformed through medical care and that medical care should be available to all
"so that the poor, as well as the rich, may avail themselves of the benefits."[2]

The first committee's recommendations went unheeded by the General Assembly. Instead, three years later the legislature appointed a new Committee

on the Penal Treatment of Inebriates, several members of which attended the American Association for the Cure of Inebriates annual meeting in 1873 to explore the subject more fully. Like their predecessors, the committee visited a number of inebriate asylums to view their rehabilitation practices: the Inebriates Home at King's County, Long Island; the New York State Inebriate Asylum; the Washingtonian Home in Boston, and the House of Correction at Deer Island (Boston). They also surveyed the annual reports of other American and British asylums and concluded that (1) intemperance was a disease and appropriately treated through medical means; (2) asylum treatment offered the most hope for curing the disease; (3) penal treatment for inebriates was useful if it differed from that directed at ordinary criminals; (4) the protection of society by diminishing crime and pauperism and the reformation of the inebriate were the goals of penal treatment; and (5) the present penal treatment of inebriates is a failure in all respects.[3] The key, according to the committee, was to change the public's mind about inebriety, to build a popular image of the condition as a disease, and to cultivate widespread support for the reform of the penal institutions currently handling the inebriate. The committee concluded that work, a disciplined daily schedule, medical care, and a supportive mutual aid atmosphere were the best means to address hopeful cases of inebriety. For those who were resistant to reform, however, the committee recommended a graded system of penal and custodial care—"workhouse care"—with longer terms of commitment and earned wages sent to inebriates' families after the state had taken its cut. While the former resembled the treatment employed at private inebriate asylums and homes of the day, the latter represented a departure from the small fines and short sentences that were regularly dispatched to repeat offenders. "We have been deeply and painfully impressed with the conviction that our present system of treatment, including the practice of the justice and police courts, and the jail system, is one continuous method of educating criminals," concluded the committee.[4]

No action was taken on the workhouse proposition, but the legislature did incorporate the Connecticut Reformatory Home for Inebriates, and it passed a law permitting the commitment of inebriates to asylums and jails for care. The following year, however, the legislature turned its attention to the penal aspects of the problem, re-naming the investigative committee the Committee on the Penal Treatment of Inebriates and Vagrants and charging them to develop a plan for the wholesale reorganization of the correctional system that fined and punished the habitual drunkard.[5] The committee's conclusions regarding penal treatment for inebriates and vagrants changed little from its initial report endorsing the workhouse plan and the "therapeutic value" of disciplined labor.[6]

By 1875, then, Connecticut was addressing public medical care for inebriates with deliberation. Yet for all the therapeutic rhetoric that emphasized the onward march of enlightened statecraft and the development of humane and rational policies toward the insane and the inebriate, it is clear that the latter still occupied a middle point on the spectrum from vice to disease. The 1875 *Report of the Committee on the State Board of Health and Vital Statistics* made plain that inebriety was hardly the only condition viewed through both moral and medical lenses in Connecticut. As they made their case before the legislature for creating a new state board of health, the advisory committee emphasized the strong ties between the vicious and the ill. Just as "social, moral, and personal agencies have been acting on the human organism, producing important structural changes and disturbing the vital elements," so too pernicious living circumstances might compromise humankind's best instincts.[7] A morbid environment, noted the committee, "brutalizes and dwarfs the intellect, corrupts the morals, breeds intemperance and sensuality, and is ever recruiting the ranks of the vile and 'the dangerous.'"[8] In Connecticut, the drive to reform inebriates may be seen as part of a larger effort afoot in the last quarter of the nineteenth century to create a healthier moral and physical environment that would further humanity's progress toward perfection. This process involved the creation of a centralized and activist state health and welfare administration that ranked inebriety among its most pressing problems. Connecticut was similar to Massachusetts, Minnesota, Iowa, and New York City in this respect; all were investigating new and rational forms of administering their public welfare services.

Yet if Connecticut legislators recognized inebriety as a social and medical problem, they still hesitated to fund the construction of a new inebriate facility. The legislature changed the name of the Connecticut Reformatory Home to the "Asylum at Walnut Hill" in 1875 after the corporators of the asylum had purchased a large tract of land for its construction in the state capital of Hartford. The choice of this urban environment reflected a rationale not unlike that of the neo-Washingtonian homes—that "an inebriate should be reformed in the midst of those temptations which he is sure to encounter when discharged."[9] The city might offer wholesome entertainments and activities for

inebriates, not just vice districts lined with saloons. Churches, libraries, muse-
ums, and institutions of higher learning were present in the metropolis too,
and they might exert elevating influences on urban-based patients. Certainly,
there was a ready supply of inebriates in the city.

An unsuccessful campaign to secure state monies, however, left Walnut Hill
with no other funding choice than to become a private inebriate asylum catering to society's privileged classes: "Beginning without resources, and depending entirely upon the income from patients, we are constantly obliged to
refuse indigent but worthy cases, thus favoring the impression that we were
designed to benefit the rich only," complained Superintendent Thomas
Crothers in 1878.[10]

As we saw in the previous chapter, Walnut Hill's image had changed little
by the end of the century. It opened as a private institution and remained so,
in spite of its regular petitions to the legislature for support. In 1886 the original act of 1874 that authorized the commitment of dipsomaniacs to an inebriate asylum was amended to permit the commitment of inebriates to either
an inebriate asylum or an individual guardian. As revised, the act allowed inebriates to be sent to private institutions such as Walnut Hill, but it did not offer
the indigent the financial assistance required for a stay.[11] Rather than lament-
ing the shortage of state facilities for treating inebriates, however, Thomas
Crothers began to court the paying patient and to expand the hospital's base
of funding. His new strategy proved successful, for the asylum at Walnut Hill,
later the Walnut Lodge Hospital, became one of the nation's premier institutions for treating the alcoholic, offering care to "many of the most prominent
men in the country."[12] Indeed, a guide to the City of Hartford, Connecticut,
published in 1912, maintained that "in all parts of the United States, Hartford
and the Walnut Lodge Hospital are synonymous."[13]

It was just about that time that the Connecticut Medical Society became interested in creating a state institution for inebriates. Although the Connecticut Temperance Union had recommended this in 1906 in its publication
*Connecticut Citizen*—a proposal reiterated by the Social Service Commission of
the Protestant Episcopal Church in 1908—it was not until 1911 that the Temperance Union and the state medical society joined forces to advise the creation of a farm for inebriates. Their proposal was based on the work begun in
1893 at the Massachusetts Hospital for Dipsomaniacs and Inebriates. In 1912
the Connecticut Medical Society formed a special Committee on the State
Farm for Inebriates. Two years later, both houses of the state legislature passed

a bill that appropriated $100,000 for the construction of a State Farm for Inebriates, but the bill failed to win the governor's executive approval. There were higher priorities as the nation entered World War I.

Much to the state medical society's surprise, however, a Senator Alcorn introduced an alternative bill in the 1915 legislative session. The revised bill authorized no special state expenditures for the inebriate farm. Instead, it stipulated that the State Farm for Inebriates be established as a satellite department of the Norwich State Hospital. The legislature and governor approved the bill, and Norwich was left to fund the new State Farm for Inebriates out of its regular operating expenses. Over the summer of 1915, a farm a few miles from the Norwich State Hospital was refurbished with inebriate labor and equipped to support treatment for sixty patients; its doors were formally opened to inebriates and dipsomaniacs that November.

Treatment for inebriety at Norwich resembled that at the state inebriate hospitals in both Massachusetts and Iowa. Patients entering the State Inebriate Farm were "given a careful physical examination" at the State Hospital.[14] Physicians took complete family and personal histories, paying particular attention to the length of time an inebriate had been addicted to alcohol and/or drugs, the number and length of jail sentences imposed, and any previous treatment. After a brief period of rest and recuperation at the main hospital facility—what we might call "detoxification"—patients were transferred to the State Farm. There the recovering inebriates were put to work preparing food, tending the furnace, maintaining the farm buildings, and managing livestock, poultry, and an assortment of crops. They were also responsible for various construction projects at both the State Hospital and the State Farm, jobs such as enlarging the hospital's dam and reservoir. In the summer months, neighboring farms hired select inebriate patients to assist them.

Henry M. Pollock, the superintendent of the Norwich State Hospital and the State Farm at its opening, had lobbied for the creation of a "hospital farm for inebriates" since 1910. Pollack felt the pressure of mounting requests for alcohol treatment at Norwich State Hospital. Moreover, many of the patients were indigent, lacking the resources to visit the Walnut Lodge Hospital, and Pollack was without sufficient legal authority to detain them. "Whether a dipsomaniac can be regarded as insane and may on this ground be committed to an insane hospital, is a question which may be debatable," asserted Pollock.[15] Even if made legal by the legislature, such a commitment was not especially helpful, the superintendent maintained, for the insane hospital had little more to offer the inebriate than a place to dry out. Once his sobriety had returned, the inebriate's sanity usually followed, disqualifying him for care at a state facility. The hospital had to release him as "not insane." Pollock's concerns echoed those raised by the Association of Medical Superintendents of American Institutions for the Insane.

Pollock raised further objections to court commitments: if the commitment is legal (accomplished through a court proceeding), little good is served by such procedure, he lamented. The stigma, inconvenience, and public spectacle of the court proceeding deterred most families from seeking admission to the state hospital early in the course of mental illness, Pollock claimed. Urging an early and voluntary intervention policy that was consistent with a growing mental hygiene movement in the United States, Pollock regarded the treatment of intemperance as another step in preventing serious mental illness.[16] In any case, he believed the physician needed to be in charge and to possess authority over the course and duration of treatment, a view he expressed to Norwich's board of trustees in 1910.[17] Reiterating his plea for an inebriate farm in 1912, Pollock argued that such an institution would reduce Norwich State Hospital's overcrowded census.[18]

When the State Inebriate Farm finally opened in the fall of 1915, Pollock assured the board that every effort was "made to make the life of the inmate at the State Farm a normal one." Inebriates would be treated differently than the insane. The doors and windows "are never locked," and patients are

> subject to simple rules that differ but little from those observed in the usual household, the inmate enters and departs as he pleases. He rises to have breakfast at six-thirty, is employed during the day, and after an evening spent in reading or employed in games or writing goes to bed at nine. On Sunday he attends church in the village where he receives a cordial welcome from both the pastor and people; or hears the services held at the Farm by its devoted chaplain. As far as possible he is strengthened physically and morally.[19]

Pollock's words are reminiscent of the superintendents of the earlier neo-Washingtonian homes. He prioritized domesticity, religion, and strong bodies and minds. The key for him was to distinguish care at the State Farm from that which inebriates might receive at the jail; to provide his patients with work that would strengthen their character, as well as their physiques; and to prepare them to resume a job in the regular workforce once they were deemed sufficiently fit. At Norwich, rehabilitating inebriates was a medico-moral project focused on the whole patient, his physical disease, his moral fiber, and his social world.

By 1920, however, the tide had changed. A new superintendent had taken the reins of Norwich State Hospital in 1918. Not only was physician Franklin S. Wilcox less enthusiastic about his hospital's "inebriate department," but he was finding it increasingly hard to justify maintaining it with the passage of Prohibition. "We have practically discontinued our Inebriate Department," wrote Wilcox in his 1920 annual report. Norwich received "about one [patient] a month in contradistinction to a dozen a month prior to the enforcement of the Volstead Act."[20]

As John Burnham has argued, Prohibition was most effective in its early years.[21] The situation changed, however, once the bootleggers and moonshiners had established extensive distribution networks. By 1928, Superintendent Elijah Burdsall reported, "Inebriates come to us in as large numbers as before prohibition and prove quite a problem."[22] With bootleggers freely selling their wares, added Burdsall, it was much too easy for recovered patients to fall back into their old habits. The hospital was forced to mix its inebriate and insane populations once again, with similar unsatisfactory results. Following Prohibition's repeal in 1933, the state legislature began another round of considerations regarding a separate inebriate hospital.[23]

## Minnesota

The "Minnesota Model" of chemical dependency treatment developed in that state in the 1940s and 1950s and spread throughout America and abroad; it continues to influence addiction medicine today.[24] Less well known are the state's early experiments in treating inebriates at the Minnesota Inebriate Asylum (renamed and reorganized in 1878 as the Minnesota Hospital for the Insane at Rochester) and at the Willmar State Hospital between 1875 and 1920.

In 1873, the Minnesota Board of Health submitted a report, "The Duty of the State in the Care and Cure of Inebriates," to the state's governor, Horace Austin. The Board of Health urged the governor to declare that "inebriation is a disease; that it demands the same public facilities for its treatment as other diseases; that it is a curable disease under proper conditions," and that it is an important "duty of the State" to assist inebriates, more than any other dependent or diseased population, because alcohol consumption is "a source of national, state and municipal revenue."[25] In other words, since the state profited from liquor tax revenues, it was complicit in the creation of drunkards. The report's chief author, Board of Health officer Dr. Charles Hewitt, was clearly taking his cue from the AACI, whose charter principles bore a strong resemblance to those expressed in the report. That same year, the legislature passed a law imposing an annual fee of ten dollars on saloon keepers throughout Minnesota, the money to be deposited in a special fund for the construction of a state inebriate asylum. In 1875 the legislature chartered the Minnesota Inebriate Asylum, and land was purchased the next year in Rochester, a city that would later host the Mayo Clinic. Construction began on the inebriate asylum in 1877.

By 1878, however, liquor dealers' opposition to the ten-dollar tax was cresting. Saloon keepers refused to pay the annual fee, and this prevented the completion of the asylum. So vocal was their resistance that the liquor lobby convinced the state legislature not only to repeal the tax but to abolish the inebriate asylum altogether. Construction continued in Rochester, but the facility opened in 1879 as the Second Minnesota Hospital for the Insane, becoming the Rochester State Hospital in 1893. All that was left of the Minnesota Inebriate Asylum was an "inebriate department," which the legislature eliminated in 1897 when responsibility for the care of inebriates devolved to the counties.

If state care for the habitual drunkard was less than enthusiastically embraced by Minnesotans in the late nineteenth century, it is not difficult to understand why, based on the opinions of alienist R. M. Phelps, the assistant superintendent of the Rochester State Hospital. Writing in 1893, Phelps lamented the secrecy and profiteering of the Keeley Cure on the one hand, and on the other "the elusive and vague nature of the various articles" published by the leading exponents of the disease theory of inebriety. Men such as Thomas Crothers, Joseph Parrish, and T. L. Wright failed to secure for their readers "a clear definition of what inebriety is considered to be," apart from their insistence on its disease status, claimed Phelps, who took his own turn at defining the term. Yet Phelps's own definition hardly approached a "careful, plainly-worded" understanding. The inebriate, he asserted,

> has a nervous susceptibility that is excited and weakened by alcohol, while his comrades, with the lifetime of habitual association with alcohol, are never drunk. One may possibly call it disease, but as purely a tendency in itself; I can hardly think it a vice or disease, but rather a defective constitutional state. So much for the tendency itself. The habitual yielding to that tendency I should call a *vice*. The

diseased condition following, be it mental, physical, or both, I should call a *secondary disease;* if it is mental, I should call it *insanity.* . . . May not inebriety be any or possibly all [vice, habit, disease, insanity]? The first drinking could hardly be a disease; it has hardly become a habit, and it is a vice only if the evil is recognized as such by the drinker. In the second stage, it is probably becoming a *vicious habit,* while in the third or fourth stages the habit seemingly grows stronger and is bound by induced physical failure which had been gradually invading the system.[26]

Phelps based his thoughts on his experience treating nearly two hundred patients in the inebriate department at the state hospital in Rochester. There he had asked the inebriates how they conceived of their condition. Did they think it was a disease? A vice? A form of insanity? Most did not consider themselves victims of a disease, Phelps reported, though he was quick to add that "these subjective views are only indirectly of considerable value." Ultimately, it was up to the experts, not the patients, to discern the nature of inebriety and to treat it, Phelps maintained. Yet the alienist's words suggest that even he was less than sure about the nature of the condition he was treating—vicious habit, secondary disease, or constitutional tendency.

At Rochester, Phelps employed the nervous stimulant made famous by Leslie Keeley: strychnine. He regarded this form of drug therapy as "the most powerful remedy we have for combating the taste of liquor and the neurasthenic condition following its rise."[27] Sedatives such as bromides, valerian, and chloral were also helpful when combined with a regular routine of enforced abstinence. He furnished no further details of care at Rochester, but he allowed that it should be "correctional, rather than penal, except in so far as actual crimes are involved," and that if inebriety were at all curable, it would be through a "generalized combination of all the hygienic, dietetic, restraining and moral influences that can be brought to bear."[28] Here again, representatives of a state facility appeared to promote a holistic treatment course that embraced the moral as well as the medical.

Yet that was not the only course endorsed at Rochester. Phelps was also a supporter of eugenics, and he praised the colony plan for inebriates suggested by the superintendent of the New York State Epileptic Colony, Frederick Peterson.[29] To Phelps's mind, the best course was prevention, which could be accomplished either through legislation prohibiting the use of alcohol or through the confinement of inebriates, a step that he and others believed might prevent them from furthering defective hereditary lines. In short, the assistant superintendent of the only Minnesota state hospital treating inebriates in 1893 hardly made a strong case for regarding inebriety as a disease or treating it in a hospital environment. This task was left to physician Haldor Sneve and members of the Ramsey County Medical Society, who lobbied successfully for the creation of a state inebriate hospital a decade later.

In 1903 the *St. Paul Medical Journal* published a lead editorial entitled "Concerning the Commitment of Inebriates." Lambasting the legislature's 1897 repeal of the law that allowed the commitment of inebriates to state hospitals, the editors maintained that "it is practically impossible to commit to a hospital a confirmed inebriate. Not until there is little hope of restoring a sound mentality and until the property of the patient has been spent in reckless and insane dissipation does the state permit intervention on the part of the relatives and friends."[30] In contrast to Phelps's earlier concerns to distinguish inebriety from insanity, the *Medical Journal* editors declared that it mattered not whether mental degeneration led to chronic alcoholism or alcoholism led to insanity, it was "the duty of the state to care for the unfortunate individual at a time when the disease may be cured, or the degeneration of the mind from alcoholic indulgence may be prevented."[31]

The growing mental hygiene movement and the overgrown censuses of state hospitals made early intervention in any mental illness attractive. Even more compelling to members of the Ramsey County Medical Society may have been the frustration they had felt since 1897, when responsibility for the inebriate had devolved to the counties in Minnesota. Physicians such as Haldor Sneve, who served as alienist and neurologist to the St. Paul and Ramsey County public hospitals, were now responsible for hundreds of inebriates—despite the lack of any provision for long-term care. Furthermore, according to the editors of the *St. Paul Medical Journal*, it was impossible to confine an inebriate in a state hospital until "the man is practically in the gutter, his property wasted and his family deprived of the necessaries of life."[32] At this point the law stepped in and punished the drinker, whose will was powerless this late in the game, and whose mind was very likely compromised by alcoholic dementia. In other cases, where the inebriate was successfully committed to the state hospital for treatment via probate court, patients who sobered up and wanted to terminate their stays might easily file a *habeas corpus* application and be released as "no longer insane." The situation, in other words, was quite similar to that in Connecticut.

It was time to take a rational, scientific, and practical approach to the inebriety problem, urged the *Medical Journal* editors. They recommended that the Minnesota State Medical Association bring a bill before the legislature to permit the commitment of inebriates for terms of up to six months to hospital facilities—whether public or private—with release at the discretion of the superintendent. Medical men, they claimed, especially those trained in neurology and psychiatry, were well equipped to address the inebriate's unique situation and to determine when these patients were sufficiently recovered to be released. "Legal restraint" was "the *sine qua non* of successful treatment" the editors insisted.

Three years later, little had changed. Ramsey County physicians continued to come in contact with scores of inebriates who seemed unable to break away from their liquor or drug habits. Reiterating their view that inebriety was a disease, the editors of the *St. Paul Medical Journal* now insisted that the confirmed inebriate was "as helpless as thistledown in a gale," always going "in the direction of least resistance."[33] Neither Keeley charlatanism nor short commitments to insane hospitals, neither the pleadings of relatives and friends nor the moral entreaties of the clergy—and certainly not short sentences to the workhouse—had proven beneficial in the treatment of inebriates, argued the editors. What they did consider "proper and scientific treatment" was a program similar to that used in the Commonwealth of Massachusetts. It had six parts: (1) removal from a vicious to a healthy environment; (2) commitment sentences of not less than one year, with a similar sentence on the first relapse; (3) healthy life in country air; (4) work of a productive and remunerative character; (5) proper mental and moral stimulus; and (6) hygienic habits, proper food, and general attention to physical upbuilding. The medical society recommended that the state construct an institution capable of furnishing such a therapeutic course, an institution distinct from both a state hospital for the insane and a workhouse. In spite of the journal editors' critique of penal solutions, their language revealed an attachment to the older, moral conceptions of inebriety that informed these punitive responses. This was apparent in their praise for the redeeming value of labor:

> All types of men fall victims to liquor and drug habits, and all would be able, free from temptation and away from the opportunity to gratify their morbid impulses, to work out their salvation in the health-giving air of the farm and away from the seductions of city life. An environment new and salutary would be substituted for

one enervating and vicious. Under these conditions the will would resume its sway, the aberrant impulses finding no outlet would lose their power, and good hard work would rebuild the shattered physique of the drunken wrecks.[34]

"Temptation," "the gratification of morbid impulses," "salvation," "the seductions of city life," a "vicious" environment—all of these terms suggest that the editors of the *Medical Journal* regarded inebriety as both a moral and medical condition—a condition for which the workhouse's remedy might not be so off the mark. Work redeemed the man as it strengthened his constitution, argued these physician-reformers, but the work needed to be paired with medical attention, for it was clear to members of the Ramsey County Medical Society that the workhouse alone was doing little good. Haldor Sneve concurred in 1907, noting that the 2,500 "drunks" committed to the workhouse of Ramsey County in 1906 were among the worst of the worst—chronic cases with little hope of reclamation.[35] Sneve's message was clear: inebriates needed medical care, and they needed it in the early stages of their disease before they became hardened alcoholic rounders.

Apparently, the pleas of Sneve and the Ramsey County Medical Society found a sympathetic audience, for the legislature of 1907 once again authorized both the construction of a new inebriate hospital farm and new legislation for the commitment of inebriates to medical facilities for treatment. A 2 percent tax on liquor licenses funded the new facility. It took several years for the hospital farm to open, much to the chagrin of the physicians who had labored long and hard to reverse the state's policy on inebriates and to provide for their care. By 1909 even the Minnesota Supreme Court had expressed its support for the disease concept of inebriety and its treatment in state institutions: "The trend . . . of legislation is to treat habitual drunkenness as a disease of mind and body, analogous to insanity, and to put in motion the power of the state, as the guardian of all of its citizens, to save the inebriate, his family, and society from the dire consequences of his pernicious habit." The Minnesota Supreme Court reasoned that "the state has the power to reclaim submerged lands, which are a menace to the public health, and make them fruitful. Has it not, also, the power to reclaim submerged men, overthrown by strong drink, and help them to regain self-control?"[36]

Nevertheless, three more years of "political machination and legal technicalities" passed before the Minnesota Hospital Farm for Inebriates at Willmar opened its doors with room for about fifty men.[37] By 1914 the patient census had doubled. However, in 1917, with the liquor tax lapsed and Prohibition looming on the horizon—indeed, with the National Prohibition Act introduced into Congress by Minnesota Congressman Andrew J. Volstead that same year—the state Board of Control began to reconsider its commitment to running the inebriate farm. They optimistically assumed that the number of habitual drunkards would decline significantly if alcohol's manufacture, sale, and distribution were outlawed. So in 1917 the hospital farm began to treat the mentally ill as well as the habitually drunk. In 1919, the year the states ratified the Prohibition amendment, the legislature approved the institution's name change to the Willmar State Asylum, and the hospital's primary focus became the insane, although it continued to maintain a separate ward for the state's inebriates. Later, in the 1940s and 1950s, staff at Willmar would be instrumental in formulating the "Minnesota Model" of addiction treatment, a therapeutic paradigm that has spread the world over.

## New York City

In 1906 Bellevue Hospital, one of New York City's public hospitals, treated 6,453 inebriates, about 25 percent of the hospital's total annual admissions. One year later little had improved. Menas S. Gregory, a resident alienist at the hospital, maintained that the insanity of 3,000 of Bellevue's patients could be traced "directly to alcohol" and believed that alcohol played an instrumental role in causing insanity in 40 percent of all patients admitted to his institution. Prior to 1910, state law did not permit Bellevue to hold inebriates beyond the time it took them to regain their sobriety and sanity. Thus, many patients were alcoholic rounders who were treated and released, only to return after another debauch.

Gregory and the hospital's medical superintendent, S. T. Armstrong, M.D., were alarmed at the alcohol-related admissions figures, and they recommended that city and state officials and charity workers reconsider their policies for managing habitual drunkenness in New York City. Observed Gregory: "It would be the greatest aid to humanity if measures might be taken to reduce the consumption of this poison to a minimum, and to provide proper curative institutions for those who have formed a habit but have not passed the curative stage into one of complete mental and physical deterioration. Such an institution should be custodial as well as educational. In such an institution many will find recovery."[38]

Although these early concerns came from physicians, charity and social workers shared them. Indeed, in 1906 the New York Charity Organization Society (NYCOS) had concluded independently that the city needed to establish "a new public institution, partaking somewhat of the nature of a hospital on the one hand, and of a penal or reformatory institution on the other."[39] The problem of inebriety was felt acutely by New York's growing cadre of scientific charity workers, who daily confronted the poverty, joblessness, disease (especially tuberculosis), and general social and medical disability that accompanied habitual drunkenness.

Secretary of the New York State Charities Aid Association (NYSCAA) Homer Folks summarized the situation gravely in *The Survey* in 1910, declaring that alcoholism was everywhere:

> The drunkard's family has ever been the insoluble problem in home relief, public and private. The drunkard's children have ever been the despair of child-caring agencies. Authorities on insanity accept alcoholism as one of the dependable sources of supply for their constituency. General hospitals . . . if public institutions and therefore obliged to receive all classes of patients, find the alcoholic ward a source of ever recurring trouble. The conscientious almshouse superintendent finds his best plans miscarried. . . . The lower courts are clogged with habitual drunkards; the upper courts are distracted by questions as to the responsibility of the criminal who is part way along on the road to an alcoholic insanity.[40]

Indeed, it was Folks and his assistant, Bailey Burritt, who took the lead in developing the plan for the New York City Board of Inebriety in 1908, after NYCOS had referred the project to them. Two of the state's most influential social workers, Folks and Burritt drew up a proposal for the establishment of a seven-member centralized city board of inebriety authorized to intercede on behalf of anyone arrested for public drunkenness, disorderly conduct, or vagrancy.

The commissioners of the New York City boards of correction and charities were to be permanent members of the board, which would investigate and evaluate individual cases of public drunkenness and determine an appropriate course of action, choosing from a graded range of paternal, medical, and punitive responses. First-time offenders, for example, were to be released from police custody without going to court. Repeat arrests early in their drinking careers might cause offenders to be committed to the proposed state farm for inebriates if they did not respond to probationary sentencing and subsequent fining. Chronic recidivists unimproved after several stays at the inebriate farm were to be sentenced to the workhouse or penitentiary. The board would hold the power to commit inebriates for one to three years to the proposed inebriate farm or to a penal institution. The chief virtues of the plan, reflected Burritt years later, were its ability to deal "systematically with the problem" through a board whose sole task was keeping track of individual inebriates and their care, and its founding assumption that "alcoholism and drug addiction" are "essentially a medical and psychiatric problem."[41]

Although Folks and Burritt's plan received the endorsement of the NYSCAA, it failed to secure legislative approval when presented to the statehouse and senate in 1909. By the time it was reintroduced as the Grady-Lee bill in 1910, the plan had secured the approval of several important New Yorkers, including prominent neurologists Charles Dana, Alexander Lambert, Menas Gregory, and Frederick Peterson; New York City's libertarian mayor, William J. Gaynor; and Republican Governor Charles E. Hughes.

To bolster public support for their plan, Burritt and Folks published two pamphlets acquainting readers with the problem of the recidivist inebriate and the inefficiency and futility of using penal measures to treat him. *Treatment of Public Intoxication and Inebriety* (1909) and *The Alcoholic 'Repeater' or Chronic Drunkard* (1910) offered the rationale for creating a new municipal board to manage habitual drunkards, developing an inebriate probation system, and establishing a new hospital farm colony.

*Treatment of Public Intoxication* set forth the history of institutions for inebriates, emphasizing the strides already made by the Commonwealth of Massachusetts since it had begun providing medical care for habitual drunkards in 1893. The Bay State had met with "considerable success" reforming alcoholics at its state hospital for inebriates, and the state had been able to stem the tide of alcoholic recidivists by revising their probation policy to release first-time offenders and permit the voluntary commitment of inebriates to the hospital. Both of these measures, claimed Burritt and Folks, helped remove the stigma of crime from the drunkard and unclog the courts that had previously processed such cases. The report also highlighted the state medical facilities recently established in Iowa and Minnesota, concluding that "institutional treatment for inebriates is practically the only possible way that anything can be accomplished."[42] As the nation's largest metropolis, New York City should follow the lead of other states and confront the inebriety problem directly and scientifically, urged Burritt and Folks.

*The Alcoholic 'Repeater'* revealed the impressive commitment and re-commitment figures for inebriates treated in Bellevue's alcoholic wards. Many patients were admitted and re-admitted dozens of times a year, and these were the individuals that the hospital *could* follow in the absence of any centralized registration system. A similar situation obtained in the city workhouse, where some inmates were admitted eighteen (and probably more) times in the course of a year. Demonstrating that alcoholic recidivists consumed significantly more than their fair share of municipal resources, Burritt and Folks urged that "even greater than this actual cost in dollars and cents, great as that is, is the indirect loss to the city involved in clogging the courts with such cases to the detriment of other cases which could otherwise...be made self-respecting, useful citizens," were they placed in an institution for inebriates that provided extended medical treatment, exercise, and wholesome food.[43] Having failed to secure the passage of their plan in 1909, the secretary and assistant secretary of the NYSCAA concluded their report with a bold-face message that they hoped would hit city residents where they felt it most—in their billfolds: "Every taxpayer should know that the city is constantly putting hundreds of persons through this enormously expensive, but wholly wasteful and useful [*sic*] process of in-and-out-of-hospital and in-and-out-of-prison, and that it costs him a very appreciable amount each year, for which he gets no return whatever."[44]

Whether out of economic urgency, humanitarian concern, or interstate rivalry, the support was there for the Grady-Lee bill when it was resubmitted to the general assembly; it won approval in May 1910. Shortly thereafter, additional legislation passed that allowed other cities within the state to establish their own boards of inebriety.

Of course, establishing a board of inebriety and revising state law were not the same thing as constructing and operating an inebriate hospital colony. Indeed, the Byzantine appropriations structure for the City of New York and interagency politics made securing the necessary farmland and constructing the hospital and dormitories a complicated, drawn-out process.[45] Not until August 1914 did the New York City inebriate farm colony, located on 800 acres of farmland in Warwick, New York, about sixty miles outside the city, open its doors to between twenty-five and fifty inebriates. These men were sent there, largely on the honor system, to work on the "construction of simple cottage buildings" that would serve as their dormitories.[46] By May 1915 there were approximately one hundred beds available to inebriates—a significant increase, but a far cry from the one thousand initially envisioned by Charles Samson, the executive secretary of the board of inebriety, in 1913.[47]

We know little of the therapeutics employed at the inebriate farm in Warwick. When Samson predicted that the farm would open to one thousand patients, he also maintained that the buildings would bear no resemblance to a reformatory or prison but would function as a hospital. Likewise, patients were to receive individualized attention, their personalities "studied carefully" so that they might be rehabilitated into "normal, family, industrial, and general community relations."[48] At least rhetorically, Samson presented a medico-moral vision of inebriety treatment that coincided with the perspectives of Folks and Burritt. It reflected recent developments in social casework and in psychiatry, whose outpatient focus was geared to helping patients adapt to their environment and fulfill their social roles. In 1915 Samson observed that the men at Warwick Farms were primarily employed in agricultural and construction work: building the facility, cultivating the surrounding fields, "dairying," and engaging in "institutional routine work, such as laundry, cleaning, etc."[49]

In 1917, when the Farm's medical director, or superintendent, left his post, he was temporarily replaced by Captain R. A. Tighe, who began the inebriate farm day at 6:30 a.m. by leading the men through a calisthenics program. Submitting his report to the Board of Inebriety, Tighe portrayed his management style as firm but compassionate; he attempted to provide a disciplined yet friendly atmosphere, where "the policy of kindness, consideration and justice, tempered with mercy" was the order of the day. Under Tighe's supervision, patients not only tended the farm, but they also repaired and maintained the colony's farm equipment and power plant. When he left the inebriate farm colony in November 1917, Tighe reported that he had "run the Colony as a health resort and not as a penal Colony." He believed "the work of the Board of Inebriety will prove invaluable to many of the patients" treated at Warwick Farms.[50]

Health resort or no health resort, Warwick Farms offered treatment to inebriates and drug addicts that caught the eye of a number of prominent individuals, including the medical director of the National Committee on Mental Hygiene, physician Thomas Salmon. At a meeting of the New York Academy of Medicine in 1917, the topic of discussion was "the modern conception of inebriety." Here Salmon called attention to the Board of Inebriety and the work being done at Warwick Farms, which he believed constituted "a new step toward dealing with a great social and medical problem" as well as "the first practical effort to deal with antisocial conduct in accordance with the latest views." Salmon, who believed the inebriate required not necessarily "medicinal but physical, reconstructive, and psychoanalytical" care, praised the man that Tighe had temporarily replaced, Rear Admiral Charles F. Stokes, the former Surgeon General of the Navy. In his capacity as director of Warwick Farms, Stokes had been making "an attempt to reconstruct the inebriate in his relations to himself and his relations to society." Salmon urged members of the Academy to show their support for the Board of Inebriety's work and that of Dr. Stokes, who had accomplished as much through his individual example to the inebriate men as others achieved with "elaborate equipment in which such personal factors are lacking."[51]

Nor was Salmon the only one who praised Stokes's work. When Stokes announced in the *New York Medical Record* in June 1917 that he had discovered a new way of weaning narcotics addicts from heroin and morphine, the journal's editors hailed it as an advance "among those of the most far-reaching importance to our country."[52] Writing confidentially to the mayor of New York City, John Purroy Mitchell, Stokes outlined his plan for the wartime reclamation of drug habitués. He hoped that conscripted addicts might be triaged into treatment through their entrance physicals, while others would arrive at the farm following court proceedings or voluntary admissions. After being treated briefly at a municipal hospital such as Bellevue (Stokes's detoxification process took three to four days), drug addicts—like inebriates—would be turned over to Warwick Farms, where they would be sorted into occupational groups for training and subsequent placement in various "war industries, such as agriculture, ammunition making, shipbuilding, and the like."[53] Stokes's plan never came to fruition, for he left his position as medical director of the inebriate farm in 1917 to assume the chairmanship of the Sub-Committee on Addiction under the Council of National Defense.

Having lost its widely acclaimed superintendent, the inebriate colony faced yet another blow later in the summer of 1917, when Commissioner of Correction Burdette Lewis attempted to dismantle the Board of Inebriety and bring the care of inebriates under the jurisdiction of the Department of Correction. Lewis's plans to have Corrections take over inebriate and drug addict treatment came at a vulnerable time, for the Board of Inebriety had yet to find a replacement for Stokes. The acerbic exchange between Lewis and Burritt, who had become the board's president, highlights both the administrative politics of managing inebriates in New York City and the competing views of addiction and its appropriate treatment in the early twentieth century. Lewis wished to address correctional problems from a neurological and psychiatric perspective and saw the Board of Inebriety as taking up important financial and personnel resources that might be better directed toward the city's prison system. He was "convinced a cosmopolitan inebriety institution is a mistake."[54] Burritt interpreted Lewis's proposal as a self-serving and ill-founded play for power, writing,

> I have no doubt that you need more psychiatrical and neurological assistance and I hope that you may get it but surely that is not reason for transferring the activities of the Board of Inebriety to the Department of Correction.... I do not know what exactly you mean by a cosmopolitan inebriety institution but I would at least counter with the opinion that an effort on the part of the city to deal seriously and comprehensively with the problem of public intoxication and drug addiction is wise and will sooner or later be forced upon the city authorities.[55]

Burritt believed that placing the Board of Inebriety's responsibilities in the hands of the Department of Corrections would prove disastrous, for he saw inebriety as a medical and psychiatric problem. Angered by Lewis's suggestions, Burritt responded that there was almost as much reason to commit "feeble minded, insane, or even tuberculous cases to the Department of Correction."[56] The precarious position of the Board of Inebriety was undeniable, however. Not only were they experiencing difficulty finding a replacement for Stokes, but the number of alcoholics admitted to the city's hospitals and correctional facilities was declining and had been dropping gradually since 1916.

Moreover, the post–Harrison Act years had ushered in a crisis in drug addict management that preoccupied physicians as well as health, charities, and corrections officers. Amidst the heated negotiations between city and state over how to interpret the Harrison Act (especially whether it permitted or banned maintenance), where to house and treat the city's addicts, whether or not addicts could be successfully treated, and what role, if any, the Board of Inebriety should play in managing these cases, the habitual drunkenness problem seemed to fade from view. Commented Bellevue's assistant superintendent, M. L. Fleming, in September 1919, "Three years ago it was considered nothing out of the ordinary for us to take in from twenty to twenty-five acute chronic alcoholic cases over night. Now we rarely have more than two to three such cases over night." Moreover, he added, this was not "a result of prohibition, for the reduction was gradual and manifest prior to July 1 when the so-called wartime prohibition was scheduled to take effect. . . . Getting drunk seemed to be getting unpopular."[57] By 1920 the Department of Corrections had established its own Municipal Farm and Drug Hospital at Riker's Island to treat male drug addicts, and the inebriate colony was in the process of being converted to the Warwick Dairy Farm and Honor Camp, where well-behaved inmates from the municipal reformatory at New Hampton might live without locks or bars.[58] Once again, Bellevue became New York City's de facto inebriate hospital, the first and last resort for inebriates without the means to secure more exclusive treatment.

## Continuity and Change in the Private and Public Treatment of Inebriety

The range of Gilded Age and Progressive Era institutions for inebriates was impressive. Relying on everything from the Chattanooga Muscular Vibrator to the farmer's hoe, superintendents embarked upon a campaign for the habitual drunkard's restoration to sobriety and self-sufficiency. Historians have chosen to emphasize the differences characterizing these early institutions. Unquestionably, there are distinctions to be made. Some were privately financed; others received public support. Some employed specific drug remedies; others emphasized a holistic regimen of rest, diet, exercise, experience-sharing, and spiritual growth; still others believed that Christian fellowship offered the most secure foundation for recovery. Some offered their patients posh, private accommodations; others put patients to work constructing their own communal cottages and dormitories. Some insisted that long-term residence was best; others favored short-term care. Finally, some were voluntary institutions, while others operated in tandem with the court system and favored legal commitment, even though they preferred to admit voluntary patients.

It is important to recognize these differences among institutions. Indeed, one could argue that there was a gradual evolution of care that in many ways paralleled the conceptual evolution from intemperance to dipsomania and inebriety to alcoholism. The neo-Washingtonian homes, with their emphasis on Christian brotherhood and mutual aid, reflected the most morally infused and spiritually oriented approach to inebriety treatment, with each home recognizing the disease concept to greater (Boston) or lesser (Philadelphia) degrees. The private asylums and proprietary cures based their treatment on a more physiologically grounded concept of inebriety, dipsomania, and alcoholism—in Jim Baumohl's terms, a more "somatic" understanding of the condition. Finally, state and municipal facilities recognized the physiological and psychological elements of inebriety but prioritized the inebriate's restoration to financial independence.[59]

Yet if we focus too much on the institutional typology outlined by historians, we run the risk of obscuring the characteristics that these institutions shared and the common elements of treatment that they offered their patients. For example, it was sometimes difficult to distinguish between public and private inebriate institutions: in Massachusetts, the Boston Washingtonian Home received state funding for a number of years despite its private origins and management. And although private, paying patients often preferred more luxurious accommodations, the public inebriate farms in Connecticut, Iowa, Massachusetts, Minnesota, and New York City all accepted voluntary patients who sometimes paid their own way. Several states also passed Keeley laws that permitted states and counties to commit their inebriate charges to Keeley Institutes for care. In addition, there was the issue of using specific drugs to treat inebriates. Although Leslie E. Keeley and Charles Towns were the best known of those employing proprietary drug treatment, Charles Stokes, medical director of the New York City Inebriate Farm, also developed his own drug remedy for treating addicts and alcoholics. Bellevue physician Alexander Lambert collaborated with Charles Towns, employing the latter's specific formula and regimen in the alcoholic wards at the city's most famous (or infamous) public hospital. With regard to "work as therapy," Towns advocated work or employment of some variety to keep inebriates of all social classes sober following his detoxification treatment. Towns also emphasized the need for inebriates to truly desire to be cured for his or any treatment to prove effective. Thus, work therapy was not the exclusive domain of state or municipal institutions, just as the focus on individual agency and free will in treatment was not unique to the neo-Washingtonian homes.

Even the issue of coercive, long-term confinement was not cut-and-dried. Although Connecticut, Iowa, Massachusetts, Minnesota, and New York City all passed laws permitting the long-term confinement of their inebriates, first commitments to their public facilities rarely lasted longer than six to twelve weeks. Such a stay was considerably longer than the time it took to complete the Keeley or Towns treatment, but it was not rare for inebriates to spend comparable periods at private residential asylums. Indeed, Connecticut permitted the commitment of individuals to private facilities such as Walnut Hill Lodge Hospital. Likewise, the freedom from restraint that inebriates enjoyed at the Boston and Chicago Washingtonian Homes and Philadelphia's Franklin Reformatory was not entirely dissimilar to the "honor system" that inebriates experienced at the publicly funded inebriate farm colonies, where they were often hired out to private businesses and farmers during the day.[60] To be sure, patients at the Washingtonian and Franklin homes did not have the threat of arrest or extended probation looming overhead, but inebriates in the state and municipal hospital farm colonies appear to have had more liberty than one would expect from reading historians' claims that these facilities "blurred any practical distinction between medical and correctional institutions."[61] Certainly, in the eyes of New York City reformers Bailey Burritt and Homer Folks, there were significant distinctions between the medical and the penal perspectives. As Bailey noted in fending off the intrusion of the Commissioner of Corrections in 1917, the New York City Board of Inebriety was created to address inebriety within a medical and psychiatric context.

What are we to make of the varieties of specialized care inebriates received between 1870 and 1920, if the distinctions among institutions are more fluid than previously acknowledged? First, we can surmise that there was a great deal of communication among the leaders of these institutions. This is not surprising when we consider the relatively small group of individuals running the approximately two hundred facilities for inebriates in the United States between 1870 and 1920. Both the Chicago Washingtonian Home and the Franklin Reformatory watched and learned from the experience of their colleagues in Boston. The same was true for the early state and municipal facilities. During Connecticut's attempt to create a state asylum in the 1870s, its Committee on [the] Penal Treatment of Inebriates and [an] Inebriate Asylum not only attended AACI meetings to explore appropriate medical interventions for habitual drunkards, but they also visited the Inebriate Asylum at Binghamton, the Inebriates' Home of Kings County (Brooklyn, N.Y.), the Boston Washingtonian Home, and the hospital at the House of Correction on Deer Island (Boston). The founders of the Minnesota Inebriate Farm watched the Massachusetts and Iowa experiments in inebriate treatment closely and corresponded with the superintendents of those institutions. New York City Board of Inebriety founders Bailey Burritt and Homer Folks were likewise in touch with the state inebriate hospitals as well as with those who ran the municipal work farms in Cleveland and St. Louis. At the state level, leaders made an attempt to learn from the experience of other institutions and to create a viable standard of care.

Indeed, all of these institutions, not just the state facilities, structured their care for patients in a remarkably consistent fashion, if we listen to the rhetoric of their founders, managers, and sometime observers. Initial short-term medical treatment—what we would call "detox"—might or might not employ a regimen of nervous and digestive stimulants, purgatives, and narcotics. After a few days to a week, doctors then began a general "building-up" regimen for body, mind, and moral sensibility. Class and financial considerations dictated whether one's constitution was rebuilt through electrical light baths, therapeutic massage, and fine dining, or by an early morning bugle-announced program of calisthenics followed by a full day of work in the fields. Yet exercise and robust health, work and the rebuilding of self-esteem were highly valued in all treatment venues. Diet, or regular and wholesome meals, the example of manhood set by the superintendent, the sharing of inebriates' experience with one another, and the sober bonds of brotherhood forged through group meetings and daily interactions—all of these elements of care were regarded as essential to recovery. Finally, not one form of treatment was without religious ties. All of these institutions either provided some form of worship service, released their charges to attend local church on Sundays, or encouraged patients to cultivate their spiritual lives. In other words, regardless of the therapeutic context, inebriety was treated holistically as a medico-moral disease.

Furthermore, this standard therapeutic program was most often encased in a highly masculine frame that emphasized the inebriate's physical and moral struggle against alcohol in a society where liquor flowed freely. Of course, the New England Home for Intemperate Women made every effort to restore the "fallen angels" in their charge, steering them toward respectable womanhood through religious exhortation and domestic chores, but the vast majority of inebriates were men, and most institutions served exclusively male populations. Reformers, in turn, made repeated reference to "rebuilding" or "regaining" one's manhood through the restraint of one's appetite for alcohol via physical conditioning, by the cultivation of self-discipline through the pursuit of good works for others (including reforming inebriates and their families), and through continuous employment. If, for many American men, "holding one's liquor" was a powerful signifier of manhood, the successful struggle for respectable sobriety was meant by physicians to replace this potentially self-destructive course. Thus, both the disease *and* the treatment of inebriety were shaped by gender and class expectations. The restoration of the inebriate was a process of restoring one's manhood and sobriety. And this manhood, for men of all social classes, was one shaped by middle-class expectations of family role and financial security.

One final observation before exploring the stories of the two longest lived, and arguably the most successful, state facilities treating inebriates in the Gilded Age and Progressive Era: although inebriety affected men and women across the country in all walks of life, the problem of habitual drunkenness was felt most acutely in cities, large and small. The early inebriate homes were all located in urban centers—Boston, Chicago, Philadelphia. Municipal courts were teeming with inebriates, whose repeat jail sentences and fines appeared to help neither them nor their families. Though inebriate homes were frequently situated in the midst of the urban landscape, thus permitting their charges to continue working within the city, inebriate asylums, state hospitals, and inebriate farms were usually positioned outside urban centers. These institutions were situated to capitalize on the perceived healing properties of nature and the moral and physical benefits of outdoor labor, while maintaining a convenient proximity to the urban world where alcohol flowed freely. Particularly in the case of public asylums, the mission was to quarantine the inebriates from the temptations of city life while protecting the city's sober inhabitants and legal institutions from the drunkard's disorderly presence. The public inebriate hospital was a tangible expression of the medical profession's growing commitment to policing and protecting the public's health (and wealth) in addition to healing individual patients.

# CHAPTER FIVE

# The "Foxborough Experiment"

On the sixth of February 1893, the Commonwealth of Massachusetts opened its Hospital for Dipsomaniacs and Inebriates at Foxborough. Situated on eighty-six acres of farmland just twenty-five miles southwest of Boston, this new medical facility was to provide a more humane and effective alternative to the unsatisfactory solutions already in place for the Bay State's habitual drunkards: the jail and the insane asylum. Sixty years had passed since Samuel Woodward, the superintendent of the Worcester State Asylum for the Insane, had urged the construction of public inebriate homes. Driven by a resilient Progressive faith in rational, science-based reform, Massachusetts had enlisted doctors to treat the victims of the country's most prevalent social pathogen, alcohol. The inebriates now had a home of their own.

The history of the Massachusetts Hospital for Dipsomaniacs and Inebriates is the history of two institutions with three names. Indeed, the hospital's identity appears as complex as the disease it attempted to treat. In 1905 escalating per capita costs at the Massachusetts Hospital for Dipsomaniacs and Inebriates led the state to add the chronic insane to the hospital's census. With this change, the institution became the Foxborough State Hospital.

Only two years later the superintendent, staff, and board of trustees at Foxborough were the subject of a month-long public investigation of patient abuse, mismanagement, and graft. The "Foxborough Scandal" ended with the wholesale dismissal of the hospital's officers. The governor then appointed a new administration, with the ultimate goal of moving the inebriates to a different campus. In 1911 the trustees purchased land for the new state hospital colony in nearby Norfolk. Construction began immediately and took three years, employing the fittest, most trustworthy patients for the work. When the Norfolk State Hospital opened in 1914, the remainder of the patients were transferred to the institution. Only five years later, however, with the passage of Prohibition, the state elected to close its only public inebriate hospital, bringing the "Foxborough Experiment," as it was called, to an end. Once again, the habitual drunkard was a criminal, not a sick person.

The brief twenty-six-year history of state inebriate hospitals at Foxborough and Norfolk affords an interesting window into the medicalization of habitual drunkenness. Concerned citizens, public health officials, legislators, court magistrates, Brahmin physicians, directors of private inebriate asylums, and patients all played influential roles in supporting, critiquing, and shaping the care offered at the state hospital for inebriates. Their opinions about the hospital were a reflection of their assumptions about the nature of inebriety—as a disease and as a moral failing—between 1870 and 1920. What emerges along the way, from 1893 to 1919, from Foxborough to Norfolk, is a dynamic process in which lay and medical authorities voiced their concerns about unemployment, poverty, individual liberty, the power of volition, state responsibility, and the nature of disease. These are the issues that framed intemperance, dipsomania, inebriety, and alcoholism. The exchange of views among the aforementioned parties, not some top-down attempt to impose medical authority, was at the heart of the medicalization process.

Like the drive to medicalize habitual drunkenness generally, Foxborough's history should be seen in three related socioeconomic contexts: industrialization, urbanization, and immigration. The development of industrial capitalism required a steady, disciplined workforce that was compromised when workers drank excessively. The growth of cities that accompanied industrial expansion and immigration brought thousands of people together in close quarters and required new codes of behavior to maintain public order and public health. Many of the newcomers to Boston and beyond were young men who hailed from drinking cultures—Ireland, England, and Scotland in particular. Separated from their families to make a living, a large portion of the immigrant population was without any sort of social support network.

Boston, and Massachusetts in general, experienced all of these changes during the last two-thirds of the nineteenth century. As a result, inebriety ascended to a privileged position among social problems. Public drunkenness and urban disorder came to be seen as two intimately related plagues upon the state. By the end of the late nineteenth century, the traditional social welfare institutions that handled the habitual drunkard—the jail and the insane asylum—were heavily overburdened. State hospital administrators and magistrates, taxed by an expanding inebriate population, grew unsympathetic to their plight. Moreover, as administrators were quick to point out, inebriates were usually not helped by treatment in either institution. The state hospital for inebriates opened its doors in 1893, a time when Boston's Irish population had crested, accounting for approximately 80,000 of the city's half million inhabitants, and a time when an economic depression had put many out of work. The inebriate hospital joined a second generation of specialized social welfare institutions—epileptic colonies, tuberculosis hospitals, juvenile reformatories, schools for the feeble-minded, and psychopathic hospitals—that were seen as offering rational sociomedical reform to contain society's deviant ranks.[1]

The story of reformers' campaigns to build and manage an inebriate hospital in Massachusetts offers an intimate view of popular and professional attitudes about habitual drunkenness at the turn of the century. Massachusetts was the first state to develop a public inebriate facility, and its example was watched closely by other states and municipalities as well as by the general medical and social welfare communities. Certainly the "Foxborough Experiment" was no "Massachusetts miracle," but its qualified success bolstered efforts to medicalize habitual drunkenness across the country.

Iowa, Minnesota, and New York City, for example, paid close attention to both the Massachusetts probation system for habitual drunkards and the Bay State's efforts to develop an integrated medical, vocational, and welfare relief program for its chronically intoxicated citizens. When physician Charles Rosenwasser, vice-president of the Dependency and Crime Commission of the State of New Jersey, made his case for the construction of a new hospital to treat habitual drunkards, for example, he observed that Massachusetts had demonstrated "that inebriety can be successfully treated and that a hospital for inebriates is a great saving and not an expense to the commonwealth."[2] As the first public hospital for the treatment of inebriates, Foxborough's influence was felt across the nation, from California to New York City.[3] The chronicle of the "Foxborough Experiment," then, not only reveals the negotiated nature of the medicalization process at the state level but is also an important chapter in the history of mental health and social deviance in America.

## Early Calls for an Inebriate Asylum in Massachusetts

Massachusetts first legally recognized inebriety as a disease in 1885, permitting the commitment of inebriates to hospitals for the insane. Not long afterward, the superintendents of these hospitals realized the less-than-optimal nature of this arrangement and protested the disruptive presence of the sobered-up inebriates.[4] With the state's asylums for the insane overcrowded at the century's end, inebriates were increasingly unwelcome guests. Under these circumstances, the state legislature in 1889 approved the establishment of the Massachusetts Hospital for Dipsomaniacs and Inebriates.

The movement to establish a hospital for inebriates had begun much earlier in the Bay State, however. Alienist Samuel Woodward had introduced the subject to readers of the *Boston Daily Mercantile Journal* in 1833, the same year that he assumed the superintendency of the first state asylum for the insane in the nation, the Hospital for the Insane at Worcester. As the state Board of Health, Lunacy and Charity embarked on their trial of offering medical care for inebriates in the winter of 1893, Woodward's success with the insane at both the Hartford Retreat and the asylum at Worcester offered a promising exemplar. In 1833 a home for "victims of intemperance" had seemed a novel and enlightened act of Christian temperance and goodwill; by 1893, however, it appeared a rational and pioneering act of scientific charity. Early interest in inebriate asylums stemmed from post–Second Great Awakening ideas of Christian stewardship, but late-nineteenth-century efforts were of a Progressive stripe, drawing upon the public's growing faith in scientific experts and their capacity to conquer socioeconomic problems, and capitalizing on the crescendo of specific concerns about drink-related illness, crime, and pauperism. This change in the cultural and intellectual framework of inebriate reform reflected changes taking place both in the state's inebriate population and in its medical community, economy, growing urban populations, and ethnic composition.

The 1830s and 1840s witnessed the rise of mill and factory towns such as

Lowell, Lawrence, Lynn, Fall River, and Holyoke, as Massachusetts led all other states in industrial and urban growth. Between 1860 and 1870, the number of manufacturing firms increased by 61.8 percent. By 1880, 42 percent of the workforce held jobs in industry, and only 10 percent remained in agriculture. Waves of immigrants—mostly Irish peasants prior to the Civil War, and French Canadian, German, and southern and eastern Europeans after the war— flooded Boston and spread across the state. By 1875, 25 percent of the Bay State's population hailed from the other side of the Atlantic, and within the greater Boston area, the figure was closer to one-third. That same year the census revealed that the majority of the Commonwealth's citizens resided in incorporated towns and cities.[5] The Bay State had urbanized quickly.

The opening of the hospital at Foxborough in 1893 was the high point of a transformed campaign Samuel Woodward had inaugurated some sixty years earlier with his essays on asylums for inebriates. Between 1845 and 1885 the state legislature, referred to as the General Assembly, considered plans for an inebriate asylum nearly two dozen times. These legislative efforts mirrored changes taking place in the popular and professional perceptions of habitual drunkenness, the means to combat it, and the responsibility of the state for its treatment. Between 1845 and 1885, the successive legislative calls for inebriate asylums in Massachusetts reveal that the highly moralized concept of intemperance gave way to dipsomania and inebriety. But just as there was persistent confusion over the "true meaning" of *inebriety* within the AACI and the larger medical profession, so too did different definitions of each term coexist in the Bay State. More significantly, however, all definitions included *both* moral and medical elements.

In 1845, a group of twenty-eight concerned, temperance-minded citizens from Norfolk County submitted a memorial to the Massachusetts House of Representatives "in relation to the founding of an Asylum for the temporary abode of that large class of misguided persons who so frequently render themselves obnoxious to the laws by the vice of intemperance." The Norfolk petitioners found the "victims of intemperance" poorly served by the jails and houses of correction where they were lodged, and they believed that many of these individuals could be restored to "usefulness and respectability" through asylum care. As support for their new approach, the citizens from Norfolk cited the success of the Washingtonian movement, with its focus on reforming the individual drinker, and the efforts of Bostonian John Augustus, the mechanic who had labored tirelessly to rescue more than two hundred drunks from Boston's jails. The memorialists' beliefs were similar to those of the directors of the Washingtonian Home that would be established twelve years later. They believed that Christian faith and fellowship, coupled with "the law of human kindness," held great restorative powers. Moreover, they added, curing the drunkard made good economic sense, for it removed not only him but also his family from the state charity rolls. So many house of correction inmates had committed crimes while inebriated that sobering up the Bay State's drunks also might reduce crime generally.[6]

Only once in their memorial was inebriety called a disease: the good citizens of Norfolk compared the inebriate's state with that of the madman: What, queried the petitioners, "is the drunkard, with his periodical fits of delirium tremens, *but* a maniac? A shattered mind can be healed only by skillful words, uttered in tones of kindness; by efficient medicines, administered by persuasion and not by force."[7] The comparison with insanity was a useful one, as we have seen: it reinforced the disease aspects of the condition and proposed segregated institutional confinement as an acceptable therapeutic solution. It was not, however, a solution the Massachusetts legislature was ready to endorse. Five years passed before the legislature appointed its own committee of experts and concerned citizens to "report upon the expediency of establishing an asylum for persons *supposed* to be confirmed inebriates; with a view to the ultimate abrogation of all laws punishing intemperance as a crime."[8]

Amasa Walker, businessman, economist, and politician, directed the committee's efforts and solicited the opinions of local physicians, lawyers, and clergy "well versed in all the phenomena of intemperance." So informed, the committee reached six unanimous conclusions: (1) habitual intemperance was indeed a disease and should be treated as such; (2) there was an ample clientele for asylum treatment and would be for the foreseeable future; (3) the state had not yet provided any satisfactory alternative to the jails and houses of correction that had historically failed the inebriate; (4) an inebriate asylum could be funded privately through donations and the labor of less affluent patients; (5) curing the inebriate would pay off economically for the state in the long run; and (6) Massachusetts, a state known for its munificence in aiding the blind, the insane, the idiot, and the juvenile offender, was equally bound to offer care to a far larger group who "by the habitual use of intoxicating liquors, have contracted the terrible disease of intemperance."[9]

The committee received the strong endorsement of physician John Collins Warren, a member of Boston's social and medical elite who had played a leading role in founding Massachusetts General Hospital. Warren's comments were telling, however, for he distinguished between "incorrigible" and "corrigible" inebriates, while admitting that both required care. The former were "as dangerous as a maniac" and needed to be confined as if mad. The latter were "rendered incorrigible...by want of the power of restraining them long enough to break up their habits." Warren's language suggests that in spite of his medical training, he remained committed to a penal framework when thinking about habitual drunkenness. Nevertheless, he closed by lambasting penal solutions for habitual drunkards. The asylum was preferable, he urged, as it would be "adapted to his wants, where ample provision is made for a profitable employment of his industry, and where he will enjoy every advantage for improvement, and restoration to health and body and mind."[10] Still, rather than recommend the establishment of such a facility, the committee cautiously advised the General Assembly to appoint a board of commissioners to determine the feasibility and direct costs of constructing an inebriate asylum. Once more, matters were left in the hands of a new advisory committee.

The 1851 committee appointed by the Massachusetts legislature reached the same conclusions as its predecessors. As the committee's report said, "The subject is not new."[11] The only unique aspect of the 1851 report was its acknowledgment of the multitude of petitions the group had received from temperance societies across the state, including "divisions" of the Sons of Temperance, who were in favor of the proposed inebriate asylum.[12] Support for anti-alcohol measures reached unprecedented levels in the summer of 1851, fueled in part by the middle class's fears of the recently immigrated Irish and the perceived threat that these indigent, hard-drinking newcomers posed to urban order. This was especially the case in Boston, where the Irish accounted for half of the immigrant population.

Between 1851 and 1855, thirteen states and territories passed the so-called Maine Law, prohibiting the sale of beverage alcohol and mandating the destruction of confiscated liquor. Massachusetts passed its Maine Law in 1852. With prohibition in effect in the state, however, sympathy for the individual drunkard waned among temperance groups, making support for an asylum less likely. As was the case in Connecticut, New York City, and Iowa, prohibition appeared to render treating alcoholics unnecessary. Temperance groups supported asylum care only when prohibition was not a possibility. Indeed, it was not until Governor John Andrew raised the issue some eleven years later, in the middle of the Civil War, that interest in establishing an inebriate asylum was rekindled. By that time it was clear that the Maine Law was a failure; it had not curtailed liquor sales. Drunkenness and alcohol were probably as visible as ever on the streets of Boston. Prohibition had not prohibited.[13]

A lawyer, Unitarian, and abolitionist, John Andrew was also a temperance supporter. This constellation of causes was not uncommon.[14] He opposed all sumptuary legislation, including the Maine Law, and endorsed an earlier meaning of *temperance*: moderation.[15] The state senate responded to Andrew's 1863 request that "the legislature initiate measures to establish an asylum for the treatment of inebriates" by appointing another committee.[16] In turn, the legislators visited several inebriate homes and asylums. The committee had nothing but praise for the Boston Washingtonian Home, its "efficient, skillful, and hardworking superintendent," and its commitment to accepting any and all who desired reformation—regardless of their social class or physical condition. The New York State Inebriate Asylum at Binghamton fared less well, however. "Spacious, handsome, well-arranged, and well-furnished," the New York facility reminded committee members of a luxury hotel serving a clientele of business and professional men, "a superior class better worth saving and more likely to be saved." The legislators concluded that such a selective admissions policy discriminated against "the most unfortunate of unfortunates" who were "left to the tender mercy of pretended justice."[17] Committee members favored the small-scale, domestic approach of the Washingtonian Home over the creation of one large institution. But in the end, their views were moot. The Civil War placed enormous demands upon the state. To proceed with a new inebriate facility, declared the committee, would be imprudent.

Not until 1868, after the Massachusetts had repealed its Maine Law, did the *Report of a Special Committee Appointed to Consider the Matter of Inebriation as a Disease, and the Expediency of Treating the Same at Rainsford Island* appear before the senate. Quoting John Andrew's 1863 address to the legislature at length, the special committee stated as its primary objective the evaluation of the disease concept of inebriety, a condition they described as a "distinct pathology ... presenting post-mortem appearances unlike those of any other disease."[18] The special committee also introduced the new term *dipsomaniac* into the inebriate asylum equation. Often hereditarily transmitted, dipsomania was a species of insanity, and it was often found among the better social classes. Yet, just as Andrew had argued five years earlier, the legislators now asserted that inebriety, even if classified as dipsomania, was both a disease *and* a sin. The dipsomaniac who acquired his condition through inheritance might be less culpable than the willful indulger who gradually established his devastating habit, but once the disease had gained the upper hand, the drunkard was no longer in control of his behavior.

With 96 out of 145 criminal cases arriving at the state attorney general's office "liquor cases in some form," the Commonwealth could no longer afford to postpone the construction of a new facility for inebriates. The committee's final recommendation, however, was that private citizens or "humane and philanthropic persons" found an inebriate asylum. If it were privately built, the state might become a "most liberal patron," the legislators claimed, for "no movement in the interest of temperance promises better results." In the end, the special committee requested permission to conduct another survey of the situation when "the interests of the State demand."[19] The issue was tabled for the current session, but a new report was commissioned for the following year.

By 1869, Massachusetts was officially "dry" once more, but the momentum behind the inebriate asylum proposal was firmly established. Prohibition proved difficult to enforce in Boston and other large cities, and this further ensured a steady supply of drunks for public spectacle. The joint committee's second report was terse and to the point: "It is a well accepted theory among leading medical men that inebriation is a disease and should be treated as such; that it is as much of a disease as any variety of insanity, and often originates in causes entirely beyond the control of the person who suffers therefrom. It therefore appears to the Committee that the State should number among its charitable institutions one institution, at least, for the exclusive treatment of this malady."[20] The 1869 report recommended establishing a commission for further study of the issue. The next commission was to be charged with designing a specific plan for an inebriate asylum.

Just as quickly as Massachusetts had reinstated prohibition, it dismantled the reform by popular fiat. In the same legislative session that they approved the sale of beer, representatives also appointed a new inebriate asylum commission that delivered its report in January 1871. There could be no mistaking the inverse relationship between prohibition and the state's interest in inebriate asylums. The new committee claimed three distinguished members: lawyer Otis Clapp, then president of the Boston Washingtonian Home; W. B. Spooner, lawyer and prohibitionist; and John E. Tyler, M.D., physician to the Insane Asylum at Somerville (later McLean Hospital). Clapp, Spooner, and Tyler made their argument in favor of a state inebriate asylum in terms of economy and thrift—a framework likely to appeal to both businessmen and

Brahmins within the legislature. The committee's rationale was not new, but unlike their predecessors, they employed some impressive statistics. They estimated, for example, that the Washingtonian Home, in its ten years of operation, had saved the state over $2 million in restored inebriate labor.

Using the Boston Washingtonian Home as their model, the committee recommended the construction of three institutions to accommodate various types of inebriates: hopeful middle-class men, custodial "prison hospital" men, and women. Private citizens might band together to create a Washingtonian-style facility for the treatment of a middle-class clientele that would balk at the diverse class and ethnic mix found at the Boston Washingtonian Home, where anyone who earnestly desired reform was admitted. The state, then, would be left with the construction of an institution for "the intemperate in prisons" or those who were serving short sentences for crimes committed while intoxicated.[21] A great percentage of these individuals—men and women—could be helped, the committee believed, by spending up to six months in an inebriate "prison hospital," where they would learn discipline and regular habits. The state had made 42,153 inebriate commitments to houses of correction and jails between 1865 and 1869, exclusive of those fined for drunkenness. Approximately a quarter of these individuals were women, whom the committee thought should have their own facility.

The 1871 report was the last study of inebriate asylums commissioned by the legislature. Nothing became of the recommendations for the state "prison hospital" or its counterpart for women, perhaps for the reason that Barbara Rosenkrantz has suggested in the context of public health in Massachusetts: "the tradition of private charity and humanitarian benevolence" often mediated against any outpouring of state funds to aid the disadvantaged.[22] After all, the privately managed Washingtonian Home was successfully serving clients with only a small annual subsidy from the state.

Certainly it did not help the state inebriate asylum cause to have Nathan Crosby, the police court judge in the city of Lowell, stridently attack the plan before the legislature. Addressing the Massachusetts General Assembly in March 1871, Crosby lambasted the disease concept of inebriety: "The American Association for the cure of inebriates say *inebriety is a disease*. The Board of Charities say *it must not be punished*, but reformed. And the Commissioners on Inebriate Asylums report the building of Asylums for detention and recovery. A remarkable triumvirate of excuses—of protection and *encouragement* for intemperance."[23] Crosby believed the drunkard chose to drink his life away.

The disease concept and the therapeutic approach to inebriety stood in contradistinction to the disciplined, rags-to-riches stories of the day, heralding the success of the self-made man. Unlike "Ragged Dick" of the Horatio Alger stories, the drunkard abused any opportunities for achievement that came his way. If Massachusetts was not yet ready for a state-supported inebriate asylum in 1871, however, sympathy for the enterprise was forthcoming from private philanthropists. By 1873, for example, the Washingtonian Home had assembled the $100,000 required for its expansion from individual donors within Boston. By this time, a new arm of the state's government had taken up the inebriate's cause: the Board of Health.

In 1875, Henry Ingersoll Bowditch, activist physician and chairman of the Board of Health, addressed his colleagues in the same organization on "the necessity for the State to establish one or more inebriate asylums or hospitals for the cure of drunkards."[24] The timing of his message seemed appropriate: Massachusetts had once again repealed its remaining prohibitory legislation. Bowditch, who opposed prohibition on the grounds that it denied the privileges of the many for the protection of the few, viewed the drunkard's habit as a crime against society—a breach of the social contract made between responsible citizen and state in the democratic republic. Like so many of his contemporaries, however, he found that punishment of drunkenness by fine and incarceration harmed the moral fiber of the inebriate, beggared his family, and did little to foster sobriety. The state, according to Bowditch, had as much responsibility to address "a vice [that] so saps the health of individuals and of their progeny" as it had to build public general hospitals.[25] In other words, inebriety was a viciously acquired condition, even if, as Bowditch believed, the habitual drunkard was a diseased individual.

Bowditch drew other distinctions within the inebriate population. He distinguished between three classes: those with hereditary taint, those led to drink through bad habit, and those suffering from the insanity dipsomania. Although Bowditch regarded the first and last groups as the most difficult to reform, much depended on the inebriate's recognition of his condition and willingness to comply with therapy. He believed, furthermore, that class considerations were important to successful treatment. All persons, according to the chairman of the Board of Health, were not equal—"physically, intellectually, morally, or in the surroundings of their birth."[26] Separate, more comfortable accommodations, and amusements such as billiards, cricket, and croquet were necessary to make inmates with better social standing feel at home. Re-

sponsibility accompanied privilege, however, and Bowditch envisioned inebriates with "better cultivated minds" assisting in the education of their less-
fortunate fellow patients. Here Bowditch's description of the asylum echoed
the published words of the anonymous patient at Binghamton who saw his institution as a microcosm of democracy and collegiality, where men of all walks
came together with a common goal. Finally, Bowditch believed that a separate
penal asylum with additional restraint would probably be necessary for all patients who resisted treatment, regardless of their social standing. Bowditch
may have given more thought than his predecessors to the details of building
and managing an inebriate asylum, but his report to the State Board of Health
seems to have gone the way of all previous reports recommending the construction of the new facilities: into a legislative cul-de-sac.[27]

In 1885 the newly reorganized State Board of Health, Lunacy and Charity
considered the proposed state inebriate asylum once more. Once again, *dipsomania* was invoked to refer to the condition of habitual drunkards, but it was
used inclusively to describe both periodically and chronically intoxicated
drinkers. The board members felt that both forms were species of insanity;
thus they wished to apply the protocol for the certifiably insane to the Commonwealth's inebriates: "*in loco parentis* and providing for the guardianship of
their persons and estates ... and requiring restraint and treatment to restore
them to usefulness, and to a more rational habit of life."[28] Instead of shelving
the report, the legislature passed new laws in 1885 permitting the commitment of "dipsomaniacs" to state lunatic hospitals *provided however, that no*
such person shall be so committed until satisfactory evidence is furnished to
the judge ... that such a person is not of bad repute or bad character, apart
from his habits of inebriety."[29]

The "good character" caveat was an interesting condition of admission. At
first glance, it would seem that the legislators had made a distinction between
the sick and the vicious. Hospital treatment was not for the undeserving inebriate leading a dissipated life. There were worthy and unworthy inebriates, as
there were worthy and unworthy poor. Yet the condition of admission was
phrased in a way that assumed that "habits of inebriety" were themselves the
only admissible evidence of bad character. Even though one could easily
argue—and many did—that the disease of inebriety or dipsomania compelled
an otherwise respectable, moral individual to drink against his or her better
judgment, the fact remained that the new inebriate commitment laws still recognized habitual drunkenness as a moral failing and a disease.

Of course, legislative permission to commit inebriates to state hospitals for the insane did not require additional financial appropriations; this may have been the key to its success in both houses. The move was really a sort of low-budget compromise of the past forty years' proposals. It was not what either the Board of Health, Lunacy, and Charity or any of the preceding investigative committees had suggested. It did, however, place inebriety and dipsomania squarely within the medical domain and authorize its treatment by the state. As we have seen, however, the history of mixing the insane with inebriates was fraught with problems, and it was not a policy that the superintendents of the state insane hospitals were likely to support, particularly when their facilities were already overcrowded.

In fact, within a year the presence of inebriates in insane asylums had become untenable. Boston Lunatic Hospital superintendent Theodore Fisher had detailed the unfortunate situation to his colleagues in the Massachusetts Medical Society years earlier in his essay "Insane Drunkards": "The difficulty . . . arises from the transient character of the prominent symptoms, which are only brought out under the paralyzing influence of alcohol. As one writer has said, the dipsomaniac is only sane while in the hospital. . . . [I]n a surprisingly short time he is on his feet, under perfect control, looking around for a lawyer."[30]

It would take three more years for Massachusetts to provide a new solution to this very old problem. In 1889 the General Court passed "An Act to Establish the Massachusetts Hospital for Dipsomaniacs and Inebriates," authorizing $150,000 to build the state's diseased drunkards a home of their own. Although the Massachusetts Hospital for Dipsomaniacs and Inebriates was originally intended to serve women as well as men, a funding shortage and the trustees' concerns that "the admission of both sexes cannot fail to produce a serious interference with the morals of the inmates," led them to focus energies on the more affected sex.[31]

## The Growing Pains of a New Institution

If the years leading up to the founding of the new state hospital for inebriates witnessed an ongoing debate about the nature of habitual drunkenness and the best modes of treating it, this lack of consensus presaged the new facility's contentious existence. Some judges balked at the prospect of removing penalties from the habitual drunkard's case, while others found the hospital an appropriate dumping ground for their worst recidivists. Patients' families likewise committed their besotted and unmanageable loved ones to the hospital to teach them a lesson, expecting to be able to retrieve them when a sufficient time had elapsed (as determined by the committing parties, not the hospital). Thus, commitment to the inebriate hospital and support for its work did not necessarily signal an endorsement of the disease concept of inebriety. The institution was a convenient depot for the detention of irascible, inveterate drunks. The superintendent, his staff, and the hospital trustees were thrust into a position of managing a population that varied greatly, many of whom were unlikely candidates for reform. After years of lobbying by concerned citizens, public health officials, governors, businessmen, and legislators, the state inebriate hospital came to life; but the detractors remained, and sometimes the patients joined their ranks.

In July 1889 the governor of Massachusetts, Republican Oliver Ames, appointed the first board of trustees for the Massachusetts Hospital for Dipsomaniacs and Inebriates. Ames selected as board chairman Francis Amasa Walker, an economist-statistician and president of the Massachusetts Institute of Technology. The rest of the new board were equally illustrious.[32] The trustees approached their task with caution. Only ten years earlier the nation's first experiment in quasi-state inebriate reform had failed: the New York Inebriate Asylum did not provide a useful model for state-based care.[33] Aware that their new hospital would likely "guide other States in doing similar work," the trustees focused their energies on two issues: first, determining the type of clientele it would serve; and second, working within the legislature's meager appropriation.[34]

Consulting with local medical, penal, and charities experts, Walker and his colleagues candidly acknowledged that "no one can tell what classes of persons are most likely to be sent to the hospital, nor in what proportion different classes will be represented."[35] Accordingly, the board concentrated on finding a qualified superintendent to supervise the construction of the new facility. They also lobbied the legislature to either provide $25,000 more in building funds or lower the hospital's assigned patient capacity. Unsurprisingly, the legislature chose the latter, cheaper course. The trustees, in turn, chose Marcello Hutchinson, M.D., as the hospital's new superintendent. Hutchinson had extensive experience dealing with the insane, "including many of the class which will come to this hospital" at the state lunatic hospital in Taunton.[36] He was joined by another alienist, William Noyes, who for five years had held the position of pathologist and assistant physician at Boston's prestigious private hospital for the insane, McLean Asylum. Both men were graduates of Harvard Medical School.[37]

No sooner had Hutchinson commenced his duties than he was urged by the trustees to embark on a tour of "many of the newest and best-appointed hospitals for the insane in the United States."[38] These travels, the trustees hoped, would provide their new superintendent with the most up-to-date knowledge of asylum management and construction to guide the state's experiment in inebriate treatment. Upon Hutchinson's return, the trustees were gratified to learn that he concurred with their own design preference, the "cottage plan."[39] With its smaller residences, the cottage system was believed to foster better, more intimate and domestic relations among patients. From a medical perspective, the smaller residences allowed the staff to classify patients and keep different groups separate. Hutchinson and the trustees believed this design would help cultivate responsible, moral behavior by allowing patients more liberty to direct themselves inside the hospital grounds.[40]

Having finalized plans for construction, the trustees found an appropriate parcel of land within striking distance of Boston that possessed an abundant supply of fresh water, arable land for agricultural pursuits, easy sewage disposal, and close proximity to rail lines. The new hospital would be built in Foxborough, Massachusetts, a town twenty-five miles southwest of Boston. Construction began in 1891 on three medium-sized dormitory cottages, each housing sixty-seven patients—a number that was not exactly conducive to "domestic relations." Originally cottages half this size were planned, but the trustees' insistence on fireproof brick raised the cost of each building and required the construction of fewer and larger "cottages." Reservations were expressed about the larger dormitories, for some trustees believed that it might be "impossible to secure the best results with so large a number of inmates in the several buildings."[41]

Nevertheless, the Massachusetts Hospital for Dipsomaniacs and Inebriates opened with some fanfare in February 1893. Hundreds of people journeyed to the new institution from the surrounding towns and cities, hoping to take a tour of the "handsome buildings." Three stories of plain brick with slate roofs and little ornamentation, the three cottages and the combined dining hall/kitchen/residence were set back in a row from their chestnut-lined street. A new spur of the Old Colony Railroad connected the hospital's power plant with the cities of Boston and Providence. By the *Boston Globe*'s estimate, 1,200 curious citizens visited the asylum in its first week, where they "were shown everything about the place and there was general satisfaction that the State's money had been spent so well."[42] Likewise, the trustees enthusiastically announced that once again Massachusetts had demonstrated its generous and far-seeing nature by leading the nation in providing the one thing that inebriates who desired reform had previously missed: the "compulsory detention of a patient in an institution under pleasant and healthful surroundings."[43] The trustees' resounding optimism would fade within a year.

The second annual report of trustees of the Massachusetts Hospital for Dipsomaniacs and Inebriates began on an apologetic note: "The first year in any new institution is necessarily an experimental one.... Difficulties are constantly arising.... The hospital at Foxborough has proved no exception to this rule."[44] Frankly admitting that their new institution was in crisis, the trustees introduced three basic issues that would occupy board members and superintendents for the next twenty-six years: (1) how to secure the "right clientele" for the hospital; (2) how to treat the variety of patients who were committed to the hospital; and (3) how to convince the general public, and the legislature in particular, that state hospital care for inebriates was a worthy enterprise.

## Obtaining the Right Clientele

During Foxborough's first three years, an average of twelve patients escaped each month. The hospital had "no walls around the buildings, no prison cells, in short, none of the apparatus of a penitentiary." To build these tools of incarceration would have compromised the hospital's mission, the trustees maintained. Yet, as the board rightly pointed out, inebriates with their wits restored were more nimble escape artists than the insane. Matters were made worse by the fact that "some police officers, physicians, and even judges are in the habit of telling patients that they can go to Foxborough for a few weeks and come away whenever they please."[45] Still other patients were committed as punishment by their family members and friends, people who expected their loved ones to be released when needed back home and who even helped the inebriates escape. Further experience showed that some patients were positively misled about the circumstances of their commitment and the nature of the institution to which they were being sent.

Marcello Hutchinson, the superintendent, found that the very nature of the inmates made their management difficult and their chances for reform slim. While admission to Foxborough required sworn testimony of the prospective patient's good character—no vicious inebriates allowed—the courts had not strictly enforced this condition. Nor could they, for the sheer volume of arrests for drunkenness, the absence of any centralized record-keeping system, and the failure of the courts to share information with one another made it difficult to obtain accurate criminal histories on any of the inebriates sent to the hospital.

In fact, all committing parties—whether individual family members or the courts—were guilty of sending intractable cases, or "rounders," to the hospital. Their rationale was simple: they sent to the new "experimental" institution at Foxborough the people for whom all manner of treatment had failed. Thus, many inebriates arriving at the hospital had criminal records and were embittered by years of impoverished living both inside and outside state institutions. Others were simply "the sickest," appearing before the magistrates with all varieties of alcohol-related illnesses: neuritis, cirrhosis, ulcers, skin problems, heart conditions. For many, the hospital offered one more means of legal restraint.

Often domestic troubles were at the heart of commitment proceedings as well. The case of one early patient reveals just how desperate some families were to find an institution where a close relative might be confined. Writing to Hutchinson in 1894 about her recently committed inebriate husband, Julia Osborne declined the superintendent's request that she visit her spouse:

> As regards going on to see him, I would be sure to receive nothing from him only gross insults so I intend not to visit him unless he should be seriously ill. Not otherwise. He has an insane hatred for me. . . . He was until *Rum* got the better of him a very good and kind husband . . . but *rum shattered* our once happy home and converted it into a *living hell*. I have to depend solely on my two sons for our support, my oldest boy is 23, the other is 19 and we are just struggling along, experiencing poverty for the first time.[46]

Mrs. Osborne did not hold her husband responsible for his behavior, for she maintained that the alcohol had transformed her mate. Still, she did not wish to see Michael Osborne because of the damage that he had done to her family life. Cases such as this validated the WCTU's assertion that the campaign against drink was a home protection issue. Other early commitment certificates of the hospital suggest that spouse abuse, public violence, uncleanliness, and hallucinations associated with the D.T.'s often preceded commitment to Foxborough. To the trustees, the net result was the same: the hospital was "continually burdened with this incorrigible, immoral, and demoralizing element."[47]

Responding to this situation, the trustees took two courses: they petitioned the legislature to amend their and the superintendent's power to discharge patients, and they divided the patient population into different classes based on their potential for successful reform. In 1895 Hutchinson and the trustees began a "leave of absence" program in which any inebriate who fully cooperated with the hospital staff for six months would be released—a year and a half before his official two-year commitment expired. Aware that this blanket policy might be "an unscientific way of treating disease," the superintendent and board of trustees nevertheless believed it would help rid the institution of a bad element and help gain the cooperation of patients.[48] In 1897 the legislature granted the trustees the power to "finally discharge" patients.

Hutchinson also devised a "grading" plan that divided the patients into three categories: intractable, somewhat trustworthy, and trustworthy. The intractable (violent and unmanageable) patients and the newly arrived patients were placed in a locked ward and allowed out only twice a day for meals, air, and exercise. The more trustworthy patients remained under supervision but were given access to most hospital buildings, often "graduating" within a week to the highest two wards, where they were placed "on parole" and able to visit all buildings on their own, living in unlocked wards and performing "a certain amount of work each day as part of the remedial treatment of the hospital."[49] Though Hutchinson and the trustees believed that maintaining patients on the graded honor system would lead to patients' improved mental and moral condition and enhanced ability to resist temptation, the language of the system, with its "paroles" and "incorrigible" patients, and the hospital's use of locked wards and heavy supervision, left no doubt that Foxborough was a hybrid institution—part asylum, part house of correction. Foxborough was treating a vicious disease whose sufferers required both moral and medical care.

Although they focused their energies primarily on controlling patients within the asylum, the trustees also tried to stem the flow of unsuitable cases to the hospital by teaching the court system about Foxborough's mission and desired clientele. In 1896 Hutchinson and the trustees held conferences with a number of local judges who were responsible for many of the commitments to the hospital. In 1897 the trustees also asked the Board of Lunacy and Charity to alter their commitment forms to include information about the prospective patient's criminal history and personal character—another means of gauging his or her "fitness" for medical treatment. Educating the courts and social welfare agencies throughout the state became an ongoing process for the life of the institution.

## Treating the Dipsomaniac

As Hutchinson acknowledged in 1895, "Few subjects are receiving more general and persistent consideration than that of inebriety, and there is none upon which more varied and opposing views are maintained."[50] His first few years as superintendent at Foxborough had made this painfully clear. In 1894 the Massachusetts legislature had authorized its Committee on Public Health to investigate the use of several "specific cures" in the treatment of inebriety: the Keeley Cure, Duncan Cure, Houstan Cure, Empire Cure, Gold Cure, German Cure, and Thompson Cure. By far the most popular version was that of Leslie E. Keeley, who by the end of the nineteenth century could boast five separate Keeley Institutes in Massachusetts.

The legislature's attraction to the Keeley Cure and its competitors is not hard to understand. First, these cures usually took no longer than a month to complete—a period roughly one-sixth the length (and a fraction of the expense) of the minimum hospital stay. Second, contrary to Foxborough's rehabilitative success rate, which hovered around 33 percent, Keeley claimed a hyperbolic success rate of 95 percent. Finally, one needn't even attend a Keeley Institute to take the cure, since the remedy might be obtained by mail for home use. Ideally, patients could continue to work while engaging in their own reform. "The undisputed advantages [of] following this treatment are too great to make it wise for the State to wholly ignore the matter," declared the special committee.[51] Popular understanding once more clashed with the views of physicians, who deemed such specific cures the essence of quackery. Hutchinson was forced to defend his lengthy, if shortened, therapeutic regimen. The superintendent's campaign to fend off the Keeley advocates reveals how the process of medicalization—at least at the state level—was hardly an imperialistic imposition of medical authority.

Indeed, shortly after the patent medicine assault, the governor's executive council, led by Lieutenant Governor Roger Wolcott, conducted an investigation of the hospital and made some suggestions of its own: (1) encourage the hospital's trustees to educate the judges who send inebriates to the hospital; (2) increase the number of activities available to patients, so that they do not become idle; (3) use more discrimination in selecting attendants for the hospital; (4) improve the hospital's food; and (5) insure a high level of morally up-lifting personal interaction between the superintendent and patients. These recommendations echoed many of the hospital administration's own concerns, but addressing most of them required funding that was not forthcoming from the legislature. The last item, too, would prove elusive until the superintendency of neuropsychiatrist Irwin Neff in 1908.

Hutchinson, however, was able to develop a therapeutic protocol that generally resembled the treatment offered at private asylums. He regarded inebriety as an overpowering craving that blunted moral sensibility, benumbed the better impulses, and destroyed self-control and willpower. The proximate cause of the disease was a poison in the form of alcohol, which affected the nervous tissue. The ultimate cause, however, was an underlying constitutional susceptibility to alcohol originating in an individual's hereditary makeup or caused by the repeated use of beverage alcohol. Treatment for the dipsomaniac required lengthy, compulsory confinement, according to Hutchinson, for it took time to reestablish the regular habits of sleep, diet, and exercise that would rehabilitate the constitution. This was the first step in reconstituting the individual's moral fiber.

Hutchinson and the hospital's trustees used the term *dipsomania* to include both periodical and chronic inebriety. As noted, the Board of Health had defined *dipsomania* similarly in its 1885 report, but by 1893 this usage was exceptional. This may have been a deliberate decision on the part of the inebriate asylum's officers, just as the hospital's title privileged "dipsomaniacs" before "inebriates." In theory, dipsomania conferred a more sympathetic identity on the drinker since dipsomania's hereditary origins meant that the drinker was less responsible for his alcoholic state. Moreover, dipsomania's connections to insanity, a condition already treated in state hospitals, helped doctors justify their legal power to retain patients at the hospital for specific periods of time. The public might endorse treatment for the innocent dipsomaniac, whereas it might hesitate to support a hospital for vicious drunkards.

What was Hutchinson's therapeutic regimen for reforming dipsomaniacs? During these early years, patients entering the hospital were bathed and examined for injuries, then placed in seclusion for a week or more to recover from their drink-related symptoms. Hospital rules prohibited the use of any alcohol during this detention period. Physicians treated withdrawal symptoms with bromides, laudanum, and other sedatives. Once these symptoms ceased, patients were put to work making brooms in the broom shop (a common occupation in a variety of institutions). This was an intermediate step for patients before they were "paroled" and given the liberty to walk the hospital grounds unsupervised. Broom shop labor was also a step toward engaging patients in the most important and salubrious of all hospital occupations: farm work.

Patient labor—grading roads, tending the farm, renovating buildings, running the laundry, helping in the kitchen and dining hall—reduced institutional costs and taught the inebriates the habits of industry that their keepers found them wanting. Hutchinson encouraged patients to continue their trades as barbers, painters, waiters, and even physicians inside the hospital. Patients also engaged in gymnastic exercise groups each day. These required exercises were not meant to prepare inebriates for particular athletic feats, but to "strengthen the patients' muscles, coordinate bodily functions and promote habits of order, application, and purpose; most of all to increase the power of the will in controlling, through its organs... all functions, movements and desires of the body."[52] Doctors encouraged patients to chart their increasing lung capacity, muscle power, and weight as tangible evidence of their reconstituted moral fiber and manhood. If the hospital's concern with exercise during Hutchinson's tenure seemed nothing short of obsessional, it bears mention that the improved physique and physiology of patients offered one very tangible measure of their moral regeneration. Indeed, highlighting patients' progress in physical culture took on increasing importance in the face of consistently low "cure" rates, escapes, high per capita expenses, and criticism from a variety of sources inside and outside the state legislature.[53] If the hospital intended to detain patients for months, it had to demonstrate some beneficial effects of this confinement.

Clergy from the various parishes of Foxborough held weekly religious services at the hospital, offering patients a more traditional style of moral uplift. The superintendent, for his part, was to do the same. Through frank discussion of the inebriate's disease and its consequences for both himself and his family, through encouraging words, and by example, Hutchinson and his assistant superintendents endeavored to convey compassion, understanding, high expectations, and above all, good values, to each inebriate on a daily basis. Advocates for the state's first inebriate hospital placed considerable emphasis on the therapeutically powerful "personal equation" between doctor and patient. In this way, treatment echoed the moral therapy of the early nineteenth century.[54]

The weekly routine at Foxborough also included entertainments of varying types. These took the form of edifying and frequently moralizing lectures on a variety of topics: "Understanding Electricity," "How Mountains are Made," "Japan and Its People," "The Earthworm and Its Work," "A Visit to Russia," "The Ethics of Daily Life," and "Reminiscences of Life in Andersonville and Other Prisons of the South." These lectures were alternated with weekly poetry readings, variety shows, and musical performances furnished by both "friends" of the hospital and the patients themselves. The trustees and the superintendent regarded these amusements as necessary adjuncts to hospital treatment. They argued that their patients were "of saner minds" than their comrades in other state institutions and in particular need of appropriate intellectual stimulation.

## The Worthiness of the Experiment: Convincing the Legislature

Designing a therapeutic program for patients was one matter; convincing the legislature of the program's effectiveness proved a more difficult assignment. In 1894, the American Association for the Cure of Inebriates praised the new hospital as "one of the most promising and practical institutions of the country . . . one of the great pioneer asylums in the world."[55] The citizens of Massachusetts, however, were not equally impressed. In 1896, nearly two years after the first investigation of Foxborough by the governor's council, the *Foxborough Reporter* observed:

> There has been a great deal of complaint made against the hospital. It has been said on the floor of the house that the methods employed there begat laziness, and that men who have been graduated from that institution have been useless as producers thereafter. Many attempts have been made in the Legislature to introduce the Keeley and gold cures, but these efforts have never been successful. Investigations have also been instigated against the institution, but nothing has ever come of such efforts to make a change in it.[56]

To a significant portion of the public, inebriety was a vicious disease in which the emphasis still fell on the immoral aspects of the inebriate's drinking. Because the inebriate's drinking often ended in his and his family's penury, Foxborough needed to demonstrate that therapy prepared the reforming drinker for gainful employment. If patients failed to earn an income when they left Foxborough, then the hospital had failed, too.[57] Foxborough was treating more than inebriety; it was addressing the social plague of poverty.

Yet if the criticisms of the Massachusetts Hospital for Dipsomaniacs and Inebriates were many, so too were the strategies the hospital trustees and superintendent used to portray their institution as one of the state's most worthy charitable endeavors. Annual reports tracked patients' physical progress through the exercise program. Patients were listed in these reports by their occupation, revealing the contribution the hospital was making to restoring part of the Bay State's workforce. Marcello Hutchinson described the recoveries of patients accomplished "under adverse, even seemingly hopeless circumstances."[58] Hutchinson began keeping outcome records for discharged patients one year after the hospital opened, and the figures between 1894 and 1899 ran between 25 percent and 42 percent cured. Finally, the trustees and the superintendent emphasized the hospital's own growing efficiency, particularly its lowered per capita expenses. Most of this enhanced efficiency was the result of patient labor. In one exceptional case, a patient who became hospital librarian not only solicited local publishers for books and periodical subscriptions but also catalogued the hospital's eight-hundred-volume collection and assisted others in selecting reading materials. The same man gave a series of six lectures on electricity to patients and staff, illustrated with equipment from the Massachusetts Institute of Technology.[59]

Such "points of light" were not enough to keep Hutchinson at the helm, however. In 1900 he resigned his post and took the superintendent's position at the Vermont State Asylum for the Insane. Having presided over the hospital during its turbulent early years, he had several accomplishments to his credit.[60] The trustees praised his efforts to educate the public, especially the local courts, about the proper type of patient for the hospital. In short, there seemed to be "a growing public confidence in the institution and a better appreciation of its work."[61]

Charles E. Woodbury, who had spent eight years as the inspector of institutions for the Massachusetts Board of Lunacy and Charity, took Hutchinson's place as superintendent. No doubt in his former position he was as familiar as anyone with Foxborough's trials and tribulations. An experienced institutional manager, Woodbury had served as superintendent of Rhode Island Hospital. Before that he had held assistant superintendent positions at Bloomingdale Asylum and at McLean Asylum, where he had sustained a near-fatal attack by a patient just prior to his promotion to superintendent, a position he was subsequently unable to assume. Woodbury took his new post in an atmosphere of guarded optimism, but this changed quickly.

Within a few months of Woodbury's arrival at Foxborough, the widely publicized *Report of the Advisory Committee on the Penal Aspects of Drunkenness*, which had been commissioned by the mayor of Boston, cast doubt on compulsory hospital confinement for inebriates:

> The proposition to catch drunkards and then submit them to a course of treatment which will forever destroy their desire for alcohol presents grave difficulties . . . . The weight of evidence seems to show that there is no royal road to recovery for those who are not eager for their own reform. . . . The difficulties of the situation are undoubtedly great, but there is no reason to doubt that a vigorous elimination of incurable cases and a liberal provision for small institutions adapted to individual treatment would give correspondingly good results in this Commonwealth.[62]

Although the report was intended as a wake-up call to the courts to systematically extend the reach of the probation system to monitor and control public drunkards, it also constituted an assault on the new inebriate hospital. The trustees rushed to defend Foxborough's mission, noting that it was unrealistic to expect the state to fund several small institutions since they barely funded the one in place. If the courts were more cooperative with the hospital and sent only the most promising patients, the hospital's success rate might be higher, they assured the public in 1900.[63] Moreover, inebriety was such a large problem that a full-scale effort was required, not small, piecemeal forays into reform.

This exchange between the authors of *Penal Aspects of Drunkenness* and the trustees set the tone for the next six years—Woodbury's tenure at Foxborough. Woodbury had his minor victories—raising the patient census and lowering per capita costs; enlisting probation officers to track patients' post-discharge progress; securing the hospital's right to admit voluntary, paying patients. Yet some of these measures had unintended consequences. With patient rolls exceeding the hospital's carrying capacity, inebriates were squirreled away in the cottage attics, corridors, and day rooms. Escape rates skyrocketed. Faced with overcrowded conditions and patients jumping ship, Foxborough's policies grew more punitive, more authoritarian. In 1900, Woodbury increased the required hours of work from four to five, adding an additional hour of gymnastic exercises as well. Patients who failed to meet these demands were locked in their wards. Six hours of labor per day became the rule in 1902, and Woodbury urged the trustees to approve an eight-hour day for patients in good health. The board approved seven hours in 1903. With 13 percent of patients eloping, Woodbury and the trustees petitioned the legislature for the power to send escapees to the State Farm in Bridgewater. It was not granted, however, for fear that the hospital would be perceived as a gateway to the State Farm, a perception that would not win favor with paying patients desiring voluntary admission to Foxborough. The hospital's inability to transfer escapees to the State Farm did not disguise the fact that the institution itself was becoming increasingly coercive in its methods of treatment.

The legislature deflected several petitions from citizens who wished to close Foxborough, petitions that rankled the trustees so greatly that they were forced in 1902 to acknowledge the difficulty of managing a hospital for a contested disease:

> The problems with which this hospital is called upon to deal... are more varied and difficult than those peculiar to any other type of hospital under state management. Its work and its patients belong to a class in regard to which public opinion is itself divided. Some hold that inebriety is a disease, and ask for distinctively hospital treatment. Others regard it as a crime, and demand penal methods for its restraint and cure. Others, considering the evil as a fruit of mental and moral degeneracy, insist on the supreme importance of educative, ethical and reformatory measures. To each of these classes the hospital necessarily fails to justify its existence by as much as it falls short of the special method of treatment regarded as appropriate and requisite. At the outset, therefore, the hospital finds itself embarrassed in the matter of public opinion with reference to the proper function of such an institution.[64]

Of course, inebriety was all three in 1902: disease, moral failing, and, at least in Massachusetts, a crime. Ironically, however, in their efforts to enforce a brand of compulsory confinement founded on inebriety's status as a mental disease, the hospital's trustees and superintendent were turning their institution into a prison.[65]

The differences between the inebriates and the insane would become all too clear in 1905, when the state ordered the transfer of one hundred chronic, mildly insane men to the Massachusetts Hospital for Dipsomaniacs and Inebriates. With the insane in residence, the legislature changed the institution's name to the Foxborough State Hospital. Though Woodbury and the trustees were upset with the state's move, fearing that it might signal an end to their work with inebriates, their concerns were made worse when several hospital attendants were indicted for abusing inebriate patients and a series of suits were brought against the superintendent and his staff. The trustees believed the charges baseless, but they acknowledged that pending litigation was hurting patient morale and the public's dubious impression of Foxborough.[66]

Woodbury made no mention of the indictments in his own report for 1906, heartily thanking "all those associated with me in what is a most perplexing and very trying work."[67] Observing that the hospital was still working out "an untried problem" for the rest of the country, Woodbury also noted that he could not point to any precedent that indicated the appropriate path to take in inebriate reform at the state hospital level. All of this was true. Iowa, Connecticut, Minnesota, and New York City each took note of Foxborough State Hospital as they considered their own plans for public inebriate institutions. Indeed, representatives from Iowa's Board of Health had corresponded for several years with Woodbury on the methods employed at Foxborough.[68]

No defense of the hospital's mission or pioneering position would suffice, however. By late 1906, Woodbury's and the trustees' worst fears were realized. If there was one thing that the citizens of Massachusetts were even less willing to tolerate than institutional inefficiency and soft-hearted largess to the vicious and undeserving, it was patient abuse and scandal. And if there was one time of year when abuse and scandal in public institutions was sure to receive as much attention as it genuinely deserved, it was election time.

## The Investigation, 1906–7

In July 1906, the governor's executive council, led by Lieutenant Governor Eben S. Draper, scheduled an investigation of the Foxborough State Hospital. Public skepticism and accusations of mismanagement and patient abuse had made the institution a persistent thorn in the state's side for nearly all of the its thirteen years. Elections were scheduled for fall, and Progressive Republican Governor Curtis Guild Jr. could wait no longer to review the results of the "Foxborough Experiment." The beleaguered trustees, too, desired an end to the perennial complaints against their hospital. "We are quite desirous that any investigation which should be made of the hospital should be exceedingly thorough . . . so as to sift out the miserable, irresponsible opposition which the hospital has had, injuring greatly its usefulness in many ways," pleaded Frederick Fosdick, the board's chairman.[69]

Fosdick and Governor Guild would both have to wait. A grand jury hearing in Norfolk County was listening to testimony from present and former patients to determine the truth of charges filed there by Joseph Hicks and James Brennan, two men who had been committed to Foxborough and found the conditions there unworthy of the state's citizens. Hicks was a sawyer by trade, committed to Foxborough from his native Charlemont, Massachusetts, a quiet Berkshire town at the opposite end of the state. Hicks claimed that he had been improperly committed to Foxborough, where his condition had deteriorated, thanks to the abuse and neglect he received from the hospital attendants and physicians. During his stay, Hicks also had seen other patients mistreated. Brennan, an East Boston native, complained that he had contracted smallpox upon his arrival at the hospital, but the staff had neglected his condition and denied him any treatment. Neither man trusted the lieutenant governor to lead a thorough investigation, so they pressed charges themselves, enlisting the counsel of Cambridge attorney Henry C. Long. They intended to follow the grand jury hearing with several civil suits against the hospital staff.

Long advertised in the local *Foxborough Reporter* and the more popular *Boston Globe* for any information regarding the state inebriate hospital. In a short time, his office became "the Mecca to which the discharged patients come from headquarters to get their first meal after they are discharged from Foxborough."[70] Long would later assure board chairman Fosdick that he had not offered cash for the interviews, paying patients no more than their cab fare between Dedham, where they might collect their grand jury fees, and his office—what usually amounted to thirty cents. In total, Long interviewed between 150 and 200 patients to assemble his laundry list of complaints against the hospital. Only the testimony of several dozen men was necessary to substantiate his case in the investigative anarchy that dominated Beacon Hill in the deep of winter 1907. Foxborough State Hospital and the disease concept of inebriety were soon on trial.

## Conditions at Foxborough—The Patients' Perspective

In early January, Draper convened his council, unbeknownst to the Foxborough trustees. By this time, much to the trustees' chagrin, the Norfolk County grand jury had found two of the three attendants accused of abusing patients guilty; their trials were pending. Draper and the council paid a surprise visit to Foxborough on 11 January to see the institution for themselves.

The visitors' impressions of the hospital, as described by the lone Democratic councilor, Edward P. Barry, were quite favorable: "I am free to say that the conditions there, in Foxboro, are not bad, there are no serious complaints and everything appears to be running smoothly."[71] Barry was soon to change his mind.

On 24 January, Draper called the long-awaited public investigation of Foxborough to order. The "trial" of the state hospital for inebriates was big news. Reporters from all the Boston papers attended, dividing their front pages between the Foxborough investigation and the equally sensational New York City trial of Harry K. Thaw, the impassioned assassin of accomplished architect and notorious womanizer Stanford White—a case where alcohol also played a role.[72] The lieutenant governor had no fixed agenda for the investigation and began the proceedings by soliciting the public audience's questions about the hospital. Ready to capitalize on this opportunity was Lyman Griswold, Joseph Hicks's elected representative from Franklin County. Hicks had written up his own sordid tale of abuse and mismanagement at Foxborough, which he placed in Griswold's able hands.

Printed in full in the *Boston Evening Transcript*, Hicks's statement detailed his commitment to Foxborough for a painful abscess in his ear. This was not a complaint that usually led to confinement at the state hospital for inebriates; however, prior to consulting his family doctor, Hicks had pursued "self-medication" in the form of two bottles of whiskey. When Hicks finally did seek the attention of Samuel Bowker, M.D., the physician, with the assistance of the district court judge in Greenfield, managed to commit the sawyer to Foxborough. At no point, Hicks claimed, was he informed that the hospital was for the treatment of inebriates; nor had he ever been arrested for drunkenness. Disoriented by the pain from his ear, Hicks found himself at Foxborough, stunned to learn that the institution was not a general hospital and that he was not able to leave for six months. He received no comfort from attendants, who ordered him to take a bath, denied him his clothes, forced him to rest on dirty sheets, and told him that Foxborough was "a _____ of a place to get treated for an abscess of the ear; that he would lose his ear for sure and that there was no doctor there who knew anything about treating ears or anything else." After voicing his outrage at the situation, Hicks was "kicked and pounded" in the head and chest by four attendants until he cooperated. Within a month, Hicks did receive some care for his ear. Dr. Clarence Bell, an assistant physician at Foxborough, brought him to Massachusetts General Hospital, where he learned that his ear drum had broken. Returning to Foxborough, Hicks had to wait for several days before he was given any medication by the hospital staff—the attendants warning him that "state paupers don't deserve anything."[73]

But this was not all. Hicks accused the attendants of attempting to poison him in April with an overdose of the expectorant *sal ammoniac*. That same month, another troubled patient, Frank Weatherhead, who had told both patients and attendants that he was depressed and contemplating suicide, eventually wandered into the back office of the administration building and "blew his head off with a gun that was in the office. He was buried with the paupers, but the head was not put in the coffin with the body. The head was found a few days later in the office and was buried in the garden."[74] Moreover, Hicks claimed to witness, during his stay at the hospital, the beatings of other patients, such as the son of the former mayor of Holyoke, who yelled "Murder!" whenever anyone wearing shoes approached his room (shoes indicated a staff member, as the patients all wore slippers). Hicks's statement was rife with at least a half dozen similar vignettes, and his audience was not unimpressed. Representative Griswold testified to Hicks's trustworthy character, and Hicks was promptly sworn in to answer questions from the councilors.

The most important testimony of the day, however, came from Hicks's counsel, Henry C. Long. Introducing himself as the attorney who had brought several civil suits against Foxborough State Hospital for Hicks and Brennan, Long offered to provide the council with his list of twenty-five to thirty charges against the institution. The list ranged in nature and severity: attacks against the superintendent, Charles Woodbury, for failing to come in contact with patients and for taking the best meat and produce for his own family's use; complaints of patient abuse suffered at the hands of poorly trained attendants, fresh from the farms of Maine; and an indictment of the hospital's "prison ward," required gymnastic exercises, poor quality food, and carelessness with firearms. Interviews with close to two hundred former patients had convinced Long that the superintendent was "wholly incompetent from every standpoint." Woodbury had neither the good of the institution nor the well-being of its inmates at heart, and he appeared to be using the hospital as if it were "his own private property."[75] Long's list of charges provided Lieutenant

Governor Draper with an agenda for organizing the investigation. The Cambridge counsel had furnished Draper with a roster of patient witnesses; and he had laid the temporal parameters for the probe, suggesting that the council go back to 1902, when they had bungled James Brennan's case of smallpox.

In the ensuing month, dozens of patients took the stand to describe the conditions at Foxborough. Long's witnesses came first, testifying for the "prosecution"; board chairman Fosdick then mustered several patients for the "defense." In what must have seemed an investigative free-for-all, testimony was interrupted regularly by questions from council members and from the public who packed the statehouse chambers. In the end, council members weighed the evidence presented by past and present patients, hospital officers, trustees of the institution, and medical and social welfare experts. The public investigation was a crucible for determining the fate of the state hospital. It also provided an excellent opportunity to see how the Massachusetts populace regarded the disease concept of inebriety. Governor Curtis Guild Jr. had ordered the investigation not only because Foxborough had from its inception received more than its fair share of criticism, but also because Guild wondered "if the theory of the Foxboro institution is right or wrong."[76] Was inebriety really a disease that was treatable at the state level?

Throughout the investigation, patient testimony cast doubt on Foxborough's hospital identity. It brought into sharp relief the persistence of the moral-medical ideology of inebriety. The councilors focused their questions on the atmosphere in the sick ward and the prison ward, discovering that the hospital came up short in its sanitary facilities and that staff frequently substituted punishment for moral suasion. So absent was moral guidance during the Woodbury years that when council member Edward Barry asked James Brennan if he had received any "spiritual advice" at the hospital, Brennan replied, "There was some whiskey there, Paul Jones XXX I think."[77] Under Woodbury's superintendency, unsanitary conditions were the order of the day: between 100 and 125 men shared the same toilet, bathed in just one bathtub, and used the same half dozen towels to dry off after washing. Anyone not physically ill when they entered the hospital had a good chance of contracting something within the first few weeks of their stay. Likewise, the prison ward lived up to its name. For "any slight infringement of the rules, any apparent unwillingness to do just what they were told and when they were told," patients were trundled off to solitary locked rooms, where they surrendered their clothes and were fed bread, water, and skim milk every other hour.[78]

So damning was the testimony that the state legislature several doors away in the state capitol contemplating the annual budget for the Foxborough State Hospital, lowered the hospital's allocations for the coming year. The chairman of the ways and means committee, a Mr. Walker from Brookline, "defended the principle of the institution and thought it would be necessary at least to continue it as a hospital for the chronic insane and inebriates," but his colleague, a Mr. Clark of Brockton, was less enthusiastic, lambasting Woodbury's excesses: why should the state fund Woodbury's "retinue of servants, his carriages, and his six horses for the use of his family?" A representative from Worcester, Hugh O'Rourke, doubted the worth of an institution where "not one percent of the patients are cured, and the treatment is worse than that in the jails."[79] It was clear that neither Foxborough nor the disease concept of inebriety had won many allies since the hospital's opening.

## Defending the Hospital

Though patient testimony against the hospital formed the most salacious, most riveting testimony of the trial, the hospital did have its supporters, and some of them were inebriates. Several took the stand in late January and reported that the food at Foxborough was not the worst they had eaten, that they had not witnessed cases of abuse, and that plaintiff Joseph Hicks had a reputation as a troublemaker. In early February, trustee Leroy S. Brown became the first hospital officer to give testimony. About to depart for Europe, Brown wanted to set the record straight on a number of issues. Brown had been accused of graft by several patients, and he wished to defend himself and the purchasing protocol used at the hospital. In the course of his testimony, however, it became clear that he and trustee Fosdick were arranging contracts between Foxborough and their own or their relatives' firms. Brown's testimony did not have the effect he desired, nor did it help the defense.

The defense's cross-examination of Joseph Hicks, however, did much to discredit his previous testimony. It emerged that Hicks did not have a blemish-free past when it came to alcohol. Indeed, the lumberman's commitment papers, filed by his sister, indicated that on more than one occasion Hicks had become insane from drink. Prior to his commitment, he had become so intoxicated that he fired a shot into the wall at an imaginary enemy. Much to Attorney Long's chagrin, a recent x-ray of Hicks, taken at the counselor's request, revealed that Hicks had lied about his broken ribs. Moreover, Clarence Bell, the assistant physician who had taken Hicks to Massachusetts General, testified that he had witnessed no abuse at the hospital and that some discipline was necessary when contending with inebriates who lacked self-control.

Bell brought along statistical evidence of the men's physical progress—due to carefully chosen forms of exercise and nourishing meals, he claimed. These data—portrayed as "hard facts" about the patients' improvement—were warmly welcomed by the council after days of weighing the contradictory recollections of witnesses. Perhaps the most powerful testimony came from the doyen of turn-of-the-century addiction medicine, Thomas Crothers. In the twilight years of his career, Crothers was still editing the *Quarterly Journal of Inebriety* and directing the Walnut Lodge Hospital in Hartford, Connecticut. He was called to testify for two reasons: first, to report that a patient he had treated at his private inebriate hospital in Hartford was injured *before* he came to Foxborough; and second, to acknowledge that even he, a widely acclaimed inebriety expert, had not diagnosed smallpox when he visited the hospital and saw James Brennan in late 1902.

Crothers was full of praise for Foxborough, recalling his visit and observing that the hydrotherapeutic treatment and the gymnastic exercises at the new hospital constituted the "most advanced treatment and one of the best appliances and institutions in the world."[80] All of this may have been true, but what became obvious in the course of Crothers's testimony—testimony that suggested it was not so easy to diagnose smallpox—was that in the course of cloistering their inebriate charges for months at a time to protect them from the temptations of the outside world, hospital physicians treating inebriates were growing more removed from general medical practice. This phenomenon was hardly new: neurologist Silas Weir Mitchell had made similar observations when he spoke before the superintendents of insane hospitals at the 1894 American Medico-Psychological Association meeting and declared: "Your hospitals are not our hospitals; your ways are not our ways. You live out of range of critical shot; you are not preceded and followed in your ward by clever rivals, or watched by able residents fresh with the learning of the schools."[81]

Asylum psychiatry had become moribund in many parts of the country, separated from the robust atmosphere of the general and teaching hospitals. Yet alcohol weakened the inebriate's resistance to contagious disease. Doctors at Foxborough were treating a myriad of complicated alcohol-related symptoms and diseases, not just "habitual drunkenness." It was essential that they be excellent clinicians as well as administrators.

Three days of testimony for the trustees and hospital brought about a change in public opinion, for much of the investigation boiled down to "a question of veracity between inmates and administration, and the latter has the larger claim upon public confidence, especially as its testimony is supported by the official records," reported the *Boston Evening Transcript*.[82] Yet the fate of the hospital was by no means secure. The editors of the *Boston Herald*, for example, suggested that the state hospital for inebriates might be turned into a reformatory for first-conviction drunkards.[83]

By the second week in February, Joseph Hicks's testimony stood on shaky ground. The attendant that Hicks had accused of abusing him had not joined the hospital's staff until four months following the date of the claimed beating. Long now turned his attentions to undermining Woodbury, whom the trustees had tried to force from the superintendency in 1902. The counselor also brought out the business transactions between Fosdick's engine firm and the hospital. Both were easy targets. It would be up to the other trustees and a handful of Bostonians engaged in philanthropic and social reform activities to defend the state's first inebriate hospital.

In the concluding two days of testimony, the executive council directed its questions to four concerns: (1) the relative importance of the "personal equation" and physical rehabilitation in the inebriate's reform; (2) the type of patient that should receive treatment; (3) the appropriate use of restraint; and (4) the institution's identity as a hospital or detention center. All of these issues pertained directly to the medicalization process.

One by one, the trustees took the stand, uniformly endorsing both moral suasion and physical education as methods for treating inebriates. To their minds, "a man's will power and moral qualities which have been destroyed have got to be built up by building up his physical system."[84] The board was also in agreement that patients needed a paternal physician and supportive staff who would encourage inebriates to reform through supportive words and example. Yet regardless of who ran the institution, the trustees insisted, its treatment program was undercut by the poor quality of patients sent there from the courthouses and jails. Judges continued to send "late-stage" and "incorrigible" inebriates, and this created problems with institutional order and lowered the hospital "cure" rate, since these cases rarely achieved reform. In short, the trustees argued, their hospital was forced to restrain these patients, for a substantial proportion would attempt to escape to procure alcohol.

Was the institution best thought of as a hospital if patients were so restrained? While the inebriate hospital had been proposed as an alternative to the insane asylum and the jail, its terms of confinement resembled those of both institutions. Inebriety remained a medico-moral disease. The executive council did not see restraint as a particularly "medical" intervention and questioned keeping the term *hospital* in the institution's title if, indeed, the institution itself were to be retained by the state.

At stake was the nature of the disease of inebriety, for even some trustees felt that "inebriety is not curable." The cure was to make the man self-sufficient again and to help him to resist temptation, but the weakness was forever a part of his constitution. Was inebriety really a disease, or was it a "social evil" that sapped the vigor of the working-class drinking man and prevented him from supporting his family? Some trustees believed "cure" at Foxborough should be understood as helping a man back on his feet and training him in skills that might help him keep a job.[85] They proposed that the hospital be renamed with the ominous and ambiguous title the "Foxborough Institution."

On the last day of testimony, Joseph Lee, father of the American playgrounds movement and founder of the Massachusetts Civic League; settlement house reformer Robert Archey Woods; Warren Spaulding, secretary of the Massachusetts Prison Association and a sculptor of Massachusetts probation policy; Charles Pickering Putnam, pediatrician and chairman of the Associated Charities Committee on Inebriety; and George Jelly, an examining physician of Boston's Department of Public Institutions and a former superintendent of Boston's exclusive McLean Asylum—all social reform luminaries—came to Foxborough's defense.[86]

Fearing that Foxborough's fate was hanging in the balance, Lee had already drafted a piece of legislation, House Bill No. 984, that required the state to "provide for a Suitable Institution for the Curative Treatment of Drunkenness."[87] Echoing the trustees, Lee asserted that the hospital had yet to receive the right class of patient, and that once this happened, the institution would succeed, "not by drugs, but by giving good wholesome living."[88]

Woods followed by suggesting that neither the executive council nor the general public was familiar with the complexity of the disease of inebriety: "A great many diseases have various stages, different in origin, different from other stages. Certain in certain stages are curable, in others not readily; in one stage not contagious, in others, very contagious. Drunkenness as far as it is a disease, is that sort of a disease." Woods believed that it was high time the state realized this and began to provide not just one institutional solution, but "hospitals, sanatoriums and other types of institution for the treatment of this disease in all respects, one of the most serious physical and moral evils that afflicts our community."[89]

Warren Spaulding urged acquiring a better, more supportive staff and improving the patient pool, which he felt was composed of mostly degenerate lower classes who were not worthy of state medical care. Spaulding clung to a class-based dipsomaniacal interpretation of inebriety and supposed that among the better social tiers, inebriety, like crime, was a form of insanity. For Spaulding, to mix the two classes—lower and upper, criminal and diseased—was a mistake.

By far the most radical of the group was Putnam, who observed that the hospital at Foxborough had never fulfilled its purpose: to provide the state's less-affluent citizens with medical treatment for their drinking problems, something only the rich could afford. Though the state would not support luxurious sanitarium care for its indigent inebriates, Foxborough might take a page from Massachusetts General Hospital, suggested Putnam. Mass General was a public facility that had offered its patients first-rate medical and hydrotherapeutic treatment for inebriety and permitted them to leave only when they were cured.

Following the testimony of Lee, Woods, and their social reform colleagues, the editors of the *Boston Evening Transcript* reported: "While at one time it was believed that a continuance of the institution on the present plan would not be warranted, the testimony of disinterested experts has been so strong in favor of such a policy that there will probably be no change unless it may be to strengthen its work." Still, the *Transcript* noted, a name change might be in order, since "a place of detention is hardly a hospital."[90]

Attorney Long remained unconvinced: "The question whether inebriety is a disease is an open one. It is no more a disease than prostitution. It is not a disease but it creates disease. . . . It is on prevention that the state should spend its money and not the cure."[91] This was a familiar argument, but not one that would effectively counter the opinion of "disinterested experts" in Progressive Era Massachusetts. The educated public placed great faith in professional expertise.

## Experts Reform the "Foxborough Experiment"

On 10 April 1907, the governor of Massachusetts, Curtis Guild Jr., received not one but two reports from the executive council investigating the charges against Foxborough State Hospital. The council was split four to one, with the majority believing that the charges of patient abuse, poor food, and institutional mismanagement had been greatly exaggerated and arguing that the institution should be continued. The minority of one—Councilor Edward P. Barry, the lone Democrat on an otherwise all-Republican council—deemed the wholesale replacement of Foxborough's administration a necessary step for its continuation, a step he hoped the state would take forthwith.

The committee's failure to agree on a resolution to Foxborough's problems reflected the ambiguity and contradiction embodied in the disease concept of inebriety itself. Although doctors and reformers were convinced that inebriety was a disease, they seemed unable to distinguish between innocent victims of the disease and those who had cultivated a vicious habit leading to their disability. Likewise, even though hospital treatment was mandated by the state, the best way to organize and manage this institution was far from obvious. A repressive "hospital" atmosphere might appear justified, for example, if the etiological emphasis were placed on the vicious habits of the inebriate that led to his diseased state. When did moral transgression end and disease begin? Could the two ever be considered separately?

In the end, Edward P. Barry's dissenting opinion carried the day. On 11 September 1907, the *Boston Evening Transcript* proclaimed: "A complete overturn in the management of the Foxboro State Hospital was announced by Governor Guild this morning after the meeting of the executive council. Every member of the former board of trustees has resigned and a new board is named with Robert A. Woods of Boston as chairman."[92]

The governor's decision had come in the wake of a barrage of calls for the abolition of Foxborough that followed the executive council's own reports. Far from disappearing from the statehouse agenda, the Foxborough situation had lingered, a festering sore eating away at the image of Massachusetts as a pioneering, Progressive social welfare state. Guild had always been an advocate of legislation that improved the health and well-being of the working classes, attaining a reputation as an effective leader, a champion of sociomedical reform, and a protector of the people. It was no coincidence that his decision should come just six weeks before the November gubernatorial elections.

Writing to the resigning board members, Guild observed that the state hospital for dipsomaniacs and inebriates had "been an experiment that Massachusetts is, as a pioneer in reform, trying out for the benefit of the entire country. It has not yet succeeded, but I, for one, am not willing to admit that

Massachusetts cannot succeed in attacking drunkenness as a disease."[93] A testimony to his faith in the new medical approach to habitual drunkenness, Guild's selections for the new board of trustees read like a veritable cavalcade of social welfare stars. The director of the Massachusetts Civic League Committee on Drunkenness, Robert Archey Woods, who had testified before the executive council as an outside expert, became the new chairman of the hospital trustees.[94] The others had ties to temperance societies, the medical profession, the court and probation systems, and the business and financial worlds.[95]

A preliminary investigation of the hospital suggested two problems: poor patient quality and poor location (too close both to the temptations of the city of Foxborough and to the rail lines for easy escapes). These problems would require time to address, but the trustees recommended the acquisition of land in a more remote part of the state, and they quickly set in motion a search for a new superintendent who might oversee the hospital's reorganization and reconstruction. After months of candidate interviews, the board found their man: Irwin Hoffman Neff, a specialist in psychiatry and neurology who had received his medical degree from the University of Maryland and trained at Johns Hopkins before opening a practice in Baltimore with his father. Neff had spent time working at two state hospitals for the insane in Michigan, gaining "practical experience in the administration of a large hospital and long training in the care of all classes of neuropathological patients."[96] Upon his arrival at Foxborough in the spring of 1908, the thirty-nine-year-old Neff began an aggressive campaign to turn the hospital into a thoroughly Progressive medical institution. Foxborough, trumpeted the editors of the *Boston Evening Transcript*, was entering "a new era of usefulness."[97]

With the cooperation of an activist board of trustees, Neff's plans met with at least partial success. Together, Neff and the trustees reorganized the hospital to treat patients for short periods of time, employ the social casework model, and provide vocational assistance and outpatient care for individuals across the state. Several years before psychiatrist Elmer E. Southard opened the Boston Psychopathic Hospital, extending medicine's reach to a broad range of society's deviants, Neff and his staff at Foxborough State Hospital had instituted what historians have considered the hallmark of Progressive reform: a flexible ideology of personalized treatment, based on an abiding faith in the reconstructive power of science and education for the individual and the public. "Conscience" had been reconfigured several times in

Foxborough's short history; it was now time to see if "convenience" would again prevail.

The new Foxborough administration—superintendent and trustees—had high ambitions; they attempted to create a wholly new system of graded care for the state's habitual drunkards, classifying patients as long-term custodial, short-term remedial, and indefinite aftercare. This system would eventually be integrated with a larger network of public and private educational, penal, probationary, and charity institutions within Massachusetts, some of which already handled inebriates. Realizing that state reform of the inebriate was a cooperative venture and that negotiation—not only with the state legislature, but also with the inebriate's family and the judicial, social welfare, and medical bureaucracies—was the only means to effect it, Neff and the trustees actively courted the assistance of these groups in identifying, referring, and treating habitual drunkards.

## Reeducation, Vocational Training, and Aftercare, 1908–13

By the early twentieth century even correctional facilities such as reformatories and prisons were drawing upon the medical model.[98] Yet the inebriate asylum was presented simply as an alternative to the ineffective imprisonment of drunkards. The public needed to view the institution as a hospital, not a "house of restraint." When Irwin Neff assumed his post in 1908, he quickly took steps to reestablish the hospital's identity as a medical institution. To this end, the new superintendent ordered staff changes to refocus therapeutic efforts and improve hospital efficiency. He hired a steward to manage the day-to-day contract work of the institution so that he could devote his energies to the patients at Foxborough. The superintendent also abandoned the use of locked wards, giving patients the right to traverse the entire hospital grounds. With the trustees' cooperation, Neff abolished the "three-months rule," which required all first admissions to remain at the hospital for a minimum of twelve weeks. The hospital now told patients that they could leave when their condition warranted their release, and no sooner, reasoning that an inebriate entering Foxborough had "no more right to be discharged at the time that another man is than two patients who are sent to a general hospital, each for his own disease, may anticipate the same length of residence."[99] This change in policy facilitated the hospital's efforts to treat patients as individuals, a slogan that figured prominently in Foxborough's new therapeutic regimen. Here also was acknowledgment that inebriety was a chronic disease that affected each patient differently. In any event, eliminating mandatory sentencing was designed to bring the hospital more in line with the practices of the larger medical community and to help it attract volunteer, paying patients who would previously have taken treatment at private sanitaria such as the nearby Newton Nervine. Trustees hoped that this "better class of patient" might lower per capita cost and boost Foxborough's success rate.

Two phrases came to dominate the annual reports of the Foxborough State Hospital in the early years of reorganization: "concern for the individual case" and "re-education." Indeed, Neff and the trustees regarded the heart of their work as the reeducation of the individual patient. The process of reeducation, claimed psychiatric experts, placed a high premium on the qualities of the physician in charge:

> These patients come with no adequate philosophy of life, no raft with which they can safely reach shore in their sea of trouble. They have narrow, distorted, perverted viewpoints. . . . They cannot be corrected by a pronunciamento, by laying down what the analyzer believes to be the law and the gospel on the different questions involved but must be slowly changed by a process of reeducation in which the personality of the physician and his attitude towards the whole situation plays a prominent part.[100]

Annual reports of Foxborough, and later Norfolk State Hospital, reiterated the importance of the doctor-patient relationship in treatment. "We must appreciate that the personality of the inebriate is an individual personality and cannot be expressed by a composite picture. . . . Those who fail to recognize that each inebriate is an individual case, each case requiring individual consideration, will often fail to benefit the patient," claimed Irwin Neff in 1911. Neff's work with his patients and their families as well as his tireless efforts to correspond with discharged inebriates made his therapeutic priorities plain.[101]

Indeed, Neff and his staff physicians helped inebriates discover substitutes for their alcoholic habits, and they endeavored to teach their charges to value industry, fiscal independence, and responsible citizenship. The hospital directed its reeducational efforts far beyond the patient, however. The new outpatient department taught patients' families how best to cope with their loved one's illness and offer support. Trustee Robert Archey Woods published an instructional pamphlet for probation officers and judges, *Drunkenness: How the*

*Local Community Can Be Brought to Do Its Part.* In an effort to attract and educate the local medical community, Neff teamed up with E. E. Southard, M.D., the director of Boston Psychopathic Hospital, to host a state-sponsored conference on the social and mental problems stemming from alcohol.[102] Finally, Neff and the trustees developed ties to several major colleges and universities to assist with vocational training and shore up some powerful institutional support for the hospital. They engaged the Massachusetts Agricultural College at Amherst, the geology departments of Harvard and MIT, and the Massachusetts Fish and Game and Forestry Commissions to offer courses to patients on scientific agriculture and forest management. The idea was to prepare inebriates to assume healthful outdoor occupations following their hospital stays.

In the eyes of the state, "occupational therapy" was particularly relevant to the inebriate's case, for the disease was portrayed as one of economic as well as alcoholic dependence. Work could restore the patient's physical and fiscal health. Indeed, according to the superintendent, the successful treatment of inebriates required hospitals to engage patients in "congenial, and in some cases remunerative work" and to possess "sufficient farm land to develop agricultural interests."[103] Many patients were given significant responsibilities for the operation and maintenance of the hospital as transitional employment until the patient returned to the labor force. By employing the reforming inebriates—not just in farm labor but in various salaried positions at the hospital—prior to their release, Neff married institutional economy and therapeutic efficacy, all the while breaking down the hierarchical relations that had obtained under his predecessor. Some patients were even "farmed out" to local industries, where they worked during the day, prior to returning to the hospital's safe, alcohol-free environment.

Once patients were released from the hospital, the outpatient department took charge of their cases. With regular and temporary offices located throughout the state, the outpatient department not only helped place reformed patients in new jobs but also provided follow-up support services (psychotherapy, clothes, food, letters of recommendation), screened incoming applicants to the inebriate hospital, and paid visits to the patients' homes in an effort to keep track of their progress. Keeping close watch on discharged patients, the hospital amassed a great deal of qualitative and quantitative information, and this data was not used exclusively to placate skeptical legislators.

As it amassed statistics on patients released from the inebriate hospital, the outpatient department staff came to several realizations about its charges.

First, patients hailing from the western half of Massachusetts generally fared better than those from the more metropolitan east. Hospital officials attributed the difference to the better circumstances of men from the west, who were more likely to possess homes, jobs, and families to return to following their hospital stays.[104] The trustees of the hospital had reached a similar conclusion in their special report of 1910, *Drunkenness in Massachusetts: Conditions and Remedies,* where they claimed that the Bay State's institutional solution for drunkenness would not end the inebriety problem: "This end will be attained only by the provision for all men of healthful homes, constructive recreation centers, and schools that teach skilled industry, train thought, and build character."[105] Outpatient physician John Horgan likewise placed a premium on the environmental circumstances of inebriates. He observed that even when discharged patients did relapse, poor supervision and "the tremendous force of environment and old associations" were usually the determining factors, not any viciousness or indifference on the part of patients. The message was clear: environmental circumstances as well as constitutional make-up were to blame for alcoholism. Accordingly, effective treatment for the alcoholic required a systemic, holistic approach.

Four years after the publication of *Drunkenness in Massachusetts,* the legislature appointed another committee to study the state's habitual drunkard problem.[106] The second committee to study drunkenness came at the request of Governor Eugene Foss, who had learned of New York City's Board of Inebriety and Inebriate Farm Colony and believed the new program might offer a more effective model for addressing problem drinking—an ironic request, since New York City had used Foxborough as a model for its inebriate program.[107] Both the 1910 and the 1914 reports made it clear that Neff and the trustees of Foxborough were not just putting together a hospital program that would successfully restore the lives of inebriates. They were also charged with the task of integrating the penal and medical systems, configuring them around a conception of habitual drunkenness that would make sense to courts, physicians, probation workers, and the general public. In essence, they had become social engineers. Both the 1910 and 1914 committees—composed largely of former and current hospital trustees—wrestled with the different ways in which the medical profession and the state's laws classified drunkenness, and they endeavored to create a new foundation for the triage and treatment of inebriate cases.

On the medical side, the reports claimed that the inebriate fell into one of three categories: "normal drunkards," who had acquired the habit of getting drunk over time, a habit that eventually became a disease over which drinkers had little if any control; dipsomaniacs, whose drinking amounted to a form of hereditary temporary insanity for which they were not responsible; and "defective degenerate drunkards," whose habitual drunkenness was directly tied to their underlying constitutional and moral inadequacy. Legally, the inebriate fell into three categories as well: accidental, occasional, and habitual. The first category included those who got drunk unwittingly, usually out of inexperience; these individuals were typical "first arrest" cases. The occasional drunkard became intoxicated infrequently, with little effect on his or her ability to work or manage a home; often the occasional drunkard had a couple of previous arrests. The habitual drunkard, however, was frequently, if not always, drunk; often he was a victim of some underlying mental or physical defect.

Ultimately, the committees concluded that the accidental and occasional drunkards—usually those arrested between one and five times—were best managed through probation. The habits of these individuals might be stopped early, and the disease of inebriety prevented. The diseased inebriate—the pathologic case who displayed no moral defect—was to be sent to the hospital, at least initially. The "morally debased . . . in whom the habit of drinking is but one phase of a lawless and vicious character" was to be sent to the state farm at Bridgewater or to prison.[108] A new institution, however, was to manage the men who were "pathetic and often permanent burdens" on their families; these individuals wished to refrain from alcohol, but they had passed beyond the point of reform. In a controlled, liquor-free environment, this group of custodial cases would be "given the opportunity" to become self-supporting; they would be given indefinite terms of confinement of up to two years. The hospital at Foxborough would oversee this group as well as patients with more sanguine prospects.

Thus, although the trustees had spent twenty years fighting to keep chronic cases out of the Foxborough State Hospital, by 1910 they had assumed the responsibility of caring for the "incurable" or "incorrigible" cases in addition to running a hospital facility for more hopeful cases. Woods and the other trustees acquired one thousand acres of farmland in Norfolk, Massachusetts—near Foxborough—and began to construct the Norfolk State Hospital, a facility that was to be exclusively for the medical care of inebriates and the detention and review of incurable cases. Creating the new facility required substantial financial backing. Irwin Neff and Robert Archey Woods made their case to the legislature, highlighting the economic and public health benefits to the state. Declaring that "the greatest economic loss to the State lies in the idleness of capable men," the neurologist and social worker continued: "Nineteen out of every 20 men imprisoned for drunkenness in Massachusetts last year were of American or British birth, unhandicapped in their occupation by difficulties with our language. Four out of every 5 men so imprisoned were between seventeen and fifty years.... These men at the prime of life, lost over 300,000 working days from imprisonment alone last year."[109]

Inebriety was an Anglo-American problem, not one that could be dismissed as the result of imported drinking habits affecting only minority populations. American families were financially undercut by the lost wages of their drunken breadwinners, and the whole state suffered. Moreover, added Neff, the time had come when "there should be a much more general recognition of the fact that a hospital for inebriates is fundamentally an agency for preventive medicine, and does not exist merely to assist those persons who are already suffering from inebriety."[110] For the Progressive Era neurologist, treating the habitual drunkard had a large payoff: it prevented alcohol-related physical and mental illness and the impoverishment that often accompanied them. It was mental hygiene work of the first order.

## Reorganizing the State Hospital for Inebriates: The Norfolk Years, 1914–20

During the summer of 1914, all of the inebriates remaining at Foxborough State Hospital were transferred to the new Norfolk State Hospital. Finally, announced the institution's first annual report, the state had "a hospital for inebriates which in situation and character is suitable for the purpose for which it is employed."[111] This was not the whole story, however. Originally, hospital facilities and dormitories for inebriate men, inebriate women, and detention colonists were planned. However, the legislature had not appropriated the required funding for the construction of all three facilities. Only the detention colony, its hospital, and a separate hospital for curable patients were finished. For whatever reason, the detention colony had received first priority. The priorities of the Norfolk physicians lay with those they were most likely to cure, however. So "hopeful" cases now occupied the hospital facilities in spite of its mandate to take both promising and potentially incurable patients. Although the state's judges and probation officers were beginning to discriminate between the two types of patients they were sending to the hospital, the hospital was only able to accommodate the hopeful cases.

Norfolk's inauspicious opening portended the six-year struggle that brought about its end. It signaled the final phase of an ongoing fight between doctors and social welfare reformers on one side, and state politicians and bureaucrats on the other. This was a fight to treat rather than punish the state's most populous group of social deviants—people with alcohol problems—a group for whom there had always been little popular sympathy. Between 1914 and 1920, Neff and the trustees would wage an ongoing battle with the legislature to complete the physical plant, construct facilities for women, and operate their hospital in the midst of a declining patient population that sent per capita expenses skyrocketing. Following the passage of the Harrison Narcotic Act in 1914, the hospital also treated a growing population of drug addicts.

By all accounts the hospital facilities at Norfolk were impressive, as was its beautiful setting along the Swift River. The hospital administration took full advantage of the surrounding farmland, turning the new facility into what annual reports called an extension colony of the Amherst Agricultural College. Irwin Neff and the board of trustees wished to make the hospital "of practical service to the farming and dairy interests of eastern Massachusetts."[112] Doing so, they believed, would make the community more inclined to contribute tax money for the support of the institution; it would give the inebriates excellent training in a healthy, wholesome vocation; and it would assist the discharged inebriates who wished to find employment in the surrounding towns. Norfolk's rehabilitative mission, like Foxborough's, was intimately tied to demonstrating its economic utility. Just as the inability of inebriates to function in their familial and breadwinning roles typically signaled the point at which they had lost control and required medical treatment, so the restoration of their social and economic functioning was the key to shoring up support for the hospital. At least at the institutional level, the staff at Norfolk appeared aware that support for the medical treatment of inebriates did not require support of the disease concept per se. Rather, it was allied with the restoration of familial, civic, and economic responsibilities.

The relationship between employment and alcoholism registered again in 1916, when the construction of several new cottages enabled the hospital to accommodate a full range of custodial and curable patients, but the size of the patient population began to decline. The First World War was to blame, maintained the trustees, for "excessive drinking diminishes directly with opportunity for continuous work." "When men are out of work, or discouraged or poor, a certain proportion turn to excessive drinking," they added.[113] Neff regarded unemployment as an etiological factor in alcoholism. The Preparedness Era forced industry to utilize all available sources of labor; as businesses relaxed their standards, men whose drinking problems would have posed an obstacle to their employment now found work. Increasingly, hospital physicians and trustees linked inebriety to the larger social and economic fabric, not just to an individual's hereditary makeup.

At the same time that the trustees reached these conclusions, the outpatient department was expanding. Patients were being released from Norfolk faster (the average stay decreased in length from seven weeks to four between 1914 and 1917). The hospital often put discharged patients in touch with Boston's Jacoby Club, whose motto was "A club for men to help themselves by helping others." Sponsored by the Emmanuel Church, the organization offered reformed drinking men fellowship and a leg up on the water wagon. Moreover, Norfolk State Hospital treated many patients on an exclusively outpatient basis. With the number of available jobs increasing statewide, outpatient care was well suited to allow inebriates to stay on the job while receiving help for their drinking problems from physicians and social workers at regular intervals. Indeed, the outpatient department frequently cooperated with local mills, placing reformed inebriates in jobs where they were desperately needed. Statistics seemed to affirm this treatment strategy. In 1914 some 27.3 percent of all patients released since 1909 were reported abstinent or improved, while that figure jumped to 35.8 percent in 1917, leveling off at 32.8 percent in 1918, the last year that the hospital treated alcoholics.

Two other populations were treated at Norfolk in its closing years—drug addicts and women. Their experience further illustrates the dynamic nature of the disease concept and its malleability in the face of social values and expectations. As the total population at Norfolk declined between 1916 and 1919, the percentage of drug inebriates and women increased markedly, although women were not admitted until 1918. The Harrison Narcotic Act went into effect in March 1915, two months after Massachusetts had passed its own drug law. As a result, many narcotic addicts unable to secure their morphine and heroin supplies applied to Norfolk to taper off the drugs. Although the hospital had treated drug cases since 1909, the case load had never risen above a dozen or so per year. Seven hundred and forty-six narcotics addicts walked through the doors at Norfolk between 1914 and 1916, literally inundating the hospital with drug cases. Repeated references to the drug addict's rejection of therapy and the hospital's reluctance to take on more of these cases suggest that these individuals posed significant problems for their hospital keepers. Indeed, by 1917 Neff reported: "Continued laboratory work has emphasized the truth that a large number of drug addicts are prenatally wrong, the drug taking being merely an expression of a prior mental defect. . . . The depravity, criminality and licentiousness seen in cases of this type are not end pictures—results of drug taking—but are to be regarded as degrees of mental depravity denoting a mental warp antecedent to the use of the drug."[114]

Confronted with relatively youthful cases who had taken morphine for far less time than most inebriates had consumed alcohol, Neff and his staff physicians expected to be able to cure drug cases with little difficulty. Instead, they experienced repeated failure and concluded that the trouble must lurk deep within the individual's hereditary endowment—quite a contrast to the largely environmental explanations they were using in the case of alcoholic inebriety. Bad heredity explained the failure of those "resistive to ordinary measures of treatment" to respond to hospital care.

Since there was no state facility for the treatment of female inebriates, women arrested for public drunkenness or as habitual drunkards were sent to the state prison, local almshouses, and insane asylum. In 1909 the legislature ruled that they should be committed exclusively to the insane asylum. Yet as of 1912, close to 3,500 women had been incarcerated in a wide range of penal institutions. Concluding in 1914 that the immorality of women inebriates was likely the result of "congenital mental defectiveness"—an incurable condition—prolonged detention was ruled a necessary step. For the few women who were regarded as pathological and not vicious cases, a hospital was deemed "clearly necessary."[115] The trustees requested money for such a hospital at Norfolk at least four times in the course of the institution's history, but it was not until the hospital population had fallen by two-thirds in 1918 that doctors began to fill the empty beds with representatives of "the fairer sex." Defective heredity loomed large on the horizon for these "fallen angels," for as we have seen, common medical wisdom held that women with drinking problems were more diseased than their male counterparts. However, experience treating the women at Norfolk and at the Massachusetts Home and Hospital (formerly the New England Home for Intemperate Women) eventually awakened the hospital staff to the panoply of familial and economic circumstances that shaped women's alcohol problems—an array of circumstances comparable to those of their male counterparts. The more contact physicians had with individual women, the more sympathy the women seemed to gain and the more likely they were to be seen as victims of circumstances—including poor medical care—rather than victims of bad heredity.

With state ratification of Prohibition proceeding apace in 1918, and the passage of the wartime Lever Food and Fuel Control Act, which restricted the use of grains in brewing and distilling alcohol, Norfolk's end was near. In 1919 the hospital was leased to the U.S. government for use as a nervous reconstruction facility for returning veterans. The final report of the Norfolk State Hospital observed that although public drunkards had all but disappeared, their experience was "a warning against the introduction of any State policy which would furnish inadequate treatment for inebriety in whatever form it may appear. Alcoholism and drug addiction are expressions of an individual defect, requiring distinctive and specialized treatment. Unquestionably, the State should furnish means for furthering such treatment."[116]

The rationale the Norfolk trustees used to justify the hospital's continued existence highlighted the flexibility of the constitutional model of disease that physicians applied to alcoholism. As we saw, the development of the outpatient department and the successful treatment of inebriates through psychotherapy and the adjustment of the inebriate's environmental circumstances had led physicians to take a far more hopeful perspective on their work. Moreover, in the late teens of the twentieth century, the environment—with its high employment rates and wartime restrictions on alcohol—appeared to slow down the production of chronic drunkards. Environment mattered. But so did heredity, claimed the trustees. In the face of the hospital's closing, they returned to the hereditarian element of disease, stressing that an innate defect predisposed inebriates and drug addicts to alter their consciousness. Alcohol might go away, but inferior nervous constitutions—constitutions that were vulnerable to changing environmental circumstances—would not. The nature of the disease of inebriety, it would seem, depended not only on changing social and economic circumstances but also on the specific political circumstances of the hospital where inebriates were treated.

## Medicalization in Massachusetts

The irony of the Massachusetts story is that the state closed the hospital after it had developed a relatively—some would say highly—successful system of inpatient and outpatient care for inebriates. When it became obvious during the 1907 public investigation of Foxborough State Hospital that the institution had not fulfilled its mission, it was still given a new lease on life, largely because Massachusetts Governor Curtis Guild Jr. regarded the program as a grand sociomedical experiment that would enhance his state's image as a leader in scientific social welfare reform. Guild, the Progressive leader, was swayed by the testimony and promises of social reform experts and medical authorities on inebriety.

In its new incarnation, Foxborough, and later Norfolk, became the centerpiece for a system of integrated care for inebriates that linked the reform of the individual drinker to industry, to education, to agriculture—to the restoration of civic responsibility and fiscal independence. Far from enforcing a narrow definition of inebriety as a medical disease, the reorganized hospital saw inebriety as a social disease, a medico-moral condition resulting in physical and psychological dependence on alcohol and financial dependence on the state.

The case study of the Massachusetts Hospital for Dipsomaniacs and Inebriates and all of its subsequent incarnations illustrates how contingent the disease concept of inebriety and the medicalization of habitual drunkenness were upon changing political fortunes, shifts in the economy, and the support of important figures within Boston's medical and social reform arenas. Fuel for the effort to treat inebriety as a disease stemmed from the state's perennial problem of dealing with Boston's hard-drinking Irish, English, Scottish, and French Canadian immigrants. Simply put, the Brahmin activists and their reform colleagues felt a pressing need to maintain order throughout Massachusetts. The state's longstanding rivalry with New York and its pride as a Progressive reform-minded state also assisted in the campaign to adopt an innovative approach to habitual drunkenness. Executing the plan was made all the easier because of the close ties among people such as Joseph Lee, Robert Archey Woods, Warren Spaulding, and Irwin Neff. Yet because the legislators controlled the purse strings and the judges influenced the selection of patients, the influence of these inebriety experts was hardly all-powerful. Indeed, early in the hospital's history, the Massachusetts General Assembly attempted to get the hospital to employ the Keeley Cure or other specifics that supposedly took much less time to work.

We must also remember that the changing fortunes of the temperance campaign within Massachusetts proved crucial to the institution's success and failure. Support for the medical approach to habitual drunkenness grew when support for Prohibition waned. The medical treatment of drunkards was but one alternative to stem the rising tide of liquor. Of course, the opposite was also true: when Prohibition became the order of the day, and the number of inebriates was already dropping—thanks to an expanding labor market—the state reasoned with millennial optimism that habitual drunkenness would altogether vanish from sight. Support for Norfolk State Hospital evaporated in this "dry" climate. Norfolk's closing did not necessarily indicate a vote of no confidence in the inebriate asylum ideal or the disease concept of inebriety; rather, it represented a reallocation of resources. The Bay State would prove no more successful at enforcing Prohibition than it had in treating those who suffered from alcoholism. Wisely, the Washingtonian Hospital and the Massachusetts Home and Hospital for Women—the places to which Norfolk's patients were transferred at its closing—remained open throughout America's "Noble Experiment."

# CHAPTER SIX

# Building a Boozatorium

From his room in the State Hospital for Inebriates at Knoxville, businessman and patient Ed Harris penned his request for "parole" to Iowa governor William Harding. It was January 1917, and Harris, who had spent several weeks in the state "jaghouse," claimed to have "finished taking the treatment." As evidence for his rapid recovery, the dry goods merchant from Salix, Iowa, offered his current position as the hospital's barber—a job that occupied him three days a week. No sick man could manage such a schedule, Harris claimed. Then he begged the governor to remember their meeting years back under happier circumstances, when Harris had traveled to the Sioux City executive offices to discuss business. Might the governor not attend to his case personally? Harris asked, for he desperately wanted to return to his store in Salix. In his absence the business "had gone to blazes." Harris also confided, "My little boy of 12 years is taking this [his absence from home] very hard and I will never take another drink for his sake alone."[1] Further reassuring the state's executive officer, Harris added that there was considerable distance between him and the "common drunkard" in residence at Knoxville.[2] As if on cue, a few weeks later the governor received another handwritten note, this one from Harris's son, closing with the query: "Know [*sic*] will you try and help get [my dad] out so he can take care of me?"[3]

Harris's letter and that of his son struck a responsive chord in Harding. Within less than a week of their receipt, the state's executive officer consulted with the hospital superintendent on Harris's behalf and learned that the superintendent and the Board of Control were the only parties responsible for "paroling" patients. Deferring to Superintendent M. C. Mackin's authority, the governor informed Harris that his case had been judged a favorable one but that "there is an element of time which is very necessary to complete recovery." Harding hoped that Harris would "feel disposed to accept and act upon the advice of those who have your case under observation and when you are really ready for release on parole or discharge, there should be no difficulty about it."[4]

Harris was not satisfied with the governor's reply though, and he quickly revamped his campaign for freedom. Having resided at the hospital for a month, the businessman pronounced himself "cured" in a subsequent letter, adding "I have thoroughly made up my mind to quit drinking and smoking not only for a few months, but the rest of my life, and I don't like to stay here when I am perfectly well and see my shop and living go to the dogs."[5] Once more, Harding was prompted to act, inquiring a second time into the possibility of Harris's early release. This time M. C. Mackin acquiesced, reasoning in a letter dated 6 March, "Were he to lose his business by reason of his detention here, it probably would be a factor in discouraging him and causing him to again take up his former habits of inebriety. I really think it would be much better for the boy to remain for a period of four or five months but owing to his worried condition it might really be of no value to keep him longer than the first of April. This is the date on which he seems to be anxious to be released and I think I would be justified in giving him a trial at home."[6]

On 11 April, Governor Harding received a final note from Salix. Ed Harris expressed his and his son's thanks for Harding's personal interest in the businessman's case. "I am on the water wagon for good," he assured the governor. "The good people here as a rule have given me the glad hand and while at first I thought the measure [commitment to Knoxville] was rather drastic, I am now pleased that I am rid of the Habit."[7]

So ends this—as far as we know—happy tale of one of Iowa's inebriates, but what are we to make of it? The story of the Salix, Iowa, businessman raises a variety of questions: How did medical and lay authorities understand the condition of inebriety—as a bad habit or as a disease? How did the conflicting priorities of the patient, his family, and inebriate hospital physicians affect the course of treatment? What distinguished treatment in the inebriate hospital from treatment in jail? And why was Iowa committing habitual drunkards to state hospitals in the first place?

This chapter examines the evolution of Iowa's eighteen-year state-sponsored medical program for inebriates—like the Foxborough and Norfolk State Hospitals, an early attempt to medicalize habitual drunkenness through the construction of new laws and new institutions for its treatment. I analyze the changing relations among the executive office, the inebriate hospitals, and other interested parties—families, physicians, the temperance lobby, legislators, the Iowa State Medical Society, and the Board of Control for State Institutions.[8] The medicalization of habitual drunkenness proved a difficult task in Iowa, not only because of the strong moral valence surrounding drinking problems—a force that made it impossible for many to view the inebriate as an innocent victim of a disease—but also because of the competing interests of those affected by the alcoholic and his actions: judges, physicians, hospital superintendents, temperance reformers, eugenicists, legislators, and the friends and family members of the inebriate.

As in Massachusetts, the attempt to medicalize habitual drunkenness in Iowa was not a top-down enterprise imposed upon the state by physicians. It was a politically and socially negotiated process. Indeed, a central irony of inebriate reform in turn-of-the-century Iowa was that the same congeries of interests that made medical treatment for the inebriate a pressing public issue also impeded its successful implementation. Though it took place at a time when the medical profession was expanding both its social and cultural authority within Iowa and across the United States, the effort to medicalize habitual drunkenness in the Hawkeye State revealed both the limitations of this authority and the difficulty inherent in labeling as a disease a protean, chronic condition with connections to a host of social problems and political causes.

For those familiar with Midwestern politics, it may come as no surprise that Iowa—a state with a rich temperance and prohibition heritage—would have attempted such an innovative therapeutic course. Iowa historian Dorothy Schwieder has remarked that temperance concerns "proved the most emotional, politically significant and tenacious of all issues in nineteenth- and twentieth-century Iowa."[9] As we shall see, however, there was no single impetus for inebriate reform; there were many. The temperance and prohibition movements of the late nineteenth and early twentieth centuries clearly nurtured related concerns about the plight of the habitual drunkard. Legal reformers and medical practitioners also promoted the effort to medicalize habitual drunkenness and provide for its cure.

The initial, and unheeded, calls for medical treatment in Iowa came in the 1850s from judges, who were tired of seeing the same alcoholic recidivists in their courts year after year. By the end of the nineteenth century, Iowa's reform-oriented State Board of Health and a growing medical profession with a meliorist bent supported medical treatment. Likewise, an increasingly centralized state administrative apparatus—with a newly created Board of Control of State Institutions—nurtured the cause through its efforts to efficiently manage Iowa's impoverished, diseased, and disabled citizens. State administrators and physicians alike defined inebriate reform in pragmatic terms, focusing on the good it might effect in the daily lives of individual drinkers, their families, their friends, and the state's economy. Indeed, the campaign to build a state system of care for inebriates was emblematic of a variety of changes taking place in turn-of-the-century Iowa.

By 1902 when the first state hospital opened its inebriate ward, Iowa was well on its way to building a network of specialized social welfare and medical institutions for the treatment of the state's defective, delinquent, and dependent classes. Like Massachusetts, the State of Iowa was consolidating its authority in the arenas of public health and social welfare. As early as 1888 Iowa had established a department for the criminally insane at the state prison in Anamosa. The revised Code of Iowa of 1897 provided for two new state hospitals for the insane—Clarinda and Cherokee, which opened in 1899 and 1902 respectively. In 1903 Iowa financed the construction of the University of Iowa Hospital in Iowa City. Oakdale, the state's tuberculosis sanitarium, opened its doors in 1908. The Perkins Act, passed in 1913, underwrote the treatment of children at the University of Iowa Hospital before the state bankrolled a separate institution for children in 1917. In 1919 the Iowa General Assembly passed a law to establish the state's first psychopathic hospital, linked to the University of Iowa Hospital.[10] In short, the state of Iowa expended more funds on medical care for its citizens between 1898 and 1919 than ever before.

If the establishment of inebriate hospitals was consistent with the state's expansionist healthcare policy, it also boasted significant symbolic value. By placing their state on the cutting edge of institutional expansion and reform across the country, Iowans could rank themselves with states such as Massachusetts and New York. To offer medical care to the inebriate said, in effect, "Everything's up-to-date in Iowa City," not to mention Des Moines. It was an act of enlightened compassion, scientific expertise, and rational administration that signified the heartland's participation in the modern world. The reform of inebriates, dipsomaniacs, and alcoholics was a classic Progressive reform.

As we have seen, the construction of specific institutions for drunkards was a key element of the campaign to medicalize habitual drunkenness. Across the country, reformers established more than two hundred private and public institutions for inebriates and dipsomaniacs between 1870 and 1920, and Iowa was no exception to this trend. Keeping a watchful eye on the successes and setbacks of the Foxborough and Norfolk State Hospitals, Iowa addressed its own inebriate problem. In turn, other states and cities watched both Iowa and Massachusetts. Indeed, in 1910 social worker Bailey Burritt pointed to both Iowa's State Hospital for Inebriates at Knoxville and the state hospital for inebriates in Foxborough, Massachusetts, as blazing the path for other cities and states.[11] If the longest-lived state hospital system for inebriates belonged to Massachusetts, Iowa was not that far behind.

## The Origins of Iowa's Inebriate Reform Campaign

At the April 1902 quarterly meeting of the Iowa Board of Control, a new and controversial issue arose. A debate took place over the nature of the inebriates treated within the state hospitals for the insane and the Home for Old Soldiers. Josiah F. Kennedy, secretary of the State Board of Health, initiated the exchange with his paper, "Inebriety and Its Management."[12] With the legislature considering several bills to provide state care for inebriates, the Board of Control had decided to study the matter. They had commissioned Kennedy, whose anti-tobacco sentiments were widely known in the state, to conduct the investigation. Kennedy regarded tobacco as what today is called a "gateway drug," one leading to alcohol consumption and ultimately to inebriety and crime. Moreover, he regarded the state's swelling inebriate population as a blot on the crest of the State Board of Health, noting:

> Here is a large class of acquired and preventable diseases patent not only to the physician and sanitarian, but to the layman as well, and as yet the State Board of Health has not discovered any effectual, if possible, way of prevention; nor has

the legislature, the press, or the forum been more successful. Moral suasion, legal suasion, education, the teaching in our public schools of the evil effects of alcohol, the daily exhibition by its unfortunate victims of its dangerous results have all been tried and are still on trial, and yet, as the ranks of the inebriates are thinned by death, there seems to be an on-coming army to take their places.[13]

Kennedy related in detail the story of inebriate hospitals across the country, quoting extensively from the annual reports of the state hospital for inebriates in Massachusetts. He concluded with several recommendations from alienists and reformers around the globe in support of the inebriate hospital idea. For Kennedy the construction of institutions for inebriates was just a beginning, however. Far-ranging in his reform vision, the Board of Health secretary declared that the battle against habitual drunkenness required not only prevention through temperance instruction at school and at home but prohibitions against drinking within "the great corporations"; termination of state employees who used intoxicants—cigarettes, tobacco, and alcohol; rigorous laws against the sale of alcohol and tobacco to minors; and the elimination of confirmed inebriates' "right to beget a tainted offspring."[14]

Nothing if not comprehensive, Kennedy's plan bore the stamp of Iowa's vigorous temperance movement and presaged the state's eugenics concerns by about a decade.[15] It represented the sort of Progressive Era activism that had characterized Iowa's medical profession, and it was of a piece with the centralizing administrative reforms taking place in Iowa's state government. As historians Amy Vogel and Lee Anderson have shown, the temperance lobby, medical profession, and eugenics movement in Iowa were far from isolated communities.[16] And during the early years of the twentieth century, all three were poised to take advantage of the agenda of the new centralized administrative agency in the state, the Board of Control of State Institutions. Indeed, the success of Iowa's experiment in state-sponsored medical treatment for inebriates depended largely upon the ties among these groups.

Given the ongoing alcohol concerns of Iowa's physicians, it was hardly surprising for the topic of medical care for inebriates to receive such prominent attention from the Board of Control in 1902. By 1871 Josiah F. Kennedy had presented several resolutions before the Iowa State Medical Society to recognize the existence and agenda of the American Association for the Cure of Inebriates, founded the previous year. Kennedy lobbied to appoint an ongoing committee on inebriety that would report on the topic "as upon any other medical subject."[17] In 1880 a temperance-minded contingent within the state medical society presented the Iowa General Assembly with a petition to ban the sale of intoxicating liquors at state, district, and county fairs.[18] And just three years later the president of the society, H. C. Huntsman, M.D., of Oskaloosa, devoted a significant portion of his inaugural speech to the liquor "traffic" in Iowa, encouraging "the medical profession to support this fearless young State in its gigantic struggle with a social disease that honeycombs society."[19]

In 1887 society president A. W. McClure, M.D., of Mt. Pleasant urged that the group focus more of its energies on "nervous and mental diseases, and the department of State Medicine and Public Hygiene."[20] Inebriety qualified as such a disease, according to George F. Jenkins, M.D., of Keokuk, who devoted most of his 1892 presidential address to the subject, referring to intemperance as "a disease and not simply a vice." For Jenkins, inebriety "more seriously and disastrously affected the moral and civil affairs of the State and Nation than any other disease that comes under the notice of the physician." Jenkins chastised moralizing temperance advocates and urged the state to consider instead a law that would recognize inebriety as a disease and provide medical treatment in a special facility for "the alcohol habit."[21]

Concerns about the alcohol problem in Iowa, and the medical treatment of inebriates in particular, reflected an increasing commitment on the part of the state medical society to serve the public's health and, like other medical personnel across the country, to consolidate their own social authority.[22] Indeed, in 1906, just a year before the State Hospital for Inebriates at Knoxville opened its doors, Sioux City's William Jepson, then president of the society, observed: "A constantly recurring necessity seems to exist of impressing upon our legislative bodies, including those charged with the moulding of public thought and directing legislative action, that we are vitally concerned in all that pertains to the health and welfare of the public. . . . From this organization must ever flow in the future that thought which will be the guiding light to our lawmakers in making matters pertaining to the betterment of the physical and mental welfare of our citizens."[23] In short, stewardship of the individual patient *and* the body politic were priorities for Iowa's leading physicians.

The medical society's great interest in the alcohol problem was not unusual in Iowa. Another prominent organization, the State Board of Health, was preoccupied throughout the 1880s with similar concerns. The founding of specific facilities for inebriates at two of the state's insane hospitals in 1902 coincided with unprecedented growth in the Board of Health's administrative authority. After its establishment in 1880, the Board of Health had served as little more than a state advisory agency, maintaining the power to recommend, but not enforce, sanitation and anti-nuisance measures it deemed in the public's interest. In 1902 smallpox struck Iowa's capital, Des Moines. The disease disrupted city life and led other major Midwestern cities to reconsider their economic ties to Iowa's capital, where officials were slow to put costly quarantines into effect. A legislative "panic" ensued, and the board was given the authority not only to recommend but also to enforce its policies.[24] Strengthening the Board of Health was very much a political expedient to secure Iowa's business status within the Midwest. In one stroke, the state relieved its trade partners' fears of contagion and took a significant step toward securing the health of the state's citizens through rational administration.

Politics, however, was nothing new for the Board of Health. The temperance lobby, and the Woman's Christian Temperance Union in particular, had courted the agency throughout its first two decades. Fighting to maintain its mantle of scientific objectivity and political neutrality, the board consistently stopped short of adopting a strong pro-temperance position. Instead, board members used their organization "to inject medical science into the temperance debate," winning the favor of an important political constituency—the WCTU—while addressing a legitimate public health concern.[25]

Nor was the Board of Health the only state organization to don the mantle of science, or at the very least, scientific or rational management. Iowa's Board of Control of State Institutions was established in 1898 to improve the efficiency and management of its charitable and correctional institutions through central governance. Prior to the board's formation, the individual institutions functioned autonomously: they were, in essence, small fiefdoms, competing with each other annually for the state's largesse. Poor relief and its attendant institutions were managed with near-autocratic authority by county judges from 1851 to 1860, when county boards of supervisors assumed these responsibilities. It is not surprising, then, that the very first pleas for a state inebriate asylum, voiced in 1863, arose from the judicial sector, long responsible for the welfare of the state's dependent paupers and insane.[26] Even after control for the county charities and corrections institutions had been ceded to the new boards of supervisors, inspections of the institutions remained the province of the county judges and their prosecuting attorneys—a system that existed until 1868.[27] Between 1870 and 1898 there were repeated calls for a central state agency to manage Iowa's welfare and correctional institutions, all unsuccessful.

By 1897, though, Iowa's emerging constellation of state institutions for the deaf, dumb, insane, orphaned, and criminal classes was shrouded in controversy. Local corruption and general mismanagement had led to "a feeling of hostility between institutions and a feeling of opposition toward them on the part of [the] public and Legislature, induced by sentiment that institutions were the vehicles of special interests, and not unselfishly representative of a beneficent purpose of government."[28] To combat this skepticism and ill will, the General Assembly appointed the Healy Investigating Committee, which delved into the administration of the state's asylums, homes, and prisons and ultimately recommended that a new, rationally and centrally organized governing board be created for their supervision. The Board of Control, appointed as a direct result of the Healy Committee's report, was charged with three basic responsibilities: (1) general, including financial, control of the state charitable, penal, and reformatory institutions; (2) financial control of the state's institutions of higher education; and (3) administrative control of the county and private institutions for the insane.[29]

Efficient management also meant that the General Assembly encouraged the Board of Control to make sure that the state's institutions were kept abreast of the latest developments in the care of their respective populations. Thus it was the board's task to undertake investigations regarding the management of "soldiers' homes, charitable, reformatory and penal institutions in this and other countries" and to encourage "scientific investigation of the treatment of epilepsy by the medical staff of insane hospitals and the institution for the feeble-minded . . . to publish from time to time bulletins and reports of the scientific and clinical work in such institutions."[30] It was in this capacity that the board asked J. F. Kennedy, secretary of the State Board of Health of Iowa, to present an overview of the management of inebriety at its quarterly meeting in April 1902.[31] The bulk of Kennedy's report was devoted to the medical measures already implemented in Europe and in the United States at institutions for inebriates.

Finally, turn-of-the-century developments within Iowa's temperance movement—specifically, the weakening of the state's prohibition policy—created a more hospitable climate for novel alcohol control measures, including medical care for inebriates. As with other states, the enthusiasm for medical approaches to alcoholism increased as prohibition faltered. In 1889 Iowans elected a Democratic governor, Horace Boies. His election spelled an end to thirty-two years of Republican rule in the state and signaled Iowans' dissatisfaction with prohibition, a staple of the Republican platform. Republicans eventually gave up their strong support for prohibition, ceding control over such issues to the legislature, and won the governorship back in 1893. One year later, in 1894, the legislature voted in the Mulct Law, which did not repeal prohibition but gave local communities the option of violating prohibition upon a favorable local vote and the payment of a certain fee. Now cities were able to exercise a certain level of control over their own wet or dry status.[32] With the state's policy on prohibition easing, legislators initiated what Dorothy Schwieder termed "an almost bewildering array" of new liquor legislation to keep the traffic and the problems associated with drinking in check. In 1909 alone, there were at least nineteen bills on liquor reform introduced at the General Assembly.[33]

Six years later, in 1915, statewide prohibition won the day again. By the time national Prohibition was enacted, "almost every known method of regulating the liquor traffic [had] been given a trial in Iowa."[34] In short, the establishment of the state inebriate hospital should be seen as but one of many new checks placed on the alcohol trade by Iowans during the Progressive Era.

## From "Nuthouse" to "Jaghouse": An Iowa Chronology

Despite Josiah F. Kennedy and the Board of Control's recommendation of an inebriate hospital, state legislators were not eager to expend the necessary funds to establish a separate institution for the treatment of inebriates, so they first imposed a less-expensive solution. In February 1902, House representative Mahlon Head of Greene County introduced a bill before the General Assembly to establish a special ward for inebriates at one of Iowa's state hospitals for the insane. The bill, approved by the House the next month and by the Senate and its Committee on Public Health in April, was significant for its originality and its placement of inebriety squarely within the medical domain. Yet there was little support for the measure among the state hospitals' executive officers. The superintendents of Mount Pleasant and Clarinda State Hospitals,

Charles Applegate and Max Witte, and the chairman of the Board of Control, Judge John Cownie, countered that the treatment of inebriates at a *separate* institution, as Kennedy's report had proposed, offered Iowans the best solution. But the economy-minded legislature disagreed. A few weeks after the Board of Control's March meeting, the bill cleared the Senate. The new inebriate law went into effect on the Fourth of July.[35]

On the 21 July, the *Cherokee Democrat* noted that the state had received its first inebriate, S. N. Bidne, a blacksmith from Norma. Bidne was "in the habit of getting drunk, and when in this condition, sometimes dangerous."[36] Most recently, while intoxicated, he had attempted to shoot a woman. As the first person to be tried under Iowa's new inebriate law, Bidne had to tough it out in the Forest City jail until the Board of Control decided which hospital would receive the state's habitual drunkards. Unlike the insane, who rarely if ever were held in jail as they awaited room at Iowa's insane asylums, inebriates were sentenced to the state hospitals under the same conditions that governed the commitment of individuals to the industrial schools of the state.[37] Similarly, district court judges determined the length of one's "sentence" to the inebriate hospital, not hospital physicians; and the governor, rather than the hospital superintendent, held the power to "parole" patients. Medical authority was far from complete. The state's policy reflected both the moral and the medical dimensions of inebriety: in order to receive medical treatment, the prospective inebriate was first detained in jail, then tried before a judge, then sent to the state hospital for a period of time designated by the magistrate.[38]

Within twenty-four hours of Bidne's trial, the Board of Control reached its decision, the *Des Moines Register and Leader* reported: "Inebriates to go to Mount Pleasant—the Board is Hostile.... The board does not approve of the new law and believes it was a mistake to make the inebriate department a part of one of the state hospitals.... It is expected there will be an ample number of patients, and that it will not be long until the new department is overtaxed."[39] So began Iowa's eighteen-year experiment in the medical management of the state's inebriates.

By September 1902, just two months after Mt. Pleasant had established its inebriate ward, referred to in official documents as "the inebriate hospital," sixty-nine habitual drunkards were receiving treatment, and the rate of admission for inebriates was outpacing that for the insane. In October the Board of Control designated Cherokee State Hospital, the newest (and emptiest) of the state hospitals, as the institution to care for inebriates from the north-western part of the state.

As in virtually all other state hospitals in the country, the mixing of inebriates and insane patients proved difficult. Opening up Cherokee to habitual drunkards might have relieved congestion, but it did not ease the tensions between the two patient populations. Upon making his monthly inspection of the Mount Pleasant State Hospital, for example, Board of Control Chairman John Cownie reported that the insane held the inebriates in utter contempt. Chatting with a patient whom he mistakenly thought an inebriate, Cownie had spoken "sympathizingly with him and consoled with him over his unfortunate habit." The patient, confined by reason of insanity, shouted in indignation, "Mr. Cownie, I want you to know I'm no drunken sot; I'm here for my health."[40] Matters had not improved by the end of the year, when a well-respected general manager of a Des Moines insurance company was sentenced to the Mount Pleasant facility for his drinking and complained bitterly about the treatment he received. In short, the inebriates, once they became sober, were insulted by their confinement with individuals who had lost their minds; the insane were offended by being housed with those they regarded as immoral and vicious in habit; and the superintendents were piqued by the resulting discord and the ease with which the inebriates, once sober, escaped from the hospital grounds. Nor did it help that Cherokee's superintendent, Max Voldeng, proclaimed after a mere six months of treating inebriates at his hospital that "caring for inebriates properly at a state hospital for the insane is as impossible as its attempt is injudicious. Besides the uselessness of keeping the inebriates, their presence is injurious to the insane patients and to the discipline of the institution. Usually, they are dirty and lazy. . . . They won't work. All they do is sit around and spit tobacco juice all over everything, making their rooms dens of filth."[41]

Responding to the complaints of both the superintendents and the patients, and to the critique and recommendations offered by the Board of Control, the state legislature in 1904 set aside over $100,000 to revamp the abandoned State Home for the Blind in Knoxville as the new State Hospital for Inebriates. The General Assembly also put the Board of Control in charge of the state's inebriate facilities, removing the authority from the governor's hands.[42] Creating the new Knoxville State Hospital took two years. Local opposition to the facility fell away as the promise of jobs became a reality for Knoxville's citizens. In January 1906 the *Knoxville Express* reported with great fanfare:

From the survey a two hours' visit to the new institution affords, we are impressed with the fact that the state has undertaken in seriousness to afford men addicted with the drink habit an opportunity to reform. . . . A special study of each patient's case will be made by the medical directors, and an earnest attempt made to combat and eradicate the disease of alcoholism. It is hoped that when patients are dismissed from the hospital that they will have been built up into the best physical condition they are capable of. As Superintendent Willhite says, the work of the hospital must necessarily be, in a large measure experimental, and if it proves to be successful in any large degree it will be the greatest thing in the world.[43]

Finally, the inebriates had a home of their own. Over the next fourteen years, five different superintendents served terms at the State Hospital for Inebriates at Knoxville. In 1913 the hospital developed a two-tiered system that separated "hopeful" inebriates from the so-called incorrigible inebriates, who, although deemed unlikely to reform, were thought to benefit from prolonged confinement within a structured farm setting.[44] Parole became largely a discretionary procedure controlled by the superintendent and Board of Control, and a pay system for patient labor was established that compensated the working inebriate, funneled money back to the hospital for his support, and sent what was left to the patient's dependents, if he had any.[45]

The inebriate hospital at Knoxville never served women, who comprised between 4 and 10 percent of the inebriates treated in the state between 1902 and 1920. In short, the situation in Iowa regarding women inebriates was remarkably similar to that in Massachusetts. Female inebriates continued to take the cure at Mount Pleasant State Hospital for the Insane, just as women in the Bay State received care from the Westborough State Hospital. Each hospital maintained an inebriate ward specifically for the care of women.[46]

In 1919, with the passage of the Volstead Act, Knoxville closed its doors to the state's alcoholics, and the state sold the facility to the federal government as a hospital for returning veterans of the First World War.[47] The dwindling numbers of inebriate men were sent to the Independence State Hospital, and the women continued to be treated at Mount Pleasant.[48]

## Recurring Desires, Conflicting Priorities: Treatment Challenges Inside and Outside the Inebriate Hospital

Treating Iowa's inebriates was a difficult, unenviable task. As with the condition itself, treatment often pitted the physical and psychological needs of the drinker against the emotional and economic needs of his or her family. The correspondence of Ed Harris and his son with the governor makes this plain. Initially, treatment also opposed the needs and desires of inebriates to those of the insane; this much is clear from the report of Chairman Cownie. Moreover, shortly after Iowa initiated its inebriate reform program in 1902, doctors at the state hospitals realized that their own therapeutic intentions were being undercut by the needs and priorities of the court system, especially the county judges who committed patients. Addressing these and other challenges posed a frustrating problem for all involved and one that, much like the disease concept of inebriety itself, was constantly renegotiated. How then did each of these parties make sense of the problem of inebriety and its treatment?

## Cultural Framing of Inebriety in Iowa

At least rhetorically, Iowa's state hospital physicians conceived of inebriety as a disease of modern civilization, somewhat akin to George Miller Beard's concept of neurasthenia.[49] Classifying the condition as "one of the most serious menaces accompanying the twentieth-century civilization," whose "direful effects seem to have been fully realized in all civilized countries," Charles Applegate, superintendent of the Mount Pleasant State Hospital, voiced an opinion shared by most of the directors of Iowa state institutions.[50] Dealing with this "defective class" was "becoming more difficult as our modern social life becomes more complex. . . . Not in the whole field of medicine is there a disease so far-reaching in its ruinous effects upon the habitué himself, home, family, and society at large," added W. S. Osborn, who became superintendent at Knoxville in 1906.[51] The comments of Applegate and Osborn highlight the perceived seriousness of the problem they confronted. They are also interesting for the connections the physicians drew between Iowa and "modern civilization."[52] Applegate made this tie more explicitly when he observed that "the statistical records of the police courts of Paris, London, New York, and Chicago, show a rapid increase in juvenile criminality, and charge this increase to alcohol. Our small towns, too, have caught the disease."[53] The problems of the metropolis had become the problems of the heartland; and it followed that Iowa should engage in reform efforts on par with nations such as Great Britain and France and states such as Massachusetts and New York, places keeping "abreast of the times by enacting restrictive laws to enable us to protect, treat, and if possible, cure this unfortunate class of citizens."[54]

If inebriety was a disease of modern civilization demanding a modern, scientific response, it was by no means clear to Board of Control members that the state had a moral or financial obligation to provide the most up-to-date care to inebriates. Even after three state hospitals had established inebriate wards, Judge L. G. Kinne, a member of the Board of Control, noted that inebriates, whether diseased through defective heredity or vicious habit, did not *deserve* the state's largess, but should receive it anyway because "the state can do no better service to society at large than to restore to health and to the ranks of the productive laborers these men and women who, without such aid become mental and physical wrecks and who tend to sap the morals and health of the people, thereby greatly adding to the vast army which is a constant public burden."[55] Kinne, who had spent many an hour considering the plight of the "defective, delinquent, and criminal" classes, articulated two related arguments in favor of state care of inebriates: inebriety was a fount of other physical, mental, and social disease; and inebriety turned productive citizens into unproductive and obsessive consumers of alcohol.

It may be tempting to dismiss Kinne's concerns about the prophylactic value of alcoholism treatment as so much rationalization; however, on average about 25 percent of patients suffered from what hospital physicians called "general disease" at the time of admission, with others suffering from injuries, circulatory problems, digestive ills, and genitourinary problems. Many also had neurological problems. Indeed, physicians labeled only an average of 50 percent of their patients in "good" condition upon their admission between 1904 and 1918. In short, many inebriates did have other physical problems that accompanied, and were sometimes exacerbated by, their drinking.

Relying on Benedict Morel's theory of degeneration, Kinne and his comrades in reform believed that one generation's inebriety could be hereditarily transmitted to the next as a defective nervous constitution that subsequently could appear in the form of inebriety, epilepsy, insanity, nervousness, moral depravity, or criminal behavior. The fact that on average 50 percent or more of admissions between 1904 and 1918 had at least one parent who drank intemperately confirmed this belief. Thus, in the minds of Board of Control members, if no effort were made to confine and treat the inebriate ranks of Iowa, this dangerous subpopulation had the potential to spawn a race of physical and moral degenerates that would tax the state and national coffers as never before. Eugenic arguments such as this were a staple of discussions of the state's duties. Inebriate reform was promoted as enlightened and scientific statecraft.

Similarly, in a mostly agricultural state that prided itself not only on its productive farms but also on its metal mining firms and Mississippi River industries, inebriety posed a particularly disturbing threat. Some reformers believed that the desire for alcoholic stimulation originated, on the one hand, in the mentally taxing work of the professional and merchant classes and, on the other hand, in the debilitating work conditions and standard of living that burdened the unskilled laborer. Certainly the patient census reflected each of these vocational groups. In other words, the demands of production (whether in the office or factory or on the farm) led individuals to search for relief in the form of drinking. However, most reformers focused on the act of consuming alcohol itself as the force that turned men, and to a lesser extent women, into unrestrained consumers. Thus, Knoxville's first superintendent, W. S. Osborn, recommended gardening and farmwork as restorative pursuits for inebriates not only because the physical activity might strengthen weakened physiques but also because these pursuits substituted "healthy activity for unhealthy activity, sober thought to produce instead of drunken craving to consume."[56] In a state as agriculture-oriented as Iowa, gardening and farmwork also could be seen as vocational training, something that was perhaps not as manifest in Massachusetts. The unchecked consumption of alcohol violated the productive ethic that Iowans held dear and foreshadowed the public's rising concern with addiction, a concern that became a staple of twentieth-century consumer society.[57] Treatment was meant to restore inebriates to productive citizenship.

If reformers in Iowa perceived inebriety as a disease of consumption run amok in modern civilization, threatening to compromise future generations, they also reluctantly regarded it as an "American" disease. This is not to say that they were ignorant of the toll habitual drunkenness took in other nations. Far from it. Inebriety was a disease that connected Iowa to the metropole, whether Boston or Paris. Yet the statistics collected by the superintendents of insane hospitals treating inebriates between 1902 and 1906 demonstrated conclusively that second- and third-generation Americans dominated patient censuses—not Germans, not the Irish, not Scots, not Britons, not Slavs, not even Scandinavians.[58] An average of 75 to 85 percent of patients admitted each year had been born in Iowa or another state, and between 50 and 60 percent were born to "American" parents. Of course, relatively speaking, Iowa did not have the burgeoning immigrant populations of the northeastern or western United States, or of Midwestern cities such as Chicago. Nevertheless, the perception was that foreigners drank more than Americans, especially Iowans,

212 Alcoholism in America

who had lived under prohibition for so long. Reformers expected to find foreigners or first-generation Americans presenting at the inebriate hospitals in disproportionately high numbers. When Charles Applegate reported his findings to the contrary—that 137 of 155 admissions to the inebriate hospital at Mount Pleasant in 1902–3 were second- or third-generation Americans—the directors of Iowa's penal, medical, and social welfare institutions were shocked: "It is really surprising that the nations of Europe where a great majority of the people indulge in intoxicating liquors should furnish so few confirmed inebriates for treatment . . . while the American, where [sic] food is better and more abundant than in Europe, and where there is less temptation for the use of intoxicating liquors on account of poverty, should lead all others with one hundred and thirty-seven."[59]

All was not well in the land of milk and honey. Board of Control Chairman John Cownie asked, "What could account for the fact that although more Europeans than Americans drank, the Americans were more likely to become inebriates?" Answering his own question, Cownie claimed it was the American character—"the persistency with which the American goes after everything he undertakes."[60] In other words, the strength of the American drive was a weakness when it came to inebriety. The commandant of the old soldiers' home suggested that it was simply Americans' "pernicious habit of treating," something not shared by foreigners, who generally paid only for themselves.[61] Superintendent M. T. Gass of the soldiers' orphans home thought that Americans recognized their drinking problems more easily and sought out treatment more frequently.[62] This was a sanguine interpretation, but one with which the matron at the same institution disagreed. Instead, Mary Hilles, who claimed familiarity with "mothers of all classes," believed simply that "the foreign mother is a better home-keeper than the mother of the same class among the Americans."[63] Foreign mothers prepared more wholesome meals than their American counterparts, claimed Hilles. Thus, foreign families were better nourished and less in need of artificial stimulants such as alcohol to help them through the day.

No explanatory consensus was ever achieved with regard to the abundance of American patients, but the fact was undeniable: inebriety in Iowa was an American disease. It was also a disease that revealed the state's caretakers' prejudices for and against the foreign presence in Iowa. Ultimately, its new nativist image may also have helped to garner support for medical reform efforts.

## *Competing Medical and Moral Visions of Inebriety*

Those who treated inebriates at Iowa's state institutions did not work in a vacuum—they relied on the courts and the public for their patients, and they found themselves dealing with these two entities at every turn. What most troubled the superintendents of Mount Pleasant, Cherokee, Independence, and eventually Knoxville, was the stream of "incorrigible" inebriates who were committed to these hospitals from the county courts. Such patients were recognizable to physicians not only by their symptoms and chronicity but also by their "moral taint." Frustrated after his first year as the director of the Knoxville State Hospital, Superintendent W. S. Osborn declared: "The indiscriminate commitment of persons because they are given to drink brings degenerates, criminals and men of low moral standing in which there is little or no hope of benefit. The last named class of patients do not want to be benefited, but prefer the life they have been leading."[64]

Two years later Osborn's successor, H. S. Miner, reported that the problem persisted, for county courts regarded the hospital as "a dumping place for all the good-for-nothing bums and petty criminals in the community. Every one who was a menace to society, whether an inebriate or not, if he indulged in intoxicants at all, ought to go to Knoxville."[65] These commitments were devastating to hospital order and efficacy. Escapes were rampant among this class of patients, who diverted the hospital staff's energies and diluted the institution's "cure" rates.

As we have seen, other states had similar problems with the treatment of "incorrigible" inebriates and with their commitment to state inebriate hospitals. Indeed, the problem of the "incorrigible" inebriate revealed much about the difficulties of medicalizing habitual drunkenness. The managerial priorities of each institution—court and hospital—were instrumental in defining who was an "inebriate" requiring medical treatment. At least initially, the courts wished to get rid of their worst recidivist cases and regarded these individuals as appropriate candidates for medical care (after all, nothing else had worked!), whereas the hospital wished to treat "hopeful" cases early in their drinking careers. Ultimately, the inebriate hospital reached an agreement with the courts and state legislators, taking the incorrigible cases if they were detained in a new branch of the facility, the inebriate "reformatory." Physicians saw the separation of these two patient classes as essential to maintaining a hopeful and uplifting atmosphere for those who might benefit from hospital confinement. The parallel with Massachusetts, where separate facilities were established for incorrigible, chronic, and hopeful cases, is unmistakable. In order to treat the promising patients, both the State Hospital for Inebriates at Knoxville and the Foxborough State Hospital compromised and agreed to treat chronic recidivist cases at a separate branch of the institution. The priorities of the state and its judicial branch were not necessarily those of the physicians, but they had to be factored into state-supported care nonetheless.

The case of the incorrigible inebriate—indeed, the term itself—further reminds us that inebriety was perceived as a hybrid medico-moral condition, one that doctors believed involved the power of the will and the power of heredity, and one that likewise addressed issues of criminal justice and medical treatment.[66] If hospital physicians saw teaching the courts how to select appropriate candidates for medical care as an important step toward effective hospital treatment, they regarded educating the public as another. The families and friends of inebriates who committed their loved ones to the hospital commonly harbored two false assumptions: first, that treatment required but a few weeks (much like the Keeley Cure); and second, that their committed relatives would be returned to them upon request.[67] Moreover, it was not clear that the Iowa public regarded inebriety as a disease in the first place, despite the state's imprimatur. Between 1902 and 1906, the superintendents of Iowa's state hospitals for the insane could pride themselves on their up-to-date understanding of the disease of inebriety, but bringing the public in line was a more difficult matter. As Mount Pleasant's Charles Applegate lamented in 1903: "There seems to be but little charity, and less sympathy, shown the poor unfortunate inebriate by the general public, and it may all be due to the fact they do not consider inebriety a disease, but the results of the victim's own sin and folly."[68]

Newspaper coverage of the inebriate hospitals' work suggested a similar reluctance on the public's part to regard inebriety as just another disease. For example, initially Knoxville's residents so vigorously protested the state's decision to place the inebriate hospital in their town that one Iowa paper concluded that "the drunkard is considered by all classes as on a lower level than lunatics or convicts."[69] It is hard to tell if public opinion had changed much by 1906 when the inebriate hospital at Knoxville opened and the *Knoxville Journal* editorialized that "those incarcerated in the institution will not be permitted to rest on flowery beds of ease, nor will they be subjected to unnecessary harsh discipline. They will be furnished good comfortable rooms, good diet, proper medical treatment and those who are physically able will be required to work."[70]

Language such as this only fed the ambiguous identities of inebriety and its institutions. Was addiction a bad habit or a disease? Was inebriate reform medical or penal? Were inebriates incarcerated or admitted? Addressing the Thirty-Fourth General Assembly in his biennial message on 9 January 1911, Governor B. F. Carroll offered his own answer: "Some of the persons sent there [Knoxville] need medical attention, perhaps when first committed most all of them do, so that it would be necessary to maintain a hospital, but a larger percent of the inmates, after the first few days or weeks, at most, are abundantly able to work and need to be thoroughly disciplined. . . . In other words, the institution should partake *both* of the nature of a hospital and a reformatory."[71] From the start, then, the mission of Iowa's new hospital for inebriates was being defined by a powerful layman as both medical and disciplinary.

## Medico-Moral Therapy in Iowa: Tonics and the "Wheelbarrow Cure"

Governor Carroll's concerns were shared by others. The superintendents of the state institutions for inebriates and the members of the Board of Control daily confronted the challenges of curing a morally loaded, chronic disease that took men and women away from their families and often compromised their financial security. Meanwhile, the taxpayers and legislators, conflicted in their attitudes toward inebriates, wanted assurance that their dollars were being put to effective use. Devising a treatment regimen that attended to the medical and psychological needs of patients, to the financial needs of their families, and to the political demands of legislators and citizens was no easy task. Searching for models of successful state programs for reforming inebriates, the Board of Control turned to other states. Of these, Massachusetts appeared to offer the most guidance.

In truth, the treatment protocol established at the Iowa institutions was remarkably similar to that offered inebriates in the Bay State.[72] Elements of this regimen were quite standard, though matters were a bit more complex than Mount Pleasant's Charles Applegate claimed when he averred that the object was to "simply confront the disease and treat it."[73] Most patients arrived in an intoxicated state. Before putting these individuals to bed, the admitting physician and his assistants made preliminary mental and physical exams. Blood and urine samples were sent to the pathologist. Once the patient had sobered up, usually twenty-four to seventy-two hours following admission, a second exam was performed. This would include the patient's own narrated history and an attempt to verify previous diagnoses. At this point the admitting physician often learned that the patient's alcohol habit had begun as an effort to relieve the pain of some underlying injury, disease, craving, or personal tragedy. Medicines, tonics, and physical therapies followed, according to the case. Strychnine, the active ingredient in *nux vomica,* was used frequently as a digestive aid and nervous tonic, especially in cases of difficult withdrawal. The same held true for tincture of cinchona and tincture of gentian, which were both used as digestives. Physicians also employed chloral, sulphonal, and bromides, all powerful nervous system depressants, in conjunction with strychnine, especially when sleep proved difficult. Hydrotherapy, electrotherapy, and massage supplemented these tonics.[74]

Once the immediate effects of alcohol and its withdrawal had passed, physicians started inebriates on a light diet of toast, oatmeal, and milk accompanied by large quantities of coffee and tea—as "stimulation without intoxication." When the patients' health improved, the diet became more varied. Patients continued to receive their daily doses of strychnine, just as they did at many private sanitaria. If, after a few weeks, patients progressed as expected, they started a program of physical culture, exercise, and employment, usually within and around the hospital grounds. Physicians also considered entertainments, lectures, and general socializing essential elements of therapy. Through these means, and the gradual awarding of liberties around the grounds for good behavior, physicians hoped to reform their patients and return them to productive citizenship.

Though the medico-moral elements of therapy were evident in the Iowa superintendents' prohibitions against card playing, the hospital lectures "along moral lines," and the emphasis on putting patients to work, the therapeutic issue that best highlighted the moralistic frame of inebriety was employment. Light occupation—vocational therapy—routinely played a part in the treatment of the insane, but the inebriates' situation was more complicated. Simply put, if the state was willing to care for its inebriated ranks, legislators believed that the taxpayers of Iowa should receive something in return. The Board of Control agreed, reasoning that inebriety might be a disease, but it was a largely self-inflicted disease. They considered it an illness whose victims' moral failings frequently were responsible not only for their condition but also for their loved ones' financial worries and the state's bloated roster of dependents, defectives, and degenerates.

The contrast between the "innocent" insane and "guilty" inebriates is clear when we consider the rehabilitative labor expected of each group. The insane might engage in gardening, farming, domestic labor, and some lighter occupational pursuits, but the Board of Control actually considered coal mining a potential form of "vocational therapy" for inebriates. Although many of the men who were admitted to Knoxville were categorized as "unskilled labor," more than their shortage of skill led the Board to suggest coal mining. According to Board member John Cownie, coal mining was ideal for several reasons. First, mining required little skill; second, a mine might supply fuel economically to all state institutions; third, the prospect of mining coal at the state hospital for inebriates was so loathsome an image that it might deter many drinkers from alcoholic excess; fourth, it was easy to keep watch on the inebriates if they were underground; fifth, after laboring in the mine, inebriates would be too tired to escape; and sixth, mining might be done all year round, as opposed to farmwork, which was seasonal. Coal mining was thinly veiled punishment for inebriates; the real appeal lay in its punitive prophylactic ability to deter drinkers from alcoholic excess and to provide for the state's economic interests. If leg irons and locks were not a part of the inebriate hospital, a mine shaft might serve instead.

The state never constructed its coal mine at the inebriate hospital. Instead, it supplied a fleet of wheelbarrows. Most patients who stayed at Knoxville for more than a few weeks ended up taking "the wheelbarrow cure." The hospital loaned its inebriate patients to local farmers at harvest time, and it employed patients to grade the land around the institution and manage the hospital farm. With Knoxville employing more than fifty men to landscape the grounds, John Cownie eagerly reported to the *Knoxville Express:* "Our wheelbarrow cure for dipsomaniacs is working wonderfully well . . . [and] is the best thing we have found yet. . . . I tell you when the men get through with that cure they will hesitate a long time before they touch whisky again and have to go back to the wheelbarrow."[75] Through his employment, the inebriate earned a wage that was split between the hospital and the patient's family, if he had one, or the hospital and the patient himself, if he had no relatives. The wheelbarrow cure was meant to appease a treatment-shy public and return dollars back to the state's coffers; it also was intended to train patients to provide for themselves and their dependents. In 1911 the state built a brick works at Knoxville to keep the men at work year-round.

## Ed Harris in Context: The Patients and Their Stories

If the Board of Control, the state legislature, the superintendents, and the newspapers helped mold the medical and penal identity of the inebriate hospital, so did patients, their families, and their friends. Indeed, the last three parties put the inebriate asylum to uses that were frequently at odds with the therapeutic goals of reformers. Iowans might concede that the state needed a medical facility for its inebriates, but they wished to put it to their own social and often personal ends.

Take, for example, the case of women seeking divorces from their habitually drunk husbands. Throughout the history of the inebriate hospitals in Iowa, an average of 50 percent of patients were married; approximately 30 percent were single, and the rest were widowed, divorced, or separated. In 1903, shortly after the passage of the new inebriate law, the *Cherokee Democrat* reported that "wives are taking advantage of the new dipsomaniac law to get divorces."[76] Habitual drunkenness had been considered grounds for divorce in Iowa for some time, but it was a difficult condition to prove. With the opening of the inebriate asylum, wives had a new way to certify that their husbands drank to excess regularly: they could have their men committed to the inebriate hospital. Within weeks of the opening of the Mount Pleasant Hospital for Inebriates, at least five women committed their husbands as inebriates and promptly filed divorce petitions.[77]

Matters had not changed much by the time Knoxville opened its doors in 1906: the *Des Moines Register* regaled readers with the story of Harvey Connor, a sometimes abusive inebriate who had "turned his wife and children out of doors and converted his house into a sort of wholesale liquor establishment." When Mrs. Connor could take no more, she threatened divorce. But in a peculiar twist of fate, Mr. Connor actually avoided the divorce proceeding by agreeing to be committed to the state inebriate hospital. In his case, the act of taking the cure indicated to civil authorities an earnest desire to put his life back in order and return to the ranks of responsible husbands. Connor received a term of "three years, or until cured."[78] In these cases, whether the condition was considered a disease or not mattered less than the legal recognition that the condition was "real" and therefore grounds for civil action.

Some inebriates were remarkably adept at enlisting not only their families but also their friends and concerned townspeople in the commitment and parole processes. Consider the case of Karl Pedersen, a horse buyer from Decorah, a small river town in northeastern Iowa. Though this Norwegian immigrant's "pecuniary status" was listed as "poor" upon his admission to the Independence State Hospital for Inebriates in September 1903, Pedersen proved rich in friends. According to his admitting history, Pedersen had bought horses successfully for seventeen of his thirty-five years, but in the previous two years he had lost a considerable sum in the horse trading business. Subsequently, Pedersen had turned to drink. A willing and cooperative patient, he was well liked by the hospital staff, one of whom noted on 30 October: "Gets along very nicely. Is quiet and well behave[d], and [works] in the dining room where he is a very good helper. Is not very profane."[79] Such comments reveal the priorities of both staff and institution: successful institutional management required compliant behavior; successful treatment meant the patient's adoption of good manners and work habits.

That same fall day, the citizens of Decorah also cast their vote in Pedersen's favor, petitioning W. P. Crumbacker, the Independence superintendent, to recommend the horse trader's parole to the state's governor, Albert Cummins.[80] Signed by Decorah's mayor (who was a physician), the city's district court clerk, sheriff, marshal, hotel manager, and several bankers and businessmen, the petition was also endorsed by W. D. Lawrence, M.D., medical director of the Lawrence Sanitorium for the cure of inebriety and the drug and tobacco habits in Minneapolis, Minnesota, where one of Pedersen's best friends had been a patient. Though Pedersen's original "term" at the inebriate hospital was listed as eighteen months, he was paroled just seven weeks after the arrival of the petition, upon his taking a pledge to avoid both drink and drinking establishments. Like his fellow parolees, Pedersen was asked to make monthly reports to the governor, approved by the clerk of the district court, certifying his abstinence. One year later a report from the Decorah sheriff indicated that Pedersen was again drinking, but significantly less than before his confinement.[81]

Pedersen was not alone in having his community rally behind him. Jan Vickers, a 26-year-old printer from Jones County, near the State Penitentiary at Anamosa, was committed to the inebriate ward of the Independence State Hospital for the Insane in January 1903 by his mother. Concerned about the "bad company" her son kept when drinking and about alcohol's pernicious influence on his behavior and ability to earn his living, Mrs. Vickers thought his case warranted medical treatment. Though its exact date is not recorded, a petition was filed with the hospital superintendent on Vickers's behalf, requesting his parole. Signed by the clerk of the district court who had processed Vickers's original commitment papers as well as several attorneys, merchants, a newspaper editor, a physician, the mayor of Anamosa, and others, the petition proclaimed that personal acquaintance with the printer had convinced residents that "if paroled . . . [he] will keep the obligations of his parole and abstain from the use of all intoxicating liquors."[82] Vickers, however, took matters into his own hands, escaping on April Fool's Day, after two and a half months of confinement at Independence. The Anamosa sheriff, W. A. Hogan, returned Vickers to Independence a month later, and his parole was granted in early June. Six months later the same sheriff reported that Vickers was serving a jail sentence of fifteen days for violating the terms of his parole by drinking. "We want him returned [to the state hospital] as soon as he is discharged," added Hogan.[83]

The discourse shaping patients' commitment, treatment, and release was hardly confined to petitions. The committing parties—family, friends, or officers of the court—played an essential role in supplying the patient's history. When Dennis Rowley, a 41-year-old Cedar Rapids railroad worker of Irish ancestry, entered the Inebriate Hospital at Cherokee in December 1902, the law firm that had helped Rowley's wife initiate commitment proceedings previously (only to be dissuaded by a number of Rowley's friends, including the Catholic priest of Cedar Rapids) reported that Rowley was a good worker in spite of his hard drinking, that he had been abusive toward family and friends because of his drunkenness, and that his family struggled to support themselves since his earnings were spent on alcohol: "We hope that you will be able to reclaim Mr. Rowley and if you are able to correct his habits he will be able by industry and application, both of which he possesses in a high degree, to make restitution for his former misconduct and mistreatment of his family. He keenly appreciates the abuses and mistreatment they have received at his hands, and if he can but be cured of the drink habit, he will become a faithful citizen. We hope you will succeed in righting him and in sending him home in complete possession of himself."[84]

The language of this letter highlights several issues. First, the inebriate's character was an important element in the treatment process. The law firm emphasized Rowley's potential for productive citizenship, should the hospital succeed in curing him of the liquor habit. Again, the institution's missions were both clinical and social. The disease of inebriety was a "habit" that required the "righting" of the individual, as well as his being "cured." Finally, it is interesting, but not surprising, that a law firm rather than a physician supplied the useful patient history. It was usually aberrant social behavior—violence, domestic abuse, squandered wages—not the clinical manifestations of alcohol consumption, that made habitual drunkenness so disturbing. The hospital treated both. And the inebriate's personal and clinical histories offered hospital physicians clues as to how well that treatment might take. These narratives also identified difficulties at work, troublesome associates, or poor family relationships. Each of these environmental factors and others could portend failure, no matter how much progress was made inside the hospital walls.

Cases such as Pedersen's, Vickers's and Rowley's reveal the socially negotiated nature of treatment and, ultimately, the way the public and physicians viewed inebriety. The cooperation between medical and lay agents—the townspeople, lawyers, the hospital superintendent, and the governor—make the political nature of treatment for the inebriate clear. Medicalization of habitual drunkenness in Iowa was neither complete nor physician-orchestrated. Inebriety was both clinical entity and social disease. Even the term *parole*, typically applied in penal contexts, confirmed inebriety's hybrid persona. In violating his physical constitution through drink, the inebriate had also transgressed Iowa's civil polity, and therefore all manner of citizens had a say in the path he followed before, during, and after treatment.

## Iowa and the Progressive Cause of Medicalization

The Progressive Era has received more attention from historians of Iowa than almost any other period in the state's history.[85] A passing glance at John Briggs's *Social Legislation in Iowa* gives a sense of the remarkable diversity of social reforms—from domestic relations to defective delinquency, from prostitution to public health—that swept across the Hawkeye State in the early twentieth century. What makes the state's experiment with medical care for habitual drunkards so integral to Iowa's history is its resonance with the full spectrum of social reform concerns. Indeed, as much as any reform passed in turn-of-the-century Iowa, the creation of inebriate hospitals embodied a diversity of elements that characterized Progressivism in America: the search for order; the rise of "issue-focused coalitions"; the secular institutionalization of Protestant moral values; the growth of an increasingly regulatory state with a well-articulated, efficiently organized, social reform mission; the maturation of the professions; and the expansion of scientific and medical authority.

Iowa's efforts to provide medical care for inebriates were part of the larger changes taking place within the state at the dawn of the twentieth century: the reform of Iowa's government and the centralization and expansion of state authority for health and social welfare institutions; the professionalization of Iowa's physicians and their attendant commitment to both clinical medicine and public health; and the cyclical tides of temperance and prohibition reform. In essence, Iowa was able to put in place an unusual medical and social reform measure thanks to its ties to these larger developments. The eugenics movement, active in Iowa from the 1890s through the 1920s, also nurtured a public socio-medico-economic discourse that placed priority on curtailing drunkenness.[86] For Iowa's turn-of-the-century medico-moral entrepreneurs, to reform the inebriate was to stem the tide of liquor-induced hereditary degeneration and its attendant disease, poverty, and crime. Inebriate hospital advocates repeatedly offered this rationale for their work. In short, the Progressive Era was an opportune moment to propose an alternative to the failed "treatment" solutions of the mental asylum and jailhouse. The inebriate hospital idea drew ideological and institutional support from a variety of important political, economic, social, and medical sources that typified Iowa's participation in the Progressive movement between 1900 and 1920.

Iowa's experiment in inebriate reform speaks more specifically to the difficulty inherent in expanding medical authority to treat social problems. This is the process of medicalization—a process too often characterized as the medical profession's heavy-handed, near-unilateral efforts to bring certain physical conditions or behaviors into their domain. Perhaps the archetypal example of medicalization is the case of madness. Medieval Europeans understood madness in theological terms as punishment for sinful behavior; early moderns regarded the mad as socially noxious, dependent, and sometimes dangerous. Not until the Enlightenment, when Britain began to require medical certification to confine the mad to asylums, did physicians become the keepers of the mad. The late-eighteenth- and early-nineteenth-century moral therapy employed at asylums in Europe, Great Britain, and the United States was heralded by the medical profession and the public as an unprecedented humanitarian and therapeutic advance. In America, efforts to build asylums for the insane were led by the "father of American psychiatry," Benjamin Rush, who took charge of the Pennsylvania Hospital in 1783. Social reformer and school teacher Dorthea Dix picked up where Rush left off in the early nineteenth century, campaigning vigorously and successfully throughout the country for the construction of new asylums. By the mid-nineteenth century, madness had become mental illness, falling squarely within the physician's domain.

In calling for new inebriate asylums, reformers from Benjamin Rush on routinely invoked the story of the medicalization of madness, arguing that the same level of humanity shown the insane should be given the inebriate. Chronic inebriety and its neurological lesions were not only thought to precipitate mental illness, but some types, "dipsomania" for example, were regarded as forms of insanity. As Knoxville superintendent W. S. Osborn remarked in 1907,

> The application of present day methods in treating inebriates is not unlike the unscientific measures resorted to in the treatment of that kindred disease, insanity, during the middle ages. In the light of such experience, in view of the great number of crimes committed, the nameless havoc wrought together with the fact that inebriety is the most fruitful and prolific source of all diseases which afflict mankind, can we say that inebriates receive just and proper consideration from their fellow men? Must not the state recognize its responsibility, and recognizing such, owe it to the safety and welfare of its people . . . to isolate and treat these unfortunates? . . . [Inebriates] are diseased individuals.[87]

Although people drew a host of distinctions as well as similarities between inebriates and the insane, the comparison is a useful one because it sheds light on the process of medicalization. The powerful position of psychiatry today may be attributed in part to the expansive disciplinary actions of Progressive psychiatrists wishing to extend their medical domain, but some territories proved more difficult to claim than others, and some proved less attractive in the course of time. It is clear that in Iowa physicians were hardly alone in advocating for the disease concept of inebriety and the medical treatment of the condition. The first cries for an inebriate asylum came not from doctors but from judges in the county court system. Indeed, some hospital superintendents actively opposed offering medical care to habitual drunkards when the issue was first raised. The legislature, however, voted the state's new medical policy into place, and the superintendents were left no alternative but to accept it.

Thus, the story of Iowa's inebriate hospital experiment makes clear that offering medical care for inebriety was hardly a top-down process. Though the policy was initiated "from above," by the state legislature and carried out by hospital physicians, court systems, and even the governor, the commitment and treatment processes involved inebriates' families, friends, and fellow citizens. The involvement of these latter parties, as we have seen, suggests limitations to both state and professional authority in the medicalization process. Although each of these agents participated in the inebriate hospital experiment, they frequently did so on their own terms, and it is difficult to say whether their participation in initiating treatment or demanding its end signaled an endorsement of the disease concept of inebriety.

Ironically, the medicalization process might have received support from parties uninterested in the medical perspective per se but interested in its particular social utility. Recall the example of frustrated women seeking divorce from their chronically drunk husbands. The women used commitment to the hospital as a means of validating their complaints against their spouses and facilitating divorce. Court systems, similarly frustrated by their worst drunkard recidivists, deemed these individuals "worthy" of medical treatment—this in spite of the protests of hospital superintendents who found such cases both "incorrigible" and "incurable." In short, both individual and institutional priorities influenced participation in the medical enterprise, and such involvement did not necessarily signal a change of perspective on the nature of inebriety from vice to disease.

Thus, if the political, professional, and institutional circumstances in Iowa were auspicious for the creation of a new medical approach to caring for the habitual drunkard, the implementation of the medical model—medicalization—proved less successful because the various nonmedical parties involved continued to pursue their own goals; and these objectives often clashed with treatment regimens, undermined the authority of hospital physicians, and sabotaged patients' chances of successful reformation. Had the hospital's success rate been more promising, matters might have been different. But the superintendents' resistance to using the term "cured"—a reasonable reluctance on their part, given the intractability of the condition they treated and the difficulty they had in keeping their patients for the desired therapeutic course—broadcast loudly the problematic nature of their medical mission.

Two other factors worked against the wholesale adoption of the medical perspective: first, the medical facilities for drunkards addressed a small percentage of the alcoholic population; that is, many more drunkards were sent to jail for their petty crime of public drunkenness than were confined at inebriate hospitals. Thus, medical care could hardly supplant the traditional criminal justice solutions to this vexing problem. Second, Prohibition and the First World War cut short the medical efforts of physicians, drying up much of the political concern for the treatment of drunks. Many legislators doubted the necessity of medical care for the inebriate when the manufacture and sale of alcoholic beverages was banned. As in Massachusetts, wartime prohibition and the tendency for down-and-out drunks to either enlist in the armed services or obtain employment in a desperate labor market diminished the patient censuses at inebriate hospitals across the land. In the end, Iowa's efforts to medicalize habitual drunkenness were unsuccessful for as wide a range of reasons as they were initiated.

CHAPTER SEVEN

# On the Vice and Disease of Inebriety

Throughout the nineteenth century, scores of habitual drunkards wrote about their experiences with alcohol and their return to sobriety through various means. Published as temperance tracts, these narratives constituted a popular literary genre. As Elaine Parsons recently observed, "All Americans, not just those who particularly cared about temperance reform, knew the standard generic elements of the story of the drunkard's decline."[1] Parsons identified six key characteristics of the narratives: first, the drinker is a fine, upstanding young man before he begins to imbibe; second, the drinker's decline transpires through outside influences; third, where the drinker is accorded some responsibility for his own decline, his defect is a passion for excitement or bad company; fourth, the desire for alcohol becomes overpowering, taking precedence above all other aspects of the drinker's life; fifth, the drinker, now a drunkard, loses control over his family life, his job, and his own body; and sixth, redemption, if achieved, comes through powerful external agencies.[2] Although scholars in literature, history, American studies, and philosophy have examined the wealth of temperance tales published throughout the nineteenth century, few if any have examined them in light of the attempt to medicalize habitual drunkenness.[3] These first-person accounts are a great resource for assessing popular ideas about inebriety, and they reveal some of the ways in which cultural assumptions about gender roles and social class framed the etiology and treatment of alcoholism.

Especially after 1900, popular autobiographical accounts began to champion the treatment of alcoholism through Freudian psychoanalysis, the mental healing techniques of the Emmanuel movement, the new psychology, and a general medical and moral course of care.[4] Some reformed drinkers borrowed medical terms, while others disparaged the patent medical "cures." Almost all accounts referenced the physical and mental aspects of inebriety as well as the condition's moral dimensions. Taken together, these narratives suggest that the disease concept was part of popular culture and that inebriates themselves played a role in its promotion.

Excitement surrounded the application of science to the social problem of habitual drunkenness, and it was no more palpable than in the September 1915 issue of *Illustrated World,* where the editors introduced Lucian Cary's "What Drives Men to Drink?":

> We've been told a hundred times that "Alcoholism is a disease." We've acquiesced in the statement, though but vaguely understanding it, believing all the time probably that in this connection "disease" means moral weakness. Psychology, powerfully bolstering up medical science, now shows us the nature of this disease. It tells us why, in many cases of alcoholism, sons of the best families "go wrong"; why too frequently the recognized "cures" are futile. To all who would lessen the liquor evil we commend this article.[5]

Bold and enthusiastic, the editors' words suggested that the public was grappling with the issue of alcoholism's identity. The introduction hints at the ability to keep both the moral and medical plainly in sight when discussing alcoholism—something the editors note is a disease, yet part of the "liquor evil."

Thus, a new drunkard's narrative began to compete with the more traditional temperance tales of the nineteenth century. Although a minority voice in relation to the larger body of temperance tale literature, these "recovery narratives"—my term for the non-temperance autobiographies of rehabilitation—offer an interesting point of comparison. This chapter focuses on the experiences of alcoholics, their families, and their friends, as they were voiced in both popular literature and the letters of patients treated at Foxborough and Norfolk State Hospitals in Massachusetts. By examining the ways in which inebriates discussed their condition and their struggle to get well, we gain an understanding of the difference that the disease concept made in their personal lives. We also gain an appreciation of the contributions inebriates (and the media) made in shaping popular and professional opinions about alcoholism.

## The Public Voice of Private Anguish: Published Narratives of Recovery

At the inaugural meeting of the American Association for the Cure of Inebriates in 1870, Joseph Parrish read a special communication from the patients of the Pennsylvania Sanitarium, an inebriate asylum that he directed. Published in the meeting's proceedings, the paper was entitled "Disabilities of Inebriates," and it began humbly: "Gentlemen:—We are aware that in offering to you our views upon the grave subjects, whose discussion has brought you together, we occupy the position of the condemned criminal, who, his case having been adjudicated, is simply *pro-forma* asked what he may have to say, ere the already determined sentence be passed." Yet, added the inebriates, they saw the members of the AACI as their advocates. With their "enlarged knowledge, and generous motives," members of this new organization for the promotion of the disease concept might change the way the public viewed inebriety, allowing the victims of this disease "a kinder hearing, and a revised judgement."[6] Theirs was a stigmatized condition, and this made it more difficult for them to find the sympathy and assistance they needed to secure their recovery.

Parrish read the inebriates' letter to the AACI immediately after completing his own speech on the philosophy of intemperance, giving the words of the Pennsylvania Sanitarium patients a privileged spot on the podium. Although it is tempting to interpret the communication as a self-congratulatory act on Parrish's part—after all, a third of the letter quoted "from the annual report of our President, Dr. Parrish"—I think that these words were more significant than that. By giving a voice to inebriates seeking recovery from their addiction to alcohol, Parrish and his AACI brethren allowed, if not encouraged, their participation in a popular and traditional literary genre—the temperance tale—but in a revolutionary way. Like the authors of the autobiographical temperance narratives that achieved great popularity in the second half of the nineteenth century, the Pennsylvania Sanitarium inebriates recognized in themselves a "full share of human weakness and sin."[7] However, defying both social and literary convention, they had not pursued a path of redemption and renewal through the gospel of temperance. Instead of publicly confessing, signing a pledge, and pursuing the reclamation of other drinkers, Parrish's inebriates sought the refuge of a medical institution for their cure, albeit one shaped by Christian principles.[8] In turn, their physician guides encouraged patients to spread the word about the disease concept and the medical treatment of inebriety.

Most of the published recovery narratives that focused on the restoration of the inebriate's health and social role through medical intervention were published in the first two decades of the twentieth century; they appeared in journals with reputations for muckraking and social reform reportage: *McClure's Magazine, The Arena, American Magazine, Everybody's Magazine.* These journals also published regularly on both scientific and social issues surrounding alcohol consumption. Periodicals addressing literature and the industrial arts also printed a few autobiographical accounts of inebriates.[9] In addition to promoting the disease concept of inebriety and the expert authority of medicine, these self-revealing articles offered validation to their writers, although there were no references to autobiography as a self-healing exercise. Most authors offered their stories with the explicit hope of helping other drinkers and their families. In other words, they continued the mutual-aid traditions of the Washingtonians and temperance fraternities that preceded them, but from a new perspective.

As they presented their life stories, inebriate authors frequently emphasized their own middle- and upper-class backgrounds as well as those of their (formerly) respectable fellow drunkards. The literary and social club at the New York Inebriate Asylum in Binghamton, for example, could boast "the names of men who, in their respective walks of life, have adorned and taught superior communities ... divines, physicians, lawyers, writers, artists, teachers, merchants, and more than one scholar," observed the anonymous patient who hailed from the same institution.[10] These men, together with those less well-off, were part of what the same man termed "a grand experiment in Social Science—an experiment to restore the status of prudent and faithful householders and worthy citizens, productive and honorable, a most interesting class of men, in whose fate is presented the impressive spectacle of powers, often noble, paralyzed, and affections and impulses, often pure and generous, perverted, by a prostrate weakness within and a potent and subtle enemy without."[11] The message was simple: inebriety did not just affect "down-and-outers." It was a disease that affected all ranks of society, including those most respected. By staking a claim for the middle- and upper-class alcoholic, the writers of the recovery narratives at once hoped to help those of a similar social caste, to garner support for the disease concept, and to validate their own lives as active participants in a new experiment in science and medicine.[12]

Published recovery narratives came in a variety of forms, each reflecting the particular worldview and experience of the author. Some aimed their arrows at the temperance and prohibition lobby, while others welcomed prohibition as the final solution to the alcohol problem. Some narratives specifically attacked the saloon "industry," criticizing it much as today's social critics condemn the cigarette industry and its targeting of underage smokers. Some were strictly "self-help" narratives, insisting that no matter how desperate a person's life had become through drinking, it was still possible to pull oneself out of the thralldom of alcohol. Others argued strongly for the disease concept of alcoholism and its many variations, suggesting a range of medical pathways to recovery. Almost uniformly, however, authors sought to distinguish themselves from the temperance movement. "This is no sermon," the anonymous author of "Is This Why You Drink?" assured readers: "Nobody ever hated a sermon, drunk or sober, more than I do."[13] Likewise, the inebriate narrator of "The Hardest Ride a Man Can Take" warned readers: "Understand me now; this is not a tract. Neither is this record the admission of any ordinary drunkard."[14]

An excellent example of the medically oriented recovery narrative is "Patient Number 24," in which a salesman fired for his chronic drunkenness is sent for a week-long stay in a private hospital for inebriates by his stern but big-hearted boss, whom he calls "the Chief." At the hospital, he is befriended by the physician-in-charge, "a large, strong man with inquisitive eyes," who, like the salesman's boss, takes an avuncular interest in him. As he is gradually weaned from alcohol, he is told by the physician that "You're going to fight this thing out yourself, in your own mind. Then you can go out and start all over again, if you think you'd like to, or—not. It will be up to you." His nurse likewise adds that the treatment "makes men out of horrible examples." Two days after his arrival, the salesman begins to feel like "a Hyde [who] was once more a Jekyll," and he experiences remorse, something that he tells readers physicians call "alcoholic depression." At this point, the salesman's "fight" really begins. On his fourth day in residence at the hospital, he happily reports, "I was a new man." His doctor then tells the salesman that the person underwriting his stay at the hospital is the Chief, who is described as "a heavy, slow-moving man with a heart like an oak."[15]

Through a series of conversations with the doctor, the salesman learns several important things. First, he comes to see the origins of his drinking in the nervous exhaustion and grief that followed the loss of his wife; his tendency to overwork is also implicated. The patient had sought alcohol as an anesthetic, an escape. His craving, claimed his physician, was the "result of habit and instinct and fear. Your instinct told you needed relief from work and improper living conditions. Your former habit of drinking got working—you remembered that alcohol drugs the nerves after a fashion. Then you were afraid of giving in to it, and that got you all worked up."[16] Second, the salesman comes to understand that the only cure comes from inside the inebriate's own head—"when he decides it doesn't pay to drink, determines not to drink, and doesn't drink." The hospital can bolster a person up physically and mentally, but nothing can "take the place of manhood," that is, the exercise of responsibility and willpower. Still, this is no easy task, and the salesman is encouraged to seek refuge at the hospital whenever he feels weak and in danger of drinking. Finally, the patient comes to an important realization, with his physician's assistance, that society's penal approach to the management of habitual drunkards is misguided: "The only time the alcoholic gets action through society's makeshift provisions is when he develops delirium tremens or becomes a public nuisance, in which case he is usually condemned out of hand and punished for his sins—when he should be more often treated for his disease."[17]

When the salesman leaves the hospital, he is immediately accosted by a fellow discharged patient, "a tall, slim youth, well-dressed and well-featured, with an amiable smile." The young man urges the salesman to have a drink with him, and the former model patient finds that "on impulse" he accepts the offer. Though the salesman courageously orders "Vichy and milk," the youth requests "a cocktail, a pineapple Bronx." With wobbling knees and nerves "unstrung," the salesman bids his new drinking partner farewell and leaves the saloon, only to remember the envelope that he was given upon his departure from the hospital. Opening it, he finds fifty dollars and a note from the Chief inviting him to return to a new sales position. In spite of threats to his sobriety at five and ten months after his discharge from the hospital, the salesman is able to recognize his weakness, seek additional help from the physician, and "correct" his "false mental valuations." If readers wish to help someone they know who is an alcoholic, concludes the salesman, "do these two things: if your subject's body is poisoned, scientifically unpoison it. If his mind is poisoned, *find out what the poison is and frame an antidote.*" Thus, both medical care and psychotherapy are indicated.[18]

The story of Patient 24 is a thoroughgoing medico-moral recovery tale, for it embraces both the clinical and volitional dimensions of inebriety, urging the inebriate to take charge of his own cure: first, by seeking the medical care thought necessary to strengthen both his constitution and his resolve; and second, by discussing his condition with his physician and coming to the realization that he must take responsibility for his habits and his life. Patient 24's success not only offers validation of the medical approach to treating habitual drunkenness, but it lambastes the state's failure to erect a system of institutions that will provide nonpenal care for the alcoholic. "That's just the point," snaps the doctor in the final stages of Patient 24's treatment, "The whole theory of laws concerning alcoholism is founded on penalization, not medical help. . . . In days to come our present attitude toward the alcoholic will be looked upon not only as barbarous, but as stupid and short-sighted." The care that the inebriate needs appears ongoing, a recognition of the chronic nature of his condition. Although he claims at the end of the story not to "fear alcohol" any longer, the salesman observes: "I know now that when I'm tired or seedy I need rest or medicine—not a drink."[19] His brief return to the hospital in a moment of weakness, when he does fear that he will take a drink, suggests that alcoholism is not only a disease, but a chronic one.

Embedded within the story of Patient 24 is a message about gender and sexuality as well. Through the expert guidance and fellowship of an avuncular physician and a supportive employer, the salesman overcomes his disability. The solace and camaraderie he formerly found in the saloon is replaced with that of two supportive men. The process of recovery is likened to a contest—a fight—to regain one's manhood. Exhaustion and poor habits related to the salesman's demanding on-the-road lifestyle pose significant threats to his sobriety, as does a liaison with another discharged patient to whom he is impulsively drawn. This recovery story, then, is one that revolves around sexuality and the patient's substitution of healthy homosocial relations (with the Chief and with his physician) for pathological homosexual ones (seen in the patient's impulsive attraction to the young man who is released from the private hospital).

Within popular literature, especially after 1910, there emerged an ongoing discussion of using psychoanalysis to treat habitual drunkenness. "Is This Why You Drink? An Anonymous Confession" was the first of a series of articles on alcohol and alcoholism published in 1917 in *McClure's*, a journal of literature and politics that amassed a reputation for its muckraking exposes. Turning its readers' attentions to the alcoholism-sexuality connection, "Is This Why You Drink?" was effectively an advertisement for psychoanalysis. The article received the endorsement of William Alanson White, who was then president of the American Psychoanalytic Association, superintendent of St. Elizabeth's Hospital for the Insane, and a determined popularizer of psychoanalysis.[20]

In the piece, the patient, a lawyer, submits himself to the care of "a distinguished alienist, an authority on insanity, a prolific worker in psychology and all mental and nervous disturbances." Readers, of course, are led to believe that the alienist may be White. The patient confesses that he is a "hard drinker," that he wants to quit, that he's tried several of the "so-called cures," and that he feels helpless before the urge to drink. "You are a neurotic. I can straighten you out and set you on your feet if you will give me the required time and will live here in this city," replies the psychiatrist. Six months to a year are required. And the treatment, much to the lawyer's surprise, is "Conversation!" Later, in their third or fourth session, the lawyer learns that this "treatment is of course psycho-analysis. It is the psychology discovered and put into practice by Dr. Sigmund Freud.... It consists of the most thorough analysis of your past experiences and emotional life through the proper interpretation of your dreams."[21] Freud decried such simplifications of his work, but that did not stop the press or the psychiatric community from making them.

As his dreams are interpreted by the psychiatrist, the lawyer is forced to confront uncomfortable thoughts and feelings from his youth: painful memories of things his mother said to him, his fiancée's infidelity and their parting, and other "train[s] of ideas" of which he was ashamed to speak and which he could not share with readers. Through psychoanalysis, the psychiatrist instructs the alcoholic lawyer: "We find what has wounded your soul and made you follow a twisted path, [and] we also give you the opportunity to re-educate yourself and to deal properly with our stubborn, every-day world of facts.... You will no longer be a moral coward. Every drunkard is a moral coward, because in getting drunk he runs away from the workaday world with all its problems."[22]

The crux of the lawyer's problems is discerned from a mysterious dream in which he purchases a suit for just thirty-six cents, only to find out that it is made of cheesecloth. The psychiatrist's questioning about this dream and his subsequent interpretation of it point out the lawyer's flawed relations with the opposite sex: "The downfall of your hopes as a future husband came about on the train numbered 36, in section number 3. It resulted in your giving up all ideas about marriage—and marriage is the conventional thing, the symbol of lasting happiness and usefulness in the world throughout life."[23]

The lawyer, now possessed with these penetrating insights, remarks to readers: "A man becomes a neurotic, a moral weakling, a coward, because of the shocks that have been given his love instinct. . . . Broadly speaking, it is as abnormal to be a bachelor as it is to be a confirmed drunkard. Man was not put here to hate women. He was not put here to ruin himself with alcohol or with drugs."[24]

With the knowledge of his own nature, and his bruised affections for women, the lawyer begins to reconsider his position on marriage. As he does, his craving for alcohol remarkably fades away. The lawyer becomes a man of action, once again taking charge of his life, having looked his demons "squarely in the face, and, with the aid of the physician . . . dissected them and thrown them away." The narrative ends with the psychoanalyst's observation that the "principal factor" in the lawyer's case, "as in all others," is an emotional shock that "has interfered with, or warped, or put an end to your love instinct for the opposite sex."[25]

"Is This Why You Drink?" demonstrates the utility of mental medicine in treating the alcoholic. However, even the interpretations of the psychoanalyst are infused with morality: the drunkard is a "moral coward." Psychoanalysis in this case is portrayed as a manly battle with the demons of one's past taking place under the supervision of a psychiatrist. Through the insights and encouragement of his psychoanalyst, the lawyer becomes strong enough to dissect and discard his flawed thought patterns. He becomes a man in charge of his life once more, no longer enslaved to the bottle. The anonymous author also insists that hard drinkers such as he will not find help from their general practitioners; instead, alcoholics should consult a psychoanalytically trained physician. And yet, remarkably, the lawyer also informs readers that *he* has analyzed "with success five tortured persons—two men and three women," four of whom were excessive drinkers. Thus, he may urge the use of a psychoanalyst (and he does testify to the failure of "the so-called cures"), but it is clear that his experience in analysis has trained him to perform lay analysis, too.

The message is not one of rigid professional promotion, but of mutual aid; the psychoanalyst has helped the lawyer, and now the lawyer wishes to help others like him. Indeed, he has been empowered to do this through his own analysis. This scenario bears a close resemblance to early- and mid-nineteenth-century temperance tales in which the "rescued" drunkard goes on to save others like him. This homosocial tradition of mutual aid was the basis of the Washingtonian temperance movement of the 1840s and the fraternal temperance orders of midcentury.[26] In this case, however, cure comes, not through any sort of secularized evangelicalism, but through male bonding and professional training. Finally, we see here the association of alcoholism and stunted heterosexuality. The psychoanalyst reminds the lawyer that marriage is the norm and the symbol of "lasting happiness and usefulness." Although we never learn if the lawyer's shameful trains of thought were explicitly homosexual, we do learn that he had a very close male friend ("He and I had a sort of Jonathan and David devotion for each other") who betrayed him, and it is clear from the psychiatrist's words that even asexuality is a perversion of the "natural" attraction the lawyer should feel toward the opposite sex.[27] Cure was tantamount to reclaiming one's heterosexual identity and therefore one's masculinity—in the lawyer's case, masculinity that had been taken from him by his mother's harsh words and his fiancée's betrayal.

Of course, the published recovery narratives did not present a unified view of the disease concept of inebriety. Such a view did not exist, for physicians recognized the diversity of symptoms and etiologies behind habitual drunkenness as well as its moral dimensions. If "Patient 24" presented alcoholism as a disease in its own right, and "Is This Why You Drink?" emphasized the psychoanalytic perspective, "Confessions of a Dipsomaniac," which appeared in *The Arena* in 1904, focused on the hereditary Jekyll and Hyde personality of the habitual drunkard while criticizing the religious zealotry that prevented the recognition of the diseased nature of the dipsomaniac.[28] "Was I insane during these attacks?" asked the dipsomaniac author. "Yes, but it was a strange and weird insanity," he responded—certainly an appropriate opening for an autobiographical study of an alcoholic with a double personality. Periodically overpowered by "a life that enters my soul and drags me down to the level of a beast," the dipsomaniac, a "well-known American novelist and essayist,"[29] would leave his desk and venture into the society's netherworld, "so comfortable, so mind free, so joyous and active, aimlessly moving from place to place, town to town, saloon to saloon, dive to dive." Following these irregular invasions by his second personality, the writer would emerge to pursue his work with refreshed mental powers. By the time he penned his account in 1904, he had been free of these attacks of "psychic epilepsy" for several years. He wrote to give hope and encouragement to "the silent and secret sufferer."[30]

The novelist had shown signs of having "uncontrollable impulses" at an early age: he was frequently bored in school and overpowered by a desire to "get away." Though he recognized in retrospect that he was "always insane during these periods," no one at the time viewed his early desire to escape the classroom and wander throughout the woods as pathological. That his true mental state went unrecognized, he claimed, was due to "that curse of Puritanism," which leads people to mistake "objective symptoms of disease" for vice and sin.

Eventually he dropped out of school and was privately tutored for college. It was at university that a medical student first suggested he take some alcohol when he felt mentally unstable. The writer found relief, but the periodic attacks of insanity grew more frequent, and he began to medicate himself regularly with alcohol. He soon grew dependent upon liquor, as it offered "rest, peace, oblivion" from his secondary self.[31] Just as the dipsomaniac was finishing his studies at college, "the cursed uncontrollable impulse suddenly overtook" him and he gave in. He fled to the end of town and "a dark, opprobrious den of shame" where there was "a foul bar over which fouler alcoholic drinks were served." He left university, eventually finding employment with a newspaper, where he worked diligently and successfully until his editor gave him six months of salary and sent him on assignment to London; this precipitated a nine-month-long debauch, where his second personality assumed control.

The dipsomaniac's recovery commenced when he realized through medical assistance that his condition was the result not of vice but of disease. The reproaches of his teachers, employers, friends, and family were powerless. Likewise, claimed the writer, drink "cures" such as the regimen at the Keeley Institute were of little use: "Taking the cure" might prove useful in "satisfying anxious mothers, or as a means of securing positions," but such treatment only addressed the physical symptoms of dipsomania.[32] Instead, it was the realization that he was the victim of a sort of "psychic epilepsy," that allowed his normal personality to gain the upper hand. "Dipsomania," declared the novelist, "is a symptom of disease, not the disease itself, and the disease being understood, the symptoms—which have ruined many a happy home, blighted many a brilliant brain, and placed the stigma of drunkenness on the undeserved—may be kept under control, and finally entirely suppressed, as the disease yields to modern scientific treatment."[33] Although dipsomania was not portrayed as a disease in its own right, it was a symptom of a serious nervous pathology. If one treated the underlying disease, one could also put to rest the dipsomaniac's temporary fits of insanity and thus curtail the related drinking binges.

Both "Confessions of a Dipsomaniac" and "Is This Why You Drink?" portray the diseased nature of the alcoholic, but neither regards alcoholism as the primary pathology. These two narratives also signal the transition from a largely hereditary, somatic orientation to an environmentally oriented psychodynamic interpretation of alcoholism. The 1904 "Confessions" uses the earlier term, "dipsomaniac," which emphasized the connections between habitual drunkenness and insanity. For late-nineteenth-century medical theorists, dipsomania was a form of alcoholism rooted in a hereditarily defective constitution. By the first decades of the twentieth century, however, psychiatry was coming to regard the alcoholic's problem as a neurosis, not a psychosis.[34] This is clearly the case in "Is This Why You Drink?" where outside trauma, not an inherently defective nervous system, is held out as the primary cause of the author's alcoholism.

The author of the 1918 "Slaves of an Invisible Command: The Soul and the Drunkard" presents a similar story, boldly asserting at the outset: "If the world is ever to be saved from drink, it will be by the drunkard—not by the prohibitionist." A recovered drunkard and writer, the narrator believes that prohibitionists are heartless, "negative being[s]," while the drunkard's openness, his willingness to acknowledge his shortcomings, and his seeming inability to do anything about them, make him the more complex and sympathetic of the two.[35] Here, as in the campaigns to establish inebriate hospitals in Massachusetts and Iowa, prohibition and the disease concept were poor bedfellows.

"Slaves of an Invisible Command" is an interesting story on a number of levels. First, the origin of the writer's drinking problem is both hereditary and environmental. He obtained a favorable impression of alcohol as a youngster, admiring his inebriate uncle, who possessed warmth and wit and refused to drink on Sundays so that he might teach church school. Later in life, the writer found that drinking with his comrades brought him friendship and good cheer—"were we not Englishmen following the ancient ways with warm hearts and strong heads—and well lined 'with jolly good ale and old'?"[36] It was not until he turned forty, however, that he realized he had become dependent upon alcohol. Although he made a commitment to go without it for one year and achieved his goal, the writer began to hear a voice urging him to drink. Referring to this voice as "my tempter" and the "little troll," he wrestled with his inner urging, but ultimately he submitted, finding that the voice was more muse than troll.

Searching for inspiration as he sat at his desk, the writer would hear the troll say, "Let me do it for you. . . . I have never failed you." Predictably, he would yield, but he would emerge from each drunken stupor to find pages of well-written prose suitable for publication under his own name. Alas, there was a price to pay: soon the newspaperman required stronger and stronger doses of whiskey to avoid withdrawal symptoms. One morning, when the writer felt especially low, hungover, and in need of a drink, he purchased a bottle of whiskey and returned home to drink it. Mysteriously, he was seized by a new voice, and he poured the whiskey down the wash basin instead. This was his turning point. The writer's "invisible angel" began to beat back the troll. Such experiences, asserted the reclaimed journalist, were by no means unusual. Many inebriates were torn between good and bad voices. Yet, added the author, the prohibitionist failed to understand the psychology behind this conflict, while the medical profession treated it.

Apart from the "mysterious angel" who could put a drinker back on track, announced the writer as he concluded his tale of alcohol "slavery," the drunkard had but three friends—"his own soul, a good woman, and a good doctor." It was not the physician's medical expertise that the writer appreciated most, however: "When I think of that doctor, I must withdraw what I said just now of man's incapacity for friendship; for, though his skill and his almost inspired understanding of my temperament and constitution have been much, it is his friendship that has been the real magic, which, with that beloved woman, has wrought with my own soul to set me free from that strange haunting in my blood which has at last been exorcised."[37]

This assertion about the power of friendship may reflect a lay view of psychotherapy as free and frank discussion among "friends." It also harks back to the friendship and support of the earlier neo-Washingtonian Homes in Boston and Chicago—institutions that offered medical care but relied heavily on moral therapy and the "universal law of human kindness" to assist their charges. It was the emotional support of the physician (and in this case, the woman) that meant more than anything else to the journalist. This was not unusual; many of the patients at Foxborough had similar experiences. In the last line of the quotation, the writer pulls in another of his three friends, his soul, by alluding to both the physical and the spiritual elements of alcoholism. The "haunting" is rooted in the blood, the physical constitution of the drinker; yet rather than his "blood" being cured, the haunting is exorcized. Indeed, at the end of the narrative, the recovered drunkard reminds readers that drugs can be helpful for the inebriate, allowing him to sleep and "recuperate his exhausted nerves," but as a general prescription, "'The Lord is my Shepherd' is beyond any written by the smartest physician."[38]

Lest readers think that the former slave of an invisible command had given up the medical side of inebriety for the metaphysical side, he offered them some important concluding advice in dealing with the habitual drunkard: "The first thing to remember about drink is that it is not merely a desire to swallow liquids, or to bemuse or stupefy oneself with alcohol. It is mainly a surface symptom of some physical disorder, or deep down spiritual or mental dissatisfaction. The drunkard's nature is usually predisposed by inheritance for his particular manifestation of nervous disease."[39]

Quotations such as these give us a glimpse into the ways in which the lay public was able to balance the medical and moral understandings of habitual drunkenness. The recovery narratives discussed here leave little doubt about their intent of publicizing the disease concept of inebriety (in its various forms) and promoting medical treatment. Yet they hardly avoid the moral aspects of the condition or the effectiveness of psychological and spiritual interventions. The same lawyer who urged his readers to think of alcoholism as a hereditary nervous disability in "Slaves of an Invisible Command" also observed: "Drinking has done me this one service—that I believe profoundly in the power of prayer."[40]

Like Parsons' "drunkard narratives," many of these first-person accounts shared certain characteristics even if they were not framed medically. For example, the drunkard's descent into drink was always unconscious, *involuntary*. The drinker might have enjoyed having a lively potation with his workmates—whether engineers, newspapermen, sales personnel, or lawyers—but he never consciously set out to develop bad habits or a hedonistic, dissolute lifestyle. Consider the words of the anonymous author of "Twelve Years with Alcohol: The Story of a Man Who Spent $70,000 Before He Quit": "My experience with alcohol began with infancy. My father drank to excess habitually

.... I was afraid of liquor.... Yet, with all that fear, by heredity I was predisposed to the use of liquor—nay to its abuse."[41] In the end, it was not a medical understanding of his condition that helped the man return to sobriety, but a spiritual awakening and the love of a woman.[42] Even when individual authors did not explicitly endorse the disease concept of alcoholism, and even when they took a critical view of the medical profession, they were likely to implicate heredity, a factor beyond their control, as a primary cause of their alcoholism.

Once a drinker had become dependent upon alcohol, his thirst for intoxicating beverages was likewise involuntary, rooted in a compromised constitution. Drunkards became "the tragic slaves of an invisible command, which, while under its influence, they are compelled to obey."[43] The inebriate's involuntary—or at the very least, unknowing—participation in his own decline was an essential element in securing the public's sympathy. Of course, whether or not the narratives emphasized it, it was also a salient characteristic of the disease concept. At some point predisposition or bad habit turned into disease; at this juncture, the inebriate had lost control.

Other characteristics tied the recovery narratives together as well. The role of women, discussed previously in terms of the restoration of heterosexuality and the cure of alcoholism, was frequently paramount—not only in the reclamation of a sober life, but in the drinker's decline. In "Slaves of an Invisible Command," the reformed drunkard observes that a "woman will give him that sympathy with which woman is mysteriously endowed, an inexhaustible patience of pity with which the gentlest man was never gifted."[44] This is an important part of his recovery. In "How I Broke Away from Alcohol," an anonymous auto salesman decided to "promise some woman I knew and respected, to keep away from alcohol for a month," adding that he made the promise to a woman "because for the first time I felt the need of the softer sympathy of the feminine nature."[45] After a year of sobriety, the salesman began to think about marriage and eventually tied the knot. His sobriety, readers learn, continued postnuptially. Similarly, the author of an anti-saloon narrative, "The Story of an Alcohol Slave As Told By Himself," concluded from his tabulations of his fallen friends and those who left the drinking life for a family life that "matrimony under the age of twenty-five years tends to check and stop incipient inebriety."[46] Thus, if the WCTU promoted an image of the drunkard as home wrecker, the drunkard provided a complementary if not contradictory image of marriage as a safeguard against inebriety. The liquor problem may have been an issue of "home protection" to the ladies in white, but to the drunkard on the mend, marriage could mean self-protection.

Not all marriages were portrayed so sanguinely, however. In "The Hardest Ride a Man Can Take," a lawyer who used to be a "moderate drinker" narrates his story to journalist Maximilian Foster. Although he never reached a state of chronic drunkenness, simply drinking moderately proved too much for the litigator. "Drink and you'll last two years!" his physician warned him. The doctor's concerns make the lawyer realize that everyone has a specific tolerance to alcohol and that his is low. The lawyer is a social drinker, however, and finds that he misses the conviviality of his drinking friends; sober individuals, he believes, are dull and irritating. The most memorable example of the tedium of a sober life in "The Hardest Ride" concerns a "well-known writer" at a dinner party: "He was in the first day's stage aboard the [water] Wagon; and a duller, more submerged persona I had rarely met." So disgusted is the writer's wife at her husband's depressed emotions that she urges him to drink: "For Heaven's sake, go take something! You may be able to stand it; but I can't. You are getting to be as much of a bore as he is!" she says, pointing to the lawyer.[47] Thus, the "good" wife might have been the drunkard's savior, but the "bad" wife was his enemy.

A similar scenario unfolds in "Coming Back from 'Booze'," where a hard drinking engineer's wife "developed a fierce maternal passion; she lived in and for the child . . . and of course Ruth [the wife] could no longer accompany me when I was out on field work. A second baby came, and a third, and a fourth . . . . She had no time for anything except her children. She could not understand why they annoyed me."[48] Curiously unaware of the fact that he had something to do with the arrival of the children, the engineer sees only that his wife's attentions are bestowed on their offspring rather than on him. Eventually he realizes that he failed to live up to his own family obligations, and he recommits to a sober lifestyle and a new career in the newspaper business. His success is crowned by his wife's return to their marriage. By this time she too has learned a lesson, confides the reformed inebriate, for while separated "she had had to think of something besides lavishing care on the children."[49] Clearly, alcoholism and its cure was a family matter, but it is also the engineer's humiliation at being arrested by a policeman and his commitment to starting a new career that bring him to the point of reconciliation with his wife. The medical cure that his wife and physician urge early in the story is of little consequence in his achieving sobriety—an interesting fact because there is talk of serious pathology underlying some cases of inebriety at the story's end: "No man in the clutches of alcohol need give up the struggle as hopeless—unless an inherited physical defect, a congenital predisposition, keeps his feet mired."[50] Once again alcoholism is portrayed in complicated terms, suggesting that more than one cause may be behind a man's descent into liquor—environmental circumstances and heredity may each play an important role.

In short, the recovery narratives represent a new twist on the old drunkard's narrative of the temperance movement. Alcohol's allure is threefold: first, drinking is linked to the comradeship and lively sociability of the saloon (necessitating breaking ties to old drinking buddies, and sometimes to previous careers, to promote recovery); second, drinking offers an escape from painful personal matters, both conscious and unconscious; and third, one's heredity may predispose one to find solace in alcohol. Indeed, the narratives highlight the physiological nature of the drunkard's plight, frequently implicating heredity in the etiology of chronic drunkenness. Inebriety may be a disease in its own right, or it may be a secondary disease, a manifestation of some variety of nervous defect or neurosis.

Many of these narratives also link inebriety to different types of medical treatment, though some treatments—particularly the patent medical cures—are not necessarily portrayed as effective. Gender relations are extremely important in the new narratives, with the inebriate's "guardian angel" often played by a wife or other important female presence. The return to sobriety is portrayed as a hard fight, a test of a person's manhood. Cure is linked not only to individuals taking responsibility for themselves, but to their jettisoning of friends and past habits associated with drink. Spirituality is not absent, nor are the moral dimensions of alcoholism neglected. Indeed, recovering drunkards appear able to maintain faith in medicine and religious faith simultaneously.

Alcoholism, in these narratives, continued to be a hybrid medico-moral condition that inebriates and the media promoted in all its complexity. As we shall see, the experiences portrayed in these narratives bear a rich resemblance to those of the patients at the state hospital for inebriates in Foxborough and Norfolk, Massachusetts.

## Private Voices on Public Treatment: Patients at Foxborough and Norfolk

How did patients treated at the Massachusetts Hospital for Dipsomaniacs and Inebriates think about their condition and their treatment? For patients treated before 1908, the year that the hospital underwent a complete reorganization, this is difficult to answer. However, Foxborough's new superintendent, Irwin Neff, considered the patient's family and his employment important factors in the recovery process and placed a premium on aftercare. In practice, this meant that both during and after an inebriate's stay at the hospital, the superintendent attempted to correspond with his family, friends, and employer as well as the patient himself. The emphasis on aftercare was intended to help the inebriate adjust to his responsibilities and sobriety. It was also an effort to garner evidence to support the hospital's claims to efficacy.

As a result of the hospital's new orientation, the files of the Foxborough and Norfolk State Hospitals (the successors of the Massachusetts Hospital for Dipsomaniacs and Inebriates), frequently contain correspondence between hospital staff, particularly Irwin Neff, and the discharged inebriates, their families, and friends. Taken at face value, this correspondence gives us a glimpse into the ways in which local communities in Massachusetts and the surrounding states understood inebriety. It also suggests the degree to which inebriates internalized the disease framework through their hospital treatment. Finally, the letters exchanged between hospital staff and patients provide a snapshot of patient-physician and patient-staff relations at the institution as well as the functions the hospital served for patients and their families.

Naturally, these records are not free from bias, and we need to keep certain caveats in mind when viewing and interpreting them. Inebriates released from the hospital usually wished to retain their freedom from the asylum, and this may have influenced what they, their loved ones, and their friends told physicians. Likewise, those patients who did well may have been more inclined to stay in touch with hospital physicians to report their continued success. On the other hand, patients in need of the many social and medical services provided by the hospital—those with weaker health status or fewer economic and social resources—may have had more to gain by staying in touch with hospital officials. Patients with superior education or well-developed writing skills similarly may have found it easier to pen letters to the superintendent at regular intervals. For many others, of course, their failure to stay sober impeded their ability to correspond with hospital physicians. Thus, the views of inebriety that we get from patient-physician correspondence may not be representative of the whole patient population treated at Foxborough and Norfolk between 1908 and 1920. On the other hand, a wide range of individuals carried on correspondence with Irwin Neff and the hospital staff, and their words are invaluable: they provide an insider's view of alcoholism and its medical

management in the first few decades of the twentieth century. The patient profiles that follow suggest the many ways in which patients participated in their
own treatment and that of others, accepted and rejected different aspects of
the disease concept, utilized the hospital to their own ends, and influenced
physicians' own understanding of alcoholism. In short, these vignettes represent the micronegotiations of medicalization, or partial medicalization.

### Nicholas Felson, Journalist

In February 1916, a 50-year-old journalist from Worcester, Massachusetts,
named Nicholas Felson entered the Norfolk State Hospital for Inebriates.[51] Felson had graduated from Worcester Technology at age 22; since then he had
been a reporter on several newspapers.[52] He was earning a sizeable weekly
wage of $75 the winter he sought treatment, though he had been idle for four
months during the previous year. Married with one grown child, Felson reported that his family life was "congenial," but his drinking had begun to interfere with both his personal relations and his work. Felson had drunk since
he entered college at the age of 18, enjoying the sense of well-being and sociability that accompanied alcohol. During the past three years, however, he
had begun "to lose control of himself" and "to indulge to excess." He was now
consuming on average "twelve glasses of whiskey and some beer" each day.[53]

Felson was more educated than most of the twelve thousand men treated
at the state hospital for inebriates in Massachusetts between 1893 and 1919.
He also entered the hospital as a voluntary patient, without any court hearing—a practice increasingly common after 1909 when the state began to finance voluntary admissions. Felson, however, was paying his own way; he was
a private patient who had decided with the help of his family that treatment
at Norfolk offered the most promise of ending his inebriety. Felson's story reveals the extent to which patients might influence both the hospital's staff and
the care that they and others received at the institution. To Felson, the inebriate hospital and the diagnosis of the disease of inebriety "made sense." It resonated with his own understanding of his condition. Moreover, it led him to
enter into several collaborative projects with the hospital staff, the first of
which was his own recovery from alcoholism. Factors such as his education
and general socioeconomic status may have made it easier for the physicians
at Norfolk to sympathize with the journalist and to blame his drinking on his
"evil associates" rather than some innate defect. The fact that the admitting
physician specified "evil associates" as the assigned cause of his inebriety,

however, suggests that hospital physicians were willing to make some moral judgments of their own about the inebriate's case, if not about the inebriate himself. It also indicates that the admitting physicians saw more at work in alcoholism than bad heredity.

Within a month of Felson's arrival at Norfolk, his son wrote a note to Neff expressing his and his mother's concern. Felson's wife, Minnie, was "very anxious to learn all the particulars" of her husband's confinement.[54] Neff's response came one month later in early May, noting that he was ready to release his patient soon.[55] Felson had done "exceedingly well while here and I am quite confident that he is mentally and physically much stronger than when he came to the hospital," he wrote.[56]

Felson, writing back to the hospital superintendent a week after his release, was inclined to agree with this assessment: "My family is delighted with the physical change, which, of course, they saw instantly, and more, I think with the mental change, realization of which is reaching them every day." Six weeks at Norfolk appear to have transformed Felson, who reported additional information about his condition: he no longer had a craving for alcohol, though he did yearn inexplicably for the foods of his childhood, which his family gladly provided. Felson's powers of concentration, too, had returned, and he once again approached his work with zeal: "Procrastination has departed. Those big troubles which I told you I dreaded are proving to be very much smaller than I thought and easily removable. The pessimist has been converted to an optimist."[57] Felson's informative letter to Neff indicated not only his willingness to comply with treatment—for aftercare was an essential part of the therapy offered by the state hospital—but also his eagerness to participate actively in his own "cure," reporting changes in his health and mental status as he saw appropriate.

Two months after his return home, the journalist again reported to Neff that he remained on the water wagon: "In fact," Felson claimed, "I have not experienced anything that can be called temptation, merely because I have no desire to drink." Offered cocktails and wine by his peers, Felson easily refused the alcohol: "Consequently, I have not had the chance to try out the 'will power' which I have heard so much about among the patients."[58] Nor was Felson alone in reporting his condition. When a physician friend offered him the opportunity to go for an all-day motorcar excursion, Felson jumped at the chance and sent his wife in his stead to Worcester City Hospital to meet with representatives from the Norfolk mobile outpatient department. Felson's and his wife's participation in these aspects of aftercare suggest a commitment to the medical model of inebriety, or at least to the supportive role played by the Norfolk State Hospital, a commitment that became clearer in two subsequent episodes of the journalist's post-institutional life.

Six months following his release from Norfolk, Felson referred the son of a well-to-do family to Norfolk, crafting a letter of introduction for the young man and writing directly to Neff to discuss the fellow's case: "This boy has been in the mercantile marine for some years, and while at sea has drunk nothing, but when ashore has been a drunkard. Now he is engaged to a nice girl, and the father will not consent to a marriage until the son has really shown that he intends to live an abstemious life. It is a true case of inebriacy, I think; if he drinks at all it must be to excess."[59] Having diagnosed the young man, Felson went on to recommend that he be placed in a job working for the hospital forester, who managed the institution's outdoor job training program with the Commonwealth Fish and Game Commission. Felson had worked in the same department at Norfolk and believed that he had benefited tremendously from the outdoor projects he was assigned. His referral of the young man to the hospital was not only a testimony to his faith in the institution's medical approach but also a way in which he could exert a degree of influence over the management of the facility. Through outpatient activities and visits to the hospital itself, patients kept alive their connections to the disease concept and their identities as recovering or recovered inebriates.

Felson continued to correspond with Superintendent Neff and the hospital physicians regularly for the next two years until Norfolk was leased to the federal government as a nervous reconstruction hospital. In 1917, however, Neff and Felson exchanged letters at a feverish pace, for the former patient had written his own autobiographical narrative of treatment at Norfolk State Hospital. He hoped to publish it in a popular journal, and he desired Neff's endorsement of his written account. In his initial note to Neff, Felson requested the support of the hospital superintendent:

> I want you, and Dr. Carlisle, too, if he would be so good, to go for the thing ruthlessly. No matter what changes you may think desirable, or what additions you may wish, they would be just what I want. My purpose is to give the public a clear exposition of the treatment from the standpoint and experience of a man who has undertaken it successfully.... In going over what I wrote in the beginning, soon after returning home, I found that I had failed, in the first place because I

lacked perspective, and secondly because it sounded something like "the story of my life" that one hears from the horrible example at a temperance meeting.... I have tried to tell just what my own people and, I trust, I myself, would have been benefitted by, had we seen it two or three years ago. It seems to me that there are a good many thousand people who would have an eager interest in such a story, if it is well and convincingly told. Perhaps a great good would come from it, directly and indirectly. I fancy that if this be published as I would like to see it, a gigantic number of men will curse the writer of it, because their wives and sisters will begin to look up the family pedigree for cases of mental disturbance, and harry the life out of husbands and brothers whose breath has the aroma of alcohol.[60]

Felson's concerns are noteworthy, for not only did he wish to express them publicly in print, but he also wished to have his story approved by medical authorities. He viewed his own tale as a potentially inspiring recovery narrative that would acquaint other inebriate men with a treatment option that they might otherwise overlook. Interesting too is the distinction Felson drew between his modern story of medical success and the life stories shared at temperance meetings. He had accomplished his recovery through medicine and psychology, not moral entreaty and confession. He wanted this distinction made plain. His closing words about the wives and sisters who would examine more closely their husbands' and brothers' pedigrees—or hereditary backgrounds—may also have signaled his belief in the constitutional and hereditary aspects of inebriety, even though his own case had been precipitated by "evil associates." These final words also make it clear that Felson conceived of inebriety as a masculine disease, one that touched women only through their relationships with men.

Neff worked tirelessly but unsuccessfully to find a publisher for Felson's piece. The *Saturday Evening Post* reviewed the article but found it outside the bounds of the journal; *McClure's Magazine* turned it down flatly: "We have so many articles on alcoholism on hand, that I cannot possibly order any more," wrote the editor.[61] In the end, the superintendent proved more successful at promoting the sobriety of inebriates than getting their articles into print.

Felson's story, disappointingly absent from his case file, would have resembled the recovery tales published in *McClure's Magazine* and other popular journals. It suggests that the stories appearing in *McClure's*, *Everybody's*, and *Illustrated World* may not have been too far off the mark. It also reveals the very active role that inebriates played in their own care and the ease with which many adopted the disease framework to explain their own drinking problems.

## Robert Godson, Telephone Draftsman

Robert Godson's case resembled Nicholas Felson's in a number of ways. Like Felson, Godson was university educated, although Godson's schooling had taken place at Dublin University and Trinity College in Ireland. Such a high level of education was unusual among patients. When Godson arrived at Foxborough in 1912, he had been treated three times at the Washingtonian Home in Boston. This again distinguished his case from the majority of patients, who had not been institutionalized for their drinking prior to arriving at the state hospital for inebriates. A draftsman, Godson was employed by the New England Telephone and Telegraph Company, maintaining the phone exchange equipment and working as a clerk. Interestingly, the phone company appears to have cut Godson a great deal of slack, allowing him to return to his job after each of his stays at the Washingtonian Home. At age 42, Godson was single and Protestant. Two characteristics in particular seem to have endeared him to hospital physicians: first, he possessed an enduring faith in his ability to recover from alcoholism with his doctors' aid; and second, he was blessed with a superior tenor voice that won him paid positions in churches throughout Boston and its suburbs.

The Irishman's trouble with alcohol had begun with the death of his father. Subsequent financial troubles had sent Godson into a deep depression, and he had begun to "drink heavily through desire for stimulation." According to hospital records, any significant excitement or depression could trigger a drinking episode for Godson. In 1910 the patient's sister had passed away, and her death "was followed by a severe attack." Likewise, in November 1911 Godson had hosted the controversial Irish Players, an Abbey Theater acting troupe led by Lady Gregory and W. B. Yeats, on their trip to Boston, and this proved "more than his nervous system could stand."[62] Following their visit, Godson had begun drinking heavily. At the time of his admission, the draftsman was consuming an average of "twelve drinks of whiskey per day," enough alcohol to interfere with his digestion and his job.

Hospital physicians were impressed with Godson's analysis of his situation, however, observing that the Irishman was "intelligent and evidently anxious to break from habit; shows a cooperative spirit and appears to be very much in earnest about himself. Is believed to be of rather neurotic temperament and it is believed that drinking is largely due to defective nervous system."[63]

Within six weeks, Godson was allowed to leave the hospital for 24-hour periods, always returning on time. By the end of August 1912, he had secured a position with his old employer, the New England Telephone and Telegraph Company, and was ready to try living on his own. Released from Foxborough on 31 August, Godson wrote to Irwin Neff and his assistant physician, Frank Carlisle, two weeks later with good news: he was "doing very well." More specifically, Godson reported, "My relations at the office are again very pleasant and they all expressed themselves very pleased to have me back with them again. I have made no secret of the place of my five months sojourn and everybody seems to feel that I did the right and proper thing in putting myself in your charge, especially since the result has been successful."[64] If true, Godson's words testify to a tangible level of tolerance, if not a clear endorsement, of medical care for inebriety—both by the draftsman and by his New England Telephone colleagues.

Nor was this an isolated response. Godson's case was a difficult one, and he feared that he might relapse in early January 1915. To prevent this from happening, he left Boston one Saturday in early January and checked himself into Foxborough State Hospital, believing that "even a few hours at the hospital would help me rather than if I had remained in my room." The following Monday, however, Godson deemed it best to return to work, and he left Norfolk at the crack of dawn to make it to New England Telephone on time. There, once again, he shared his story of visiting Foxborough, for which he "received nothing but praise and every encouragement."[65] For Godson, the value of traveling to Foxborough was twofold. First, he could receive care from hospital physicians and stave off another attack; and second, he could garner support from his colleagues by suggesting that he was taking responsibility for his condition and pursuing an effective course of treatment.

Indirectly, Godson's case supplies additional information about both the general public's reception of the disease concept of inebriety and the ways in which doctors at Foxborough thought about the condition. Godson's colleagues apparently approved of his taking precautions and seeking medical help for his condition when necessary. Likewise, Godson was welcomed back to the hospital on this occasion and on subsequent visits—another indication that hospital physicians saw alcoholism as a chronic disease requiring ongoing care.

One of the most striking aspects of his correspondence with both superintendent Neff and assistant physician Carlisle is that the draftsman made distinctions between "good" and "bad" former patients. It was not just a matter of being "cured" or being "sick." If Godson received kudos from his telephone company colleagues for taking an active role in his own reform, he was embarrassed by the visits of former "deadbeat" patients to his workplace: "I am sorry to say though, that I have had no less than three visits, at my office, from men I met at the hospital who wanted financial help and who evidently were doing nothing to help themselves. How they ascertained my address is a mystery. In one case, I told the janitor not to allow him round the premises again, because I had myself to consider. You will be glad to know that I am trying to continue to be a credit to your institution and my affairs are all going very smoothly.[66]

Just as the hospital wished to obtain patients who were "of good character apart from the habit of inebriety," so patients such as Godson realized liabilities when they saw them. Men who could not bring themselves to abandon alcohol and fend for themselves cast a pall on both Godson's reputation and that of Foxborough State Hospital. Godson's attitude was different with patients and inebriate friends he thought earnest in their desire to reform. In 1914, for example, Godson reported: "I met John Pines (the singer) who has been to Foxboro twice—he wanted my help, and accordingly, I offered to take him to Pondville [location of Norfolk State Hospital] tomorrow, provided he will meet me. If he keeps his promise I will have him there early."[67] In this way, Godson became an extension of the hospital staff, rounding up former patients who were having difficulty staying dry. Pines apparently followed through and met Godson, for months later Godson reported that he had seen him in Boston following his release and that "he appeared in excellent health and condition, not drinking a drop; he desired to be remembered to you and the doctors."[68] By staying in touch with the hospital, then, inebriates kept a door and a privilege open and perpetuated a tradition of mutual aid.

Successful treatment was a joint venture between patient and physician. Neff made this plain when he responded to a Christmas note Godson sent to the hospital physicians and their families in 1915. "In the first place," wrote the superintendent, "I want to congratulate you upon your success which I feel sure will be permanent. I know that you feel that your success in readjusting yourself is, as I believe I told you, due entirely to your own efforts. I want you, however, to feel that you are cooperating with us and that we are ready to do everything we possibly can for your welfare."[69]

Apparently, Godson took Neff's message to heart, for in the course of the next seven years, he routinely and voluntarily checked himself into the hospital both when he feared a relapse and when he began to drink again—a total of six times. Moreover, when Godson desired a new post as a tenor in the choir of the Episcopal Church in Melrose, Massachusetts—where Lowell Wentworth, a member of the State Board of Insanity, attended services—he requested a letter of support from Neff, which the superintendent gladly wrote.

Godson seems to have felt little or no stigma in his association with the state hospital for inebriates, for he revealed to Neff that he was not ashamed to have "Dr. Wentworth know how I happen to be acquainted with you and I shall be only too glad to have him know what you have done for me."[70] Indeed, in 1915, following several relapses that appear finally to have cost him his job at the telephone company, Godson relocated to Foxborough to be close to the hospital and to secure employment. Neff assisted Godson in getting a job at Foxborough State Hospital, which had been treating the mentally ill since the inebriates had been transferred to the new facility at Norfolk in 1914. Godson took his new responsibilities at the hospital laundry seriously, but the work proved exhausting, and the new Foxborough State Hospital superintendent was far from sympathetic to the former inebriate patients who had remained on staff at his institution.

Although offered fare back to Ireland by a man he referred to as his friend (possibly a social worker assigned to his case or a member of the Jacoby Club of Boston), Godson decided to remain in the United States—a lucky thing, for he was offered passage on the *Lusitania* just before it was sunk by a German U-boat off the Irish coastline. Sadly, by July 1915, two months after the sinking of the great Cunard liner, Godson was dismissed from his laundryman's post for drinking. Calling upon his remaining friends, he secured a temporary spot at a farm in Walpole, close to Norfolk State Hospital. By the end of the year, the indefatigable Godson was once more on his feet, writing to Neff to say that he had a new job as a draftsman at a large machine manufacturing firm in Woonsocket, Rhode Island.

With Godson's move, his attitude toward his fellow former patients had changed. "In coming to a new city and a new state, and moving in a locality removed from old associations and where I am practically unknown; I believe I have given myself an opportunity to make another fresh beginning," he wrote to Neff in early December 1915. To help insure a new start, he asked the superintendent not to forward his address to any former patients inquiring about his new location. Godson told Neff that he wished to avoid "as far as possible having anything leak out here about my recent experiences in Foxboro; until my footing is more assured at least."[71] Though he expressed his gratitude to one and all at Foxborough and Norfolk, he could not be sure how his current employer or friends would respond to the knowledge that the new draftsman in town was a reformed inebriate. In certain contexts, alcoholism still had the potential to stigmatize its sufferers. Neff complied with Godson's request and urged him to write "from time to time," adding, "I think that you should thoroughly understand that reconstruction comes slowly and in order to be permanent it is necessary that it be so."[72]

Godson progressed from working as draftsman to running the machines in the shop, which he declared was "dandy work." In the wartime economy, Godson earned more money than ever before, making munitions for "the English and Russian governments." He began to pay off his debts, and Dr. Frank Carlisle was one of the first to receive the dollar that he had loaned Godson some years before. "In a very short time," the former patient asserted, "there won't be a soul who can say 'Bob Godson does not pay his debts.' I cannot tell you what in intense satisfaction this gives me; even though I have denied myself absolute necessities to achieve this end."[73] Two years later, however, Godson sang a less optimistic song. The war had ended and so too had the labor shortage: Godson had been laid off.

Discouraged, he once more resorted to alcohol, and drinking, he once more found his way back into the care of Irwin Neff. In early March 1919, Neff was supervising patients at the Washingtonian Home because the Commonwealth had leased Norfolk to the United States government as a nervous reconstruction hospital. After a brief stay at the home, Godson found a position working as an orderly in a private hospital in Lynn, Massachusetts; he was gainfully employed once again. When last heard from, Godson had found "a much better position" at a private hospital in the Boston suburb of Hyde Park, where he wrote to Irwin Neff, asking the superintendent to send him a current or former patient for an opening as an orderly. Neff referred the Irishman's request to the head of the Emmanuel Church–sponsored Jacoby Club for reformed drinkers in Boston, sure that D. H. McFeeters, the group's secretary, would "be able to recommend a man."

## Charles Winchester, Manager

The story of Charles Winchester, a grammar school–educated manager of a spring water business, is less sanguine. Winchester was forty-four when he was first admitted to the Foxborough State Hospital as a voluntary patient in the summer of 1911. He had been drinking excessively for about eight years. A friend and physician who was also president of the Associated Charities of Haverhill, Massachusetts, where Winchester lived and worked, suggested that he seek treatment at Foxborough. There was no history of intemperance in Winchester's family. He had always drunk wine and beer, but not to excess until he experienced a reversal of fortune in the depression of 1893: he had lost his business, become despondent, and gone on a "spree" for a week. Several years later, after separating from his wife, he binged on alcohol again. Since the late 1890s, Winchester had been subject to "periodical drinking bouts." Upon his arrival at Foxborough in 1911, the admitting physician observed that Winchester had "good insight and analyses [his] case in an intelligent manner. Has thought at times that mind was not as clear and active as formerly, and believes that this was due to excessive use of intoxicants. Believed to be a favorable case for treatment."[74] As a single Protestant male in his forties with a common school education, Manchester was typical of patients at Foxborough.

We know little of Winchester's treatment at the state hospital for inebriates, but an important and perhaps influential feature of his care was his friendships with individuals outside the hospital who were concerned about his case. The first was his physician friend, Francis W. Anthony. The second was James Donald, a prominent businessman with the Standard Oil Company of New York. While the former advocated on Winchester's behalf and helped secure a spot for him at the inebriate hospital, the latter supplied monthly checks for sundries to make Winchester's stay at the hospital more comfortable. Donald was "willing to do this for the sake of old times, as [Winchester] was once a very bright and promising young man." He wrote to Neff to make sure that such financial support was permitted and to learn "whether [Winchester's] trouble is liquor or drugs."[75]

Prior to Winchester's discharge, a staff physician noted that he had cooperated "fairly well" in spite of a tendency to complain and make critical judgments. Like Godson, Winchester was allowed to leave the hospital for several trial overnight stays prior to his discharge, and he returned on schedule each time. Released after six months of treatment, Winchester assumed "a good position" at the Granite State Spring Water Company, whose offices were situated in Haverhill and Atkinson Depot, New Hampshire. Within less than a week, Winchester wrote to Irwin Neff: "Am back in the harness again, after my rest of nearly seven months. Am feeling fine and intend that my habits will be such that I will continue to feel so. Please remember me to Drs. Carlisle and Moore and to the fellows in the office. I will let you hear from me from time to time and would appreciate a line from you at any time. Thanking you for your interest in my welfare, I am very truly yours, C. J. Winchester."[76] Like many patients who corresponded regularly with the hospital's superintendent and physicians, Winchester seemed genuinely grateful for the medical and social assistance he received at Foxborough. But the ties he established at the hospital were not exclusive to the physicians and staff. Like so many of the corresponding patients, Winchester continued to stay in touch with other patients at the hospital and assist them through his ties to Irwin Neff and Frank Carlisle.

Only a month after returning to work, Winchester wrote Neff to say that he had an opening at the bottling operation in Atkinson Depot and that he had a specific patient in mind to fill the position: John Berwin, who was "one of the fellows who was going to try to do better and deserved some encouragement."[77] Winchester asked the superintendent to assist the patient, giving him train fare from Foxborough to the bottling plant. Neff complied with Winchester's request, adding: "I really think that this encouragement to others will be of decided assistance to you in living up to your resolution. I had a talk today with John and I think without doubt that he will accept the position. If he decides to accept I will finance him as well as I can and will write to you when he will report to you."[78] In a few weeks, the patient left the hospital for New Hampshire, where he arrived safely, and according to Winchester, "seemed to appreciate very much what you was able to do for him and I think he realizes that this is his chance to make good."[79]

Unfortunately, John Berwin did not pursue sobriety with the same conviction as Winchester, and Winchester was left to report this disappointing news to Neff. The superintendent's response was revealing, for it highlighted the importance of patient cooperation with the hospital not only for their own benefit but also for the hospital's successful treatment of others:

I am glad to hear of your success, but sorry to hear of John Berwin's condition. I sincerely hope that he will be able to get hold of himself and do the proper thing. I am writing to him by this mail, simply asking him why he has not written directly to me, and reminding him of his duties toward himself and the employers. I will be glad to hear in the course of a week whether or not my letter has seemingly been effectual.... With best wishes for your continued success and again thanking you for writing to me about John's condition, I am Sincerely yours, Irwin Neff.[80]

The informal ties that patients developed among themselves at Foxborough and Norfolk offered them a new way to demonstrate their solidarity—by providing assistance to one another in the form of jobs. These ties were clearly valuable to the hospital too, allowing them to keep tabs on released patients. Here was a respectable form of "treating"—not by buying a round of drinks for the local boys at the neighborhood saloon, but by offering employment to mates who needed a job. This was a significant act. It was a "gift" that the receiving patient could repay through honorable work. Equally important, it was a way in which successful patients might "purchase" freedom for their reformed but still institutionalized friends, for hospital physicians were reluctant to release a man until he had suitable employment awaiting him.

For five years and six months, Charles Winchester kept sober as he kept the books at the Granite State Spring Water Company, but in midsummer 1916, he went for a ride in a friend's automobile and was offered a drink to "brace" himself up for the excursion. He blamed "exhaustion on account of close application to business" as the cause of his fall, but the fact remained that he had taken a drink and it had been his undoing.[81] Winchester was admitted to the Norfolk State Hospital on 27 August 1916, after spending eight days at the private Hale Hospital in Haverhill. At the time of his admission, he was hallucinating about driving a car over stone walls and through houses in the company of a woman who had turned herself into a bug. Released after a week, Winchester suggested that he had strong employment prospects. When these did not materialize, he established his own business, the North East Mercantile Agency—Collections and Adjustments Everywhere. In December 1916, he wrote to Irwin Neff to say that he was confident the business would succeed. A little over three months later, Charles Winchester was dead.

Winchester's final encounter with Irwin Neff and Norfolk State Hospital came upon his readmission in March 1917. He came to the hospital voluntarily, "in a nervous condition, which was persistent." After a few days at the hospital, his condition grew worse, and he began to hallucinate. Less than twelve hours later, Winchester became cyanotic; his respiratory rate increased markedly, and in a short time he died. The diagnosis was acute alcoholic toxemia.

Though Winchester's was ultimately not a story of success, it reveals much about the relationships that patients developed with one another, with helpful friends outside Foxborough and Norfolk, and with the hospital staff. How much Winchester believed in the disease concept of inebriety is unclear. What is clear, however, is that he trusted the medical staff at Foxborough and Norfolk, as did his referring physician and friend, Frank Anthony. Whether or not he absorbed the medical framework the hospital promoted, he agreed with the treatment it provided, and he viewed the hospital as a vehicle for managing his own and his inebriate friends' lives. Arguably, the course of treatment, although based on a medical conception of alcoholism, was as much social as medical. If there was a key element to treatment, it was its comprehensive and enduring nature. Neff and his staff relied on in-house hospital care to sober up patients and cultivate their interest in helping themselves. The social relations the superintendent and other staff forged with the men who came under their charge appear to have been their most powerful weapon in the fight for sobriety.

There was no mistaking the heavy hand of paternalism that infused therapy at the state hospital for inebriates, but patients such as Felson, Godson, and Winchester seemed nonetheless to have appreciated the hospital's efforts. Perhaps an essential piece of their appreciation was the middle-class value system they shared with their reformers. These inebriate men appeared to believe that a job, financial independence, community participation, and a stable life that might involve a family or friends were desirable goals. If they cooperated with the hospital staff during and after their confinement, the institution might go a long way toward helping them achieve these ends—offering financial assistance, job referrals, medical care, advice, and a community of people sympathetic to their particular situation. As the cases of Nicholas Felson, John Godson, and Charles Winchester make plain, the hospital treated alcoholism as a chronic disease that inebriates needed to fight for a lifetime. Far from removing the responsibility for this fight from the inebriate's shoulders, the hospital reminded patients that the fight was theirs to make but that they had allies they could call upon for assistance.

## The Rand Brothers, Henry, Philip, and Thomas

The Rand brothers offer a comparable vision of the mutually dependent and dynamic relationship between patients and hospital staff. In 1903 Philip Rand was committed to the Foxborough State Hospital, the first of three siblings who turned to Foxborough, and later Norfolk, for help with their drinking habits. The Rands' father, Thomas Sr., had been a "true inebriate" before his death at the age of seventy-three. Philip was not yet thirty when he entered the hospital. A teamster, Philip Rand had been drinking constantly since January 1900. The physicians who signed his commitment papers recorded that Rand showed nervous symptoms, was repeatedly drunk, and had asked several times to be "committed to some dipsomaniac institution."[82] The details of Philip Rand's first visit to Foxborough were not recorded, but it must have started off relatively well, for in ten days he was given free reign on the hospital grounds, and in another three weeks he was placed on a work crew of painters at the hospital. After six months he left Foxborough, but he returned voluntarily four months later. Rand was in and out of the hospital twice more until he reached the time limit of his initial confinement (two years) in 1905. It would be eleven years until he again returned to the state hospital for inebriates.

In the meantime, both of Philip Rand's brothers arrived at Foxborough as patients. Henry came first, in September of 1912. Like his brother Philip, Henry was a single Catholic man of Irish descent. Unlike Philip, however, Henry had attended high school. Henry had been drinking excessively on and off for some fifteen years. He had been arrested six times, and he had served one short sentence at the Deer Island Jail in 1907. After making a pledge of temperance in 1909, Henry had remained abstinent for nine months. At his admission in 1912, the physician noted: "Is a rather chronic case and much doubt is felt as to permanency of good obtained at hospital treatment. Promises well, however, and agrees to cooperate. Insight is rather poor."[83]

As was the case with Philip, Henry's first place of employment was a shoe shop. From there, he went on to play semiprofessional baseball and later served as an umpire. For the six months prior to his arrival at Foxborough, however, Henry had worked as an attendant in various hospitals—a situation where life was highly structured and where room and board were often included in one's pay. Allowed to leave Foxborough after six months, Henry assumed a job at the state hospital for the insane in Howard, Rhode Island.

Like his brother, Henry was back at the hospital after two months. He returned voluntarily in an intoxicated condition. Released again (his employer having allowed him to return after treatment), and later returning for a third stay, Henry was given a final discharge in August 1913, the point at which he refused to stay longer at the hospital. Leaving against the recommendations of Neff and his staff, Henry was deemed unable to benefit from further treatment.

Neff may have been surprised, then, when he received a letter from Henry Rand the following day in which the former patient confessed his real reason for returning to his job so swiftly: a young woman (a nurse) whom Rand admired "above all the rest," a woman who "seemed to be my shining light, and I worshiped for five months at her shrine, and even went so far as to pray for her during my stay in Foxboro." Convinced that his love had left him, Rand wrote to Neff, "compelled by all the laws of self-protection, to seek sympathy from a congenial friend when all the world has turned against me." Appreciative of his hospital care, Rand added: "You have one grand institution there, and with that Prince, who is your colleague (I refer to Dr. Carlisle), you are compelled to fight against an unsympathizing public who see no help for the unfortunates who have worshipped at the shrine of Bacchus."[84] Closing with a poem about unrequited love, Henry Rand begged Neff to write to him in Providence with encouragement as soon as possible.[85]

Rand's letter is revealing in several regards. First, he appears to view the hospital quite favorably. Although we do not know if he subscribed to the disease theory of inebriety on which his care was founded, Henry Rand held a positive opinion of that care and further felt that the hospital physicians and select staff members were his friends: "Oh! Doctor," he wrote, "I need a friend in my extremity, and I know of no one that I would prefer to ask sympathy from than your own self—with the possible exception of Dr. Carlisle."[86] He believed that Neff and Carlisle took a genuine interest in his case, and when he felt in need of their support, he wrote to them. Moreover, Rand correctly perceived that the hospital's position was far from secure. In spite of the reorganization of the state hospital for inebriates, the public frequently remained unsympathetic to "worshipers of Bacchus" and their medical treatment. Through his own hospital experience, however, Rand had been "won over" to Neff's cause, identifying Foxborough as a "grand institution."

In May 1915, after a lengthy hiatus in their correspondence, Neff and Henry Rand began to communicate once more. Meanwhile, Rand's younger brother Thomas had resided at Foxborough for almost a year, from May 1912 to March 1913—his visit overlapping with Henry's own stay. In 1915, however, both Henry and Thomas were working as nurses at an exclusive hospital for nervous and physical disabilities, the the Long Island Home in Amityville, New York. Through Thomas, Henry learned that Neff was making inquiries about his health and well-being, and he decided that he had best write to the superintendent directly. In May 1915, Henry told Neff that he was doing well and had "renounced my vows of subjection to the Queen of Bacchante and am no longer a worshiper at her shrine." With his usual melodramatic flair, Henry wrote "If I were to unbosom myself to you I could write page after page of personal experiences and, believe me, some of it would be mighty interesting reading; I think that I can safely state without presumption that I have seen more of the deadly effects of that sub-submarine R.U.M. than any patient you ever had there."[87] Even more interesting, however, was Neff's reply: not only did he convey his delight at hearing from Rand, but he also encouraged Rand to tell him of his personal experiences. "I am sure that they will be of assistance to me in my work," he added.[88]

In subsequent correspondence, Henry Rand did "unbosom" himself to Neff. With his customary grandiloquence, Rand confessed that Neff's letter had so moved him that he was "unable for some time to remove the misty veil which covered my eyes as my thoughts wandered back and recalled to me the baseness of some of my past actions." He had "suffered too many relapses to think [Neff] could have any confidence left in me." His drinking, Henry Rand confessed to the superintendent, stemmed from an erroneous (and now corrected) belief that the problems that befell the inebriates he cared for as a nurse could never be his own: "I am of the opinion that if a large gathering of the disciples of Bacchus should convene and if each were to offer an excuse in extenuation of his over indulgence in the mythical pleasures derived from John Barleycorn that I, were I present, would be compelled to present the weakest and lamest plea of any in the assemblage to justify my obedience to the call of the trumpeter." It was the "moral kick" he received from his former employer—who had placed a letter in his civil service file stating that Rand had been dismissed for drinking and that anyone who could not stop drinking altogether would amount to nothing—that had finally permitted the nurse to see himself "as others see me" and foreswear alcohol.[89] Rand's choice of words—always interesting—suggested that he viewed his own condition as a medico-moral one that might benefit from factors such as Neff's moral encouragement and Rand's own desire to prove his condemning boss wrong. At the same time, he referred to his alcoholic episodes as "relapses," a medical term suggesting that inebriety was a chronic condition that required his constant monitoring.

Rand would come under Neff's immediate care one more time. Later that summer, after starting to drink again, he left the Long Island Home for Norfolk State Hospital. Rand was hospitalized for a period of five days for observation before returning to Amityville, where the Long Island Home staff greeted him with open arms. "As I told you when you left here," Neff wrote to his patient, "I feel quite sure that your unexpected and unanticipated relapse will in reality react in your favor, as it will show you that at an unguarded moment your weakness may re-appear without any premonition. The fact that you are living a sober life is sufficient reward to us for any good we may have done you."[90]

The two men continued their correspondence over the next year as Henry recounted tales of asylum life, caring for the Long Island Home's rich and famous clients, penned his predictions for the baseball season, and told of his brother Thomas's wedding. The correspondence between Neff and Henry's younger brother Thomas indicates that Henry fell off the wagon and wound up at Bellevue Hospital in the spring of 1917. He was able to return to hospital work at Amityville following this last recovery. By this time, however, Thomas Rand was the only brother who corresponded regularly with the doctors at Norfolk State Hospital.

Thomas was the youngest of the three brothers, and he had completed one year of high school before taking a position in a local shoe shop. He had been a shoe worker until 1909, when he began working in restaurants, both as a waiter and as a "kitchen man." A nonpracticing Catholic who upon admission "promises to do better," Thomas also claimed that his drinking was periodical—he consumed alcohol heavily for one to four weeks and then abstained for months. He had no craving for drink, except when "with companions and liquor is suggested or taken by others."[91]

Thomas Rand was single and living in Boston at the time of his admission to Foxborough in 1912. He was referred by his uncle, a physician at Massachusetts General Hospital, after being arrested four times for public drunkenness and spending ten days in the city jail at Deer Island. Upon his arrival at the hospital, doctors noted that Rand was "not prepossessing in appearance but conversation denotes some strength of character and shows some true insight. History shows that patient has successfully broken from habit on several occasions, and he believes that he can do so permanently with help. Agrees to cooperate."[92] Indeed, Thomas cooperated more than either of his brothers, remaining in contact with the hospital regularly for five years as his employment took him from a rubber plant near the hospital to the Rhode Island State Hospital for the Insane, to a state hospital for the insane in New Jersey, and finally to the Long Island Home.

Rand's first visit to Foxborough was a long one: almost ten months to the date of his arrival. Released after he had secured a position at a rubber plant in Walpole, Massachusetts, in March 1908, Thomas quickly found another job that was more to his liking. When he returned to the hospital seven weeks later, his employer was the Rhode Island State Hospital, where he served as an attendant. No reason was given for his fall from sobriety in Rhode Island. Instead, he was treated for another six weeks and released to assume a new job at the state hospital for the insane in Morris Plains, New Jersey. According to his outpatient records, he enjoyed his work there. For about a year, Thomas Rand fell out of touch with hospital officials, but he resurfaced in the fall of 1913, working at the McAuley Water Street Mission. By spring 1914, he had become a nurse with his brother Henry at the Long Island Home. Between September 1914 and September 1917, Thomas was constantly in touch with hospital officials, regaling them with stories of his own and his brothers' health, personal lives, and work.

Thomas Rand's letters indicate that he understood his own condition in both moral and medical terms. Writing to Neff in 1915, several years after his second stay at Foxborough, Rand made it clear what a battle he had waged: "One year and eight months ago today I started out to make a fight against my besetting sin, determined that come what may I would stand firm and trust to God to carry me through. . . . When (as I do) [I] sometimes look back at the misery and suffering I endured all through that "fatal cup" it spurs me on. . . . No one knows better than you do, Dr., how hard it is for a fellow to overcome that terrible thirst which had me in its grip."[93]

Irwin Neff's approval was clearly an important part of Rand's recovery. "I was deeply touched by your kind words and the confidence you have in me that I will never go back to the old life," Rand reported, "and I mean never to betray that confidence and prove to you that you made no mistake when you said I was a reconstructed man and would never drink again."[94] In a subsequent letter, he added: "Your letters have been a great help to me and my earnest desire is to show you that I value your interest in me."[95] Heavily steeped in paternalism, the therapeutic relationship between Neff and Rand appeared to work. In this case, it may have helped Rand to entertain the hospital's vision of his condition at the same time that he interpreted his disease as a "besetting sin."

Just as the published recovery tales framed the inebriate's retreat from drink as a manly struggle to regain sobriety, so Thomas Rand interpreted his own situation as the fight of his life, and he explained in a letter to Neff how he was holding fast against his addiction:

> I have made many attempts before to break away from misfortune but never succeeded but this time I decided to make one last stand and come what might I was through drinking rum; when you stop to consider the odds that were against me a stranger in this big city that has no heart and the temtations [sic], discouragements, and hardluck [sic] I was forced to overcome, I am sure you will feel that I done nobly. . . . I am in hopes that my stand will be an inspiration to John Rowan to help him on the way and if I can do anything to help him I will gladly do my best.[96]

Thomas's words tell us just how much responsibility he believed he was taking for his condition and how important it was to him to view his behavior as an aid to his inebriate comrades at Foxborough.

Neff's reply was prompt and affirmative, and he urged Thomas to write to John Rowan, for the superintendent was sure that "it offers him considerable encouragement and will be a help to him in the fight which he is now making."[97] That fall, Rand was laid off from his work at the state hospital in New Jersey and was seeking a new position. In the meantime, he worked at the McAuley Water Street Mission under Samuel Hadley. Neff put Thomas in touch with Bailey Burritt, the assistant secretary of the New York State Charities Aid Association and a New York City Board of Inebriety member, to help Rand find employment before the winter came. At the mission, Rand found that he was able to make new connections on his own, and he eventually secured a night watchman's position to tide him over until spring. At the Water Street Mission, Rand also found "quite a few of the Foxboro boys . . . and they all seemed surprised to see me changed as I am." Rand urged them to do as he had done and to "trust the man from Galilee. . . . It is one year one month and five days ago tonight since I took my last drink and I know if I keep on trusting Him who has kept me day by day, He will keep me to the end."[98] Rand's faith and the support he received from Neff, Carlisle, and the Foxborough State Hospital proved a successful combination.

By April 1915 Rand was working at the posh private Long Island Home—a job he had secured through his brother Henry. He continued to correspond with Neff and, like many of the other "successful" patients, was asked to help the hospital keep tabs on released alcoholics. "I am sorry to say that John Rowan relapsed and left the institution on March 31st," Neff wrote to Thomas: "I should advise you to write to him at his Springfield address. . . .When he answers you I would be very glad to hear from you regarding him."[99] Thomas reported back to the hospital about a number of patients, but one in particular was of great concern: his brother Henry.

"As you are aware my brother suffered a relapse and made a mess of everything, just like he has done many times before," wrote Rand to Irwin Neff in early August 1915. "I advised him to go see you and see if you would be kind enough to let him stay a few days until he braced up again," he continued.[100] Henry had taken his brother's advice, and Thomas wrote to report that his older brother was now back on the job in Amityville; Thomas regarded Neff's willingness to take Henry in as a personal favor for which he was grateful.

Neff returned Thomas's letter promptly and enlisted him further in his brother's care: "I wish you would impress upon him that our weaknesses are more pronounced when we lose sight of our limitations. In other words, they crop up unexpectedly and suddenly. Human nature is the same the world over and if we do not remember what we can do and what we cannot do, we must suffer the consequences."[101] Thomas did as asked, and he reported on his brother in every subsequent letter to Neff. Indeed, a year later, Thomas asked Neff to take his eldest brother Philip in for treatment, and he wrote to Neff regularly to inquire about Philip's health, always grateful for the treatment his family received from the hospital.

Whether Thomas fully embraced the disease concept of alcoholism or not is difficult to say; he certainly described his own drinking as both a sin and a disgrace. What is clear, however, is that the social and emotional support he received through his correspondence with the hospital staff was very important to him. Moreover, this support was extended to other members of his family who were less successful at remaining sober. As was the case with other patients, a strong relationship with the superintendent meant access to both medical care and a job referral network that could prove extremely useful to those on society's margins.

Thomas Rand's case is remarkable in one other regard as well, and here he is typical of other patients: he framed his battle for sobriety in masculine terms as a quest for manhood. "I am what you told me I could be and what I have tried my best the past twenty-two months to become, and that is 'a man.'"[102] He became "a man" by remaining sober, responsible, and fiscally independent; by continuing his association with the hospital physicians and staff; and finally, by getting married. In February 1916 Rand shared with Neff the news that "I am to become a benedict the first of next month. . . . I am getting for a life's partner a charming little girl here I know will cheer and brighten my sometimes rather lonesome existence and I know we will both be happy."[103] Marriage was a fitting end to Rand's quest for manhood, and one that matched well the published recovery narratives of his day. Neff promptly responded, sharing his congratulations with Thomas. He later enlisted Mrs. Rand in her husband's cause, asking her to be of assistance in reminding her husband to write to the hospital on a regular basis.[104]

## Frank Casey, Farmer

Frank Casey, a farmer from rural Berkshire County, some hundred and fifty miles northwest of Boston, was first admitted to Foxborough in July 1897, when the institution was still known as the Massachusetts Hospital for Dipsomaniacs and Inebriates. Casey was twenty-seven years old, Protestant, and had not completed a common school education. Committed by his family through the district court of Pittsfield, Massachusetts, Frank was "subject to epileptic fits after several days of intoxication. Attacks are severe enough to cause loss of memory. At times, when intoxicated, has threatened to commit suicide."[105] After three months of care and an authorized visit back home to see his family, Casey "eloped." He returned to the hospital fifteen years later in 1912, still suffering from epileptic seizures that were worsened by his alcoholism.

Casey's second commitment to the state hospital for inebriates did not come about in haste. By 1912 he had taken the Keeley Cure in Springfield, Massachusetts once; been sent twice to the Brattleboro Sanitarium in Vermont; and spent two months in New York City at the Oppenheimer Institute, another proprietary liquor cure. His family had been managing his case with diligence and with the assistance of his physician, William Tucker. Several years earlier, in 1908, Casey's sister Harriet had written to Irwin Neff suggesting that her brother would soon be under his care but asking the superintendent to send along an annual report of Foxborough and his recommendations to the most recent legislative session. Harriet hoped that she might be of assistance to the hospital, "feeling very strongly the necessity of some provision for the treatment of such cases as disease, mental, physical or both."[106] She suggested that Neff send the requested materials to her and to William Tucker, as he would be quite interested in learning more about Foxborough. Neff complied, but Frank Casey never arrived at the hospital.

A year later Harriet penned another letter to Irwin Neff that indicated why Casey's relatives might have hesitated to commit him to the state hospital for inebriates: "I find among some leading lawyers here great doubt as to the value of your hospital. I understand that some years ago it was simply a house of detention, carelessly managed, but that there has been an entire change and it is now in charge of men qualified and capable of treating alcoholism as a disease, and cases sent there are so treated."[107]

In managing her brother's case, Harriet had become something of an inebriety expert herself, and she asked Neff to furnish details of the course of treatment pursued at Foxborough, the medicines used, the occupations provided patients, and "the general influence and *morale* of the establishment." Harriet was interested in the use of suggestion and hypnotism in cases of alcoholism, methods that had received great acclaim within Boston's Emmanuel movement. Reiterating that "men of high character and attainment in medical circles should be interested in the treatment and care and cure, if possible, of the *disease* of inebriety," Harriet Casey requested a speedy reply.[108] Once again, Neff responded, sending additional information on Foxborough and its methods of care. The Casey family, however, waited another three years before committing Frank to the state hospital for inebriates.

Casey's case was a particularly complicated one. His seizure problem was chronic and was exacerbated by his tendency to drink. Between 1912 and 1916, Irwin Neff corresponded regularly with various members of the Casey family, including Frank himself, but there were other parties involved too: Casey; his sisters Harriet and Clara; and his wife, Georgia; his physician, William Tucker; and representatives of the district court. The superintendents of the Monson State Hospital for Epileptics, the Northampton State Hospital for the Insane, and the state hospital for inebriates all had a say in Frank Casey's treatment too. These parties each emphasized a different aspect of Casey's health, attempting (it seemed) to pass the farmer on from one institution to the next, preferably not their own. Several questions were on the table.

First, were the seizures caused by the alcoholism, or was the alcoholism an effort on Casey's part to medicate himself? Second, was Frank mentally competent or mentally unsound? Third, had the bromides prescribed for his epilepsy weakened his physical and mental powers to the point of incapacity, or had his alcoholism progressed beyond the possibility of cure? And fourth, was Casey most appropriately regarded as an alcoholic, an epileptic, or a homicidal maniac (due to his behavior toward his wife)? Frank Casey's story illustrates the active and important roles that families and courts could play in determining the inebriate's fate. Indeed, far from portraying the state hospital for inebriates as an all-powerful institution, the historical record reveals the failure of any one individual or institution to decide the inebriate's fate.

Within hours of committing her husband to the inebriate hospital in September 1912, Georgia Casey sent a trunk of clothes to Frank and enclosed a letter for Irwin Neff to familiarize him with some of Casey's idiosyncrasies. Frank Casey had bowel and heart trouble, which she feared had become chronic, and he "seem[ed] to think he must take medicine no matter how simple, and seems better for taking it."[109] Used to taking care of her husband, Mrs. Casey wished Neff to keep her updated on Frank's condition. In response, Neff wrote that it was too early to tell "anything definite about his condition," but he believed that Casey was "a confirmed case of alcoholism" and he had no doubt Casey's physical symptoms were related to this.[110] Initially, then, Neff saw Frank Casey as an inebriate first and an epileptic second.

Days later, Casey's sister Clara provided "additional insights" on the origins of Frank's alcohol problems. Assuring Neff that Frank Casey was of a strong old New England family "with no inherited tendencies toward dipsomania," Clara suggested that "the cause might be found in his large head—which was quite abnormal in childhood." Clara Casey was a graduate nurse, who confessed that she had always been interested in institutional life. Her recent work as a visiting nurse had convinced her that "inebriety should be cared for by the state as well as tubercular troubles."[111] Each of the members of Casey's family helped manage his case, and in so doing expressed their support for the hospital, for treatment founded on the disease concept, and for the indefinite commitment of Frank Casey to some medical institution.

After seven weeks of care, during which Irwin Neff revised his diagnosis of Casey's trouble as epilepsy complicated by alcoholism, Irwin Neff discharged Frank from Foxborough. Neff could not legally keep the Berkshire farmer, for he was not a "true inebriate." Responding to the pleas of Casey's family, however, he kept Frank a full month longer than he initially had wished. The Caseys' physician, William Tucker, had arranged for Frank to go to the State Hospital for Epileptics at Monson, but his family insisted on the day of his release that he be sent to the private Attleboro Sanitarium. Casey arrived at Attleboro and left the next day, bound for home by way of Boston and a meeting with Irwin Neff in the outpatient offices there. Neff concluded that home was the best place for the man and encouraged him to write to him regularly.

Postcards and letters from Frank Casey to Irwin Neff and Frank Carlisle suggest that Casey had bonded with the two men and appreciated what they had done for him. "You seem to take a special interest in my case for which I'm truly thankful. Am doing alright as far as liquors are concerned and *think* I can stick it out, and am much better in the way of nerves and dizzy headaches, but am far from well." Casey felt he had no strength and wondered if the bromides he had been prescribed by Frank Carlisle might be the cause. Could he try another medication, "something that would help my strength and ambition"? he asked. He added that he had seen Neff quoted in the *Boston Herald* on the disease of dipsomania and was very impressed.[112]

Neff responded, saying that bromides could easily be the cause of Casey's problems, but that Frank would have to see his local physician to verify this. By July, however, Casey's health had deteriorated; he was drinking again, and his wife wrote to Neff requesting that he take her husband in for a lengthy stay: "The combination of epilepsy and alcoholism seem [sic] to be a bad one. The fact that he has been sick since babyhood & taken a great amount of drugs don't [sic] seem to have helped the general mental condition. . . . Could he [Frank] return to Foxboro for say six weeks? Of course, we can have him taken care of by asking the court, but it is so hard and unpleasant. Am awfully sorry to trouble you, but we are so dependent having no man in the family and not having good health ourself [sic].[113]

At Georgia's request, Neff wrote to Frank to inquire about his health. He learned from the farmer that he had been taking "some stimulant as I found it necessary to keep a going. Now that the rush [haying season] is over I think I can get along on 'total abstinence' as I find it is far the best for me."[114] In response, Neff warned Casey that alcohol would worsen his epileptic symptoms, but he did not recommend that his former patient return to the hospital.

By September matters had gone from bad to worse, however. Lightning had struck the Caseys' big barn, razing it to the ground. Frank was distraught and found solace in drinking, his wife reported. Again, Neff recommended a private, moderately priced sanitarium, but the Caseys wanted their loved one to come to the state hospital for inebriates. Georgia wrote a letter for her husband that told Neff he needed to come to Foxborough immediately; Frank signed the letter and added in his own script, "I want to come at once."[115] Consuming a quart of whiskey a day as well as Jamaica ginger, Casey appeared in late October at the state hospital. He left within a day, homesick and convinced that he could do as well back at home. Apparently, Casey sobered up and remained dry for the entire winter, but in the spring his old problems returned.

On 29 July, Harriet Casey, Frank's sister, took matters into her own hands. After thanking Neff copiously for the attention he had given her brother and reiterating how much good it had done, she reported: "Now the trouble is on again, very, very bad, and conditions getting unendurable for the wife and child." Harriet petitioned Neff to keep her brother for a couple of months, "as long as you can, if we can get Judge Hibbard to commit him to your care again. I know you don't want him, that it is a great favor that we are asking, but in the past it has done so much good. And for my mother's sake, who is over eighty, we can't bear to have him sent to the state farm. He is the last of what has been a prominent family in the community."[116] Harriet had pulled out all the stops—terrorized wife and child, eighty-year-old mother, and her family's high social (if not economic) standing. She ended her letter by chastising the state for not offering more support to the new facility at Norfolk. When Neff offered in response only to write to Frank and not to admit him, the Casey family eventually turned to district court judge Charles E. Hibbard, who committed Frank Casey to the new hospital facility at Norfolk. Casey spent three weeks at Norfolk until he became "antagonistic to persuasion on our part."[117] Neff recommended to a distraught Georgia Casey that she commit her husband to the Monson State Hospital for Epileptics.

Monson superintendent Everett Flood, however, discouraged Georgia Casey from sending her husband to his institution: "He says that is no place for a patient like Frank," reported Georgia in a subsequent letter to Neff. Likewise, the private Brattleboro Sanitarium that had treated Casey three times by 1915 refused to take the farmer unless he were accompanied by a private nurse. Frustrated beyond measure, Georgia Casey wrote Neff: "Now the question is: Is there any place in the state, where he can be put and cared for?" Instead of her usual glowing remarks about Neff's care and concern for her husband, the strained Mrs. Casey commented: "As it is, except for the rest it has given the family to know where Frank was and that he was well cared for, I can see no benefit from his stay at the hospital."[118]

Mrs. Casey's comments reveal the social utility of the state hospital for inebriates in desperate cases where the breadwinner was incapable of working and yet very capable of terrorizing his family. Foxborough, and later Norfolk, gave beleaguered families a respite from the chaos, violence, and fear that frequently accompanied life with an alcoholic. In a sense, whether or not the family believed in the disease concept was immaterial. What the Casey family knew was that Frank needed help; they needed, as Georgia put it, "a rest"; and Irwin Neff and his staff treated patients with respect. This was more than could be said for many an overcrowded hospital for the insane or the quasi-penal state farm. The state hospital for inebriates appealed as well because it was cheaper than private facilities. After the hospital's reorganization in 1908, the Casey family and others could commit their loved ones to Foxborough with a clear conscience, knowing that their relatives would be treated humanely.

Ultimately, Georgia Casey attempted to circumvent Neff and have the outpatient department refer her husband to Norfolk, telling Thomas Foss, the outpatient department director, that "I feel that he can be cured but I believe it will take time." Mrs. Casey's plans hit a snag, however, when Foss showed Neff her letter. The superintendent recommended that she commit her husband to Northampton State Hospital for the Insane. Although it took them five months, the Caseys did commit Frank to this institution in early May 1916. Neff wrote a summary of Frank Casey's history for the Northampton superintendent and emphasized that his epilepsy really made the Monson State Hospital for Epileptics the most appropriate treatment venue. Whether or not Casey made it to Monson remains unclear.

When Judge Hibbard thanked Irwin Neff for his summary of Casey, he agreed that the farmer belonged at the epileptic hospital. Georgia Casey's last note to Neff begged him to recommend to the superintendent at Northampton that her husband be allowed to remain beyond the thirty-day observation period. "It is only toward me that he has shown this homicidal mania and there comes such a peculiar light in his eyes that it almost paralyzes one. I have good reasons to fear him."[119] In his reply, Neff assured her that Superintendent Houston at Northampton would likely decide in her favor upon reading his summary of the case.

## Patients and Medicalization

When I began this study of the history of alcoholism in America, I wondered if thinking about habitual drunkenness as a disease had made any difference in the lives of people affected by drinking problems. I am convinced now that it did, but in a variety of ways that I had not imagined. Both the published recovery tales and the histories of patients at Foxborough and Norfolk make it clear that physicians could improve the lives of their alcoholic patients substantially when patients were willing allies in the fight for sobriety. Ironically, patients did not have to believe in the disease concept of alcoholism to reap the rewards of hospital care. Equally clear is the fact that patients who did believe in the disease concept of inebriety did not necessarily abandon their moral interpretations of their condition, and it is impossible to say in these cases which intellectual framework was more helpful.

The disease framework in which physicians treated alcoholism was useful for some inebriates, whether these men thought of their condition as a symptom of an underlying constitutional defect, an expression of a neurosis, or a disease sui generis. Why was this the case? According to some patients who published their tales of recovery, to think of their battle with alcohol in disease terms was liberating. Only when the writer in "Confessions of a Dipsomaniac" realized that he was suffering from a mental disease was he able to break away from his overbearing secondary personality. Likewise, through psychoanalysis the neurotic lawyer in "Is This Why You Drink?" was able to come to terms with the unresolved emotional conflicts that had steered him into alcoholism, finally freeing himself from drink. In Massachusetts, Nicholas Felson was so impressed with the medical treatment that he received at the state hospital for inebriates that he promoted the institution in his own unpublished recovery story. For Felson, his talks with the Foxborough physicians and the outdoor work he performed while a patient had made all the difference—he was able to escape his troubles, rebuild his health working in the outdoors, and learn to think about his situation in a way that made "big troubles . . . very much smaller."

Another segment of the recovered inebriate population who took medical treatment were not so impressed with the disease framework and their newfound status as sick persons; these patients ultimately found the social support of their physician more helpful than any particular medical regimen. Time and again, inebriates at the state hospital insisted that the friendship of Irwin Neff and his fellow Foxborough physicians helped them "live up to" their commitment to sobriety. If we are to believe their writings, patients were deeply affected by the fact that physicians took an active and enduring interest in their cases. The same holds true for the "slave of an invisible command," who confided that both the love of a good woman and the friendship of a trained physician were essential elements of his own cure. At Foxborough, this "friendship" extended beyond medical treatment and emotional support: it included small personal loans, job training and placement, continued access to medical care, and negotiations with family members about the management of their inebriate relatives.

Thus, one could say that the disease concept had both direct and indirect effects on habitual drunkards. In the first case, the disease framework itself affected the way individuals thought about their situations. In the second case, the institutional treatment and social interventions that were based on the disease concept had a positive effect even if they were not medical per se. Medicalization, then, is a process in which medical ideas may provide the rationale for adopting a new therapeutic approach to a social or moral problem. This approach may include altering social attitudes and medical and judicial policies—even the creation of brand-new institutions for the treatment. Yet even when such changes have taken place, the benefits reaped by the affected parties—patients and their loved ones—may have little to do with anything unique to the medical profession.

The Rand family is a case in point. Following their release, both Thomas and Henry Rand retained a positive image of the state hospital for inebriates in Massachusetts. However, neither retained a vision of their condition as an exclusively medical one. Instead, the two brothers more frequently referred to their "besetting sin" of drink, the moral transgressions that they had made while under the influence, and the fight they bravely waged (with the help of their Christian faith) against alcohol. Both wrote to Irwin Neff and used the relationship they enjoyed with the superintendent as a means of staying sober. When weak or sick, they knew that they could return to the hospital for additional sympathy, encouragement, job placement services—anything to prevent a relapse. Was the hospital a useful institution for these men? Undoubtedly, but not because it changed the way these men thought about alcoholism or delivered a "magic bullet" for the condition. Instead, it changed the way Thomas and Henry thought and went about acquiring assistance in their struggle for sober lives.

The cooperative attitude of the Rand family, Charles Winchester, Nicholas Felson, and John Godson reminds us that medicalization proceeds, not just through the efforts of physicians, but through the work of patients and their families, and through the kindness and aid that they extended to their fellow reforming drunkards. The Rands, Winchester, Felson, and Godson all played a major role in keeping the hospital in touch with other patients and promoting the services of the institution. Indeed, one could argue that the hospital's social services were expanded by the offers of employment and comradeship that successful patients extended to one another through Irwin Neff and Frank Carlisle. The same, of course, may be said for the patients who published articles about their successful fight against alcohol; they extended a helping hand to inebriate readers.

Because alcoholism's disease status remained contested at the turn of the century, patients diagnosed and treated in medical institutions for inebriates played an influential role in swaying public opinion. Not only could they educate the public through their own autobiographical exposés, but their participation in the treatment programs affected the hospital staff's ability to perform their own jobs. At Foxborough and Norfolk, cooperative patients stood a better chance of securing the long-term support of the hospital staff—something that rarely hurt their chances of maintaining long-term sobriety. As these institutions struggled to prove their worth to the state and to validate the disease concept on which care was founded, higher rates of "material improvement" and abstinence were valuable statistics. Cooperative discharged patients were good ambassadors for the hospital—and by implication, the disease concept—especially when they were able to stay sober. As former patients made connections to other social service agencies, returned to the workforce, and reunited with their families, they offered living proof that hospital treatment for habitual drunkenness was a social good worth pursuing.

Of course, even when members of the public were convinced that medical care for the inebriate was a worthy endeavor, they could, like Georgia Casey, insist on institutionalizing individuals that physicians did not want to treat. In the end, the most important matter for Mrs. Casey was not the validity of the disease concept, but finding an institution that would treat her husband humanely and free her and her family from the terror of living with a violent alcoholic. Public support for the inebriate hospital came at a price: the custodial care of late-stage alcoholics with little hope of recovery. The medicalization of habitual drunkenness has always been a bargaining process.

In 1945, a quarter of a century after the Norfolk State Hospital had closed its doors and a dozen years after the repeal of national Prohibition, the Special Commission to Investigate the Problem of Drunkenness in Massachusetts reported: "It may seem ironical that at the conclusion of this study our thoughts and recommendations finally have become crystallized along the same general trends which repeatedly have been called to the attention of the General Court during the past thirty-five years through the media of special messages or special reports. . . . We see no other alternative to curb drunkenness in its broadest aspects on a state-wide basis, except those measures that the State has tried and failed in."[1]

Having concluded a two-year study of the habitual drunkenness problem in the Bay State, the commission observed that inebriety needed to be addressed as a "medico-social-penological problem."[2] The Special Commission dismissed the previous system-building and treatment efforts of the state hospital for inebriates as "failed" but recommended that they be tried again, for the holistic sociomedical approach Norfolk had put into practice appeared to offer the most promise for reforming the Commonwealth's alcoholic population. The recommendation was eerily similar to that of the governor's executive council, which had advised a wholesale reorganization of the Foxborough State Hospital in 1907 but had insisted that an institution like it should be continued. Over the ensuing years, Foxborough and Norfolk had succeeded in many ways, but Norfolk's last years were plagued by dwindling census figures and rising per capita expenditures that hurt its image as a useful state institution.

Ironically, as the Special Commission acknowledged, their views *were* similar to those voiced in the annual reports from Foxborough and Norfolk and the legislature's investigative reports of 1909 and 1914. Under the leadership of Judge Joseph T. Zottoli (the report was known informally as the "Zottoli Report"), the commission had divided the inebriates of Massachusetts into five categories: (1) *acute alcoholics,* or occasional drunks; (2) *early and moderate alcoholics,* or those who had not been confirmed alcoholics for more than five years; (3) *chronic deteriorating alcoholics,* or those who had been confirmed inebriates for many years; (4) *alcoholics of various classes* who have physical but not mental disabilities; and (5) *alcoholics of various classes* who, regardless of their physical state, are mentally ill.[3]

This classification system emphasized that persons with drinking problems came in many varieties, each of which required a distinctive brand of care aimed at physical, mental, and social rehabilitation. Indeed, this was a central tenet of the modern alcoholism movement, reiterated by Yale physiologists E. M. Jellinek and Howard Haggard in their 1942 *Alcohol Explored:* "In dealing with this problem [alcoholism], no order can be expected until it is generally recognized that it is not one of inebriety but of inebrieties. This distinction is not only of theoretical significance but of practical significance in the treatment and prevention of inebriety. The prevalent failure to make distinctions among the categories of inebrieties has obscured many phases of the research in this field, and, in tracing these researches, we shall point out this fact again and again even at the peril of repetition."[4] Haggard and Jellinek recognized that their concept of alcoholism as a pluralistic disease was out of step with the unitary and reductionistic notion of disease that had come to dominate modern medicine; it did not resemble the reigning "ontological" model that assumed that a disease and its mechanism were the same in each afflicted person.[5]

A pluralistic understanding of alcoholism demanded a broad therapeutic approach. Thus, the Special Commission—composed of businessmen, lawyers, educators, a physician, and a judge—advised a range of treatments to address each variety of inebriety. State hospitals for the mentally ill were to care for the last group identified in the Zottoli Report; group four was to receive care along with other patients suffering from a variety of chronic diseases at the Tewksbury State Hospital until the state could fund alcoholism clinics and wards with their own staffs inside local general hospitals. The commission also advised that chronic deteriorating alcoholics be "sentenced" to long stays at the state farm, where they would be treated, if necessary, and made as self-sufficient as possible through "farming and other occupational outlets to minimize their inevitable deterioration."[6] The acute alcoholics (occasional drunks) were to receive care in clinics that would be established in conjunction with Alcoholics Anonymous, medical schools, and local and state hospitals throughout the Commonwealth. Finally, for the early and moderate alcoholics—referred to as "the salvageable ones"—the commission recommended the construction of a new state inebriate hospital, "preferably in the country, with outlets for industrial and occupational therapy, under the care of a staff of experts in the medical and ancillary fields, and with adequate laboratory and other research facilities. We envision such selected patients going to this future hospital only under proper medical commitments through the courts, to be released only when, in the opinion of the medical authorities, it is proper for them to be released."[7] The authors of the Zottoli Report added that they had in mind the Pondville State Hospital (for cancer patients), should it be abandoned, because it had originally been built for this purpose. The Norfolk State Hospital, it turns out, had been renamed the Pondville State Hospital when it ceased to be a facility for inebriates and began its life as a cancer hospital.

History does not repeat itself, but at times it appears to come awfully close: once again, inebriates were divided into categories, depending on the nature and duration of their drinking problems; and once again, a group of professionals outside and within the medical community suggested various types of institutional treatment and probation work aimed at treating the inebriate's physical, mental, and social symptoms. Indeed, the sociomedical approach was being advocated by an expert committee with a judge, not a physician, at the helm—testimony to the powerful influence of the lay community in furthering the re-medicalization process.

Prohibition had extinguished America's collective memory of the early movement to medicalize alcoholism. Even before the ratification of the Eighteenth Amendment, states made plans to close their specialized facilities for inebriates (whose patient or inmate censuses were declining from 1916 onward), and most private sanitariums treating alcoholics either shut their doors or expanded their diagnostic catchment area to treat other nervous disabilities. With Prohibition, the alcoholic was again viewed as a moral weakling whose only options were the jail cell or, if mentally impaired, the psychiatric hospital. The knowledge base of inebriety experts was likewise discarded, dismissed, or forgotten by most of the researchers who would investigate alcoholism following Repeal. Chauncey Leake was one of the few post-Repeal alcoholism experts to recognize this, observing in 1957 that even though some of the "best American physiologists, pharmacologists, and clinicians"—men such as John Shaw Billings, Russell Chittenden, and William Welch—had scientifically studied alcohol's actions on body and mind and published their two-volume report, *Physiological Aspects of the Liquor Problem,* for a general audience, it was "astonishing how little appreciation has been given to this important work, and it is amazing how little it is known. Indeed, the general history of alcoholism has hardly been explored. Much might still be learned from the early clinical accounts."[8] In their haste to distance their efforts to combat alcoholism from the wet-dry politics of the previous generation, most alcohol researchers in the 1940s and 1950s dismissed the research findings of their predecessors. Unfortunately, this meant that alcohol and alcoholism specialists of the mid-twentieth century spent a great deal of time reinventing the proverbial wheel.

## The Modern Alcoholism Movement, 1933–70

### *An Essential Tension and an Elusive Quest*

The modern alcoholism movement rekindled public interest in treating alcoholic persons as diseased individuals. As historians of alcohol have observed, between 1933 and 1960 the alcoholic's identity was transformed from "a morally deformed perpetrator of harm to a sick person worthy of sympathy."[9] As was true with the early effort to medicalize habitual drunkenness, the number and variety of parties engaged in promoting a disease concept of inebriety and institutional treatment for drinking problems was impressive. Lay groups such as Alcoholics Anonymous; research groups such as the Research Council on Problems of Alcohol; and scientific groups such as the Yale Center of Alcohol Studies and the New York–based Public Affairs Committee, Inc.—

groups dedicated to the dissemination of scientific literature for larger audiences—played active roles in the modern alcoholism movement. They were joined by municipal and state governments, and finally, in 1970, by the federal government, when Congress passed the Comprehensive Alcohol Abuse and Alcoholism Prevention, Treatment and Rehabilitation Act (the "Hughes Act") endorsing the medical approach to alcohol problems, if not the disease concept itself. This piece of legislation established the National Institute for Alcohol Abuse and Alcoholism (NIAAA), the first federal governmental agency devoted to the eradication of problem drinking. It also signaled the high tide of support for the disease concept of alcoholism. Clearly, the process of "remedicalization" was (again) not a top-down hegemonic process spearheaded by the medical community.

Like the early alcoholism movement of the Gilded Age and the Progressive Era, the modern alcoholism movement was characterized, first, by an essential tension between medical and social (often moral) conceptions of inebriety, and second, by an elusive quest for a mechanism to explain the origins of alcoholism in each of its victims. The two were related, of course, for it was the hope of many researchers in both the early and the modern movements that discovering the holy grail of alcoholism's pathogenic mechanism would give the medical explanation of the disease the upper hand and allow for neat and tidy interventions in a complex medical and social condition. But until this day came, early and modern alcoholism specialists believed a holistic approach that recognized both the medical and the social dimensions of the disease was essential to successful treatment.

In the 1870s, Joseph Parrish had argued that one's bad drinking habits might slowly lapse into the disease of inebriety, a condition over which one had little control. Parrish also believed that a conception of inebriety as a medical condition did not exclude the afflicted drinker from moral responsibility for seeking help or for scrupulously avoiding alcohol. In the same vein, the leaders of the modern alcoholism movement recognized both the medical and the moral sides of the condition. Yale's Haggard and Jellinek regarded the modern tendency "of defining the alcohol problem as a medical one . . . [as] an essentially correct approach." Yet, they added, "when we understand the part played by the behavior of the intoxicated person in giving the alcohol problem its rank of capital importance among social issues, it becomes clear that a moral issue is also involved. It is desirable to lay stress on the medical nature of the alcohol problem and to disseminate the recognition of this in the widest possible circles, but it would be a mistake to disregard the moral aspect, for this furnishes a greater incentive for the co-operation by the citizen which is essential to the solution of the problem of alcohol."[10]

Researchers such as Haggard and Jellinek recognized that morality still played an important role in framing alcoholism in the 1940s, and they hoped that acknowledging this role would assist them in garnering support for the disease concept. Moreover, by 1960 anthropologists' comparative work on alcoholism in different cultures and sociologists' studies of distinctive class and ethnic drinking patterns in America had convinced Jellinek that "social and economic factors not only greatly influence the drinking patterns and the magnitude of the alcohol problems, but also leave their stamp on the process of alcoholism and even on some aspects of its clinical picture."[11]

The problem for Jellinek and the modern alcoholism movement was much as it had been for the early effort to medicalize intemperance—to preserve a holistic perspective on alcoholism as a sociomedical affliction, while promoting the disease concept and medical treatment for alcoholism. Jellinek was well aware of this tension, observing that "social and economic factors," while important, did not "change physiology as one sociologist or sociologically oriented psychologist would make it appear in a hypothesis which is nearer to Yoga than to sociology."[12] Thus, alcoholism experts acknowledged the importance of determining alcoholism's etiological mechanism(s), at the same time observing that much might be done to prevent and treat the condition without knowing exactly why certain people developed it. Indeed, implied in Jellinek's insistence on the existence of "inebrieties," rather than on a unified disease concept of inebriety, was a criticism of the one disease/one mechanism construct that had come to dominate modern medicine. What mattered more, he argued, was physicians' opinion of the condition. *"A disease is what the medical profession recognizes as such,"* Jellinek asserted in 1960. "The fact that they are not able to explain the nature of a condition does not constitute proof that it is not an illness. There are many instances in the history of medicine of diseases whose nature was unknown for many years. The nature of some is still unknown, but they are nevertheless unquestionably medical problems."[13]

Like the earlier movement, the fate of the mid-twentieth-century alcoholism movement hinged not only on the image of the disease of alcoholism but also on the related image of alcohol itself. Here social and political context drove the medicalization effort. The "failure" of the early alcoholism movement could be traced in part to the inability of reformers to control the governing image of alcohol. In the years leading up to the nation's great "Dry Experiment," temperance and prohibition advocates managed to control the public's impression of ethanol, portraying it as a universally corrupting poison that held only negative consequences for its users. Although the saloon was the focus for much of the anti-wet sentiment, alcohol itself was demonized, and so too its users. The public's image of the habitual drunkard was thus tainted by his intimate association with this "sinful" substance. And physicians' and reformers' embrace of a holistic model of the disease of alcoholism, which encompassed the moral as well as the medical sides of the condition, did little to change this powerful impression.

Much had changed post-Repeal, however, and the acceptance of alcohol as a normal part of American life meant that the individual drinker, not alcohol itself, was the determining factor in alcoholism. When advocates for the disease concept pressed forward their campaign in the mid-twentieth century, the politics of the issue were significantly different than they had been in 1920. Alcohol was regarded in a more neutral light; science commanded greater public respect; and alcohol researchers and treatment experts enthusiastically joined forces with lay groups to promote their scientific ideas.

## *Many Parties, One Goal: The Rise of the Disease Concept*

The drive to promote the disease concept of alcoholism arose from several sources, much as it had in the late nineteenth and early twentieth centuries. Among the most important advocates was Alcoholics Anonymous. In 1935 William "Bill" Wilson, a stockbroker, and Robert "Dr. Bob" Smith, M.D., formed this legendary modern mutual aid association for alcoholics. An idea that had impressed Wilson during a recent stay at the Charles B. Towns Hospital, one of the few inebriate facilities that had remained open during Prohibition, was the concept of alcoholism as an allergy—a concept that a hospital physician, William Silkworth, had shared with him. Although no allergy to alcohol was or ever has been recognized within science, Wilson found the idea useful in assuaging feelings of guilt about his condition and giving him a rationale to avoid drinking. Several other prominent psychiatrists and alcoholism specialists also used this metaphor in writing for a popular audience.[14] The term "disease" was not formally part of the early AA literature, which emphasized spiritual and social treatment for alcohol problems, but support for the disease concept grew informally within the AA and Al-Anon memberships in the 1940s and 1950s.[15]

More specific advocacy for the disease of alcoholism originated within the Research Council on Problems of Alcohol (RCPA), founded in 1937. Originally the RCPA was a scientific organization devoted to the study of alcohol problems; its scientific orientation was shaped by post-Prohibition politics and the economic climate of the Great Depression. Funding resources were scarce in Depression-era America, so in 1939 the RCPA turned to the alcohol beverage industry for financial support. Still smarting from the lean years of Prohibition, the alcohol beverage industry did not wish to fund research that highlighted the dangers of alcohol and risk refueling the wet-dry debate. The compromise they struck with the RCPA in 1939 was to support research that focused on "alcoholism," a condition that affected a minority of drinkers and consequently downplayed the image of alcohol as a universal poison. Depoliticizing alcohol problems was a priority for the liquor industry, and the alcoholism orientation facilitated this approach by emphasizing the alcoholic's special susceptibility to alcohol, a susceptibility not shared by the majority of the population.

Over time, the research mission of the RCPA was eclipsed by its focus on public relations, or public health awareness of alcoholism. Under the guidance of Dwight Anderson, a public relations consultant, the RCPA developed a plan for the treatment of alcoholism that Anderson hoped would be funded by taxes on both retail alcohol sales and the beverage industry. The plan included the establishment of centers for information on alcoholism; the creation of clinics and the use of general hospitals, university-linked hospitals, and state psychiatric facilities for alcoholism treatment; and the movement of social authority for the alcoholic from the penal system to the public health system. The 1945 Zottoli Report clearly bore the stamp of the RCPA as well as a sense of the history that had preceded Prohibition. This agenda for the reform of the alcoholic drew inspiration in the 1940s and 1950s from the Yale Plan Clinics, model clinics established by researchers and treatment specialists at the Yale Center of Alcohol Studies, which was founded in 1943. The clinics were grounded in a psychodynamic interpretation of alcoholism and offered a new format for outpatient care: they emphasized group therapy for alcoholics, promoted the disease concept of alcoholism, employed lay counselors, and cooperated fully with lay groups such as AA.

As seen in the work of Haggard and Jellinek, the Yale Center took a multifaceted approach to the study of alcoholism, but it also sought ways to gather, synthesize, and disseminate vast quantities of literature on the condition. Indeed, it worked in tandem with organizations such as the RCPA and the Public Affairs Committee, Inc. to promote the message that alcoholism was a disease. Pamphlets such as *Alcoholism Is a Sickness*, printed in 1946, explicitly referenced the Yale Plan clinics and emphasized that alcoholics were a diverse group of people who shared two characteristics: they could not stop drinking, and their consumption of alcohol prevented them from leading normal, healthy lives. "Until recently," continued the pamphlet's author Herbert Yahraes, "alcoholics were people to be reproached, or perhaps to be made fun of. Their troubles were laid to a weak will, or bad morals, or perversity. They were referred to as 'soaks' and 'sots'—and even as sinners who had deliberately chosen to go to hell and were lighting up the way with a bottle. Enlightened doctors now know, however, that an alcoholic is a sick person, as truly sick as a person who has diabetes or tuberculosis. And they now know that many, if not most, of them can be rehabilitated."[16] Thus, medical authority was invoked in the problem of alcoholism as it had been prior to Prohibition.

By 1946, in the post-antibiotic age, the cultural authority of medicine was significantly greater than it had been two generations earlier. As was the case with the early alcoholism movement, however, disease concept advocates regarded more than just medical authority as important in the alcoholic's recovery. As Johns Hopkins psychiatrist Robert V. Seliger wrote in his 1945 book, *Alcoholics Are Sick People*, "physicians, psychiatrists, psychotherapists, nurses, social workers, clergymen, educators, patients, and relatives" all had a role to play in the rehabilitation process.[17]

The multifaceted approach of the Yale Plan clinics set the template for reform during the modern alcoholism movement's ascent in the 1940s and 1950s. However, there was another way in which the Yale group figured importantly: E. M. Jellinek's development of a multiple "species" model of alcoholism, based on a wide review of the alcoholism literature and a study of the drinking trajectories of a group of AA members.[18] Certainly, as with the Yale clinics, the institutional locus of an Ivy League university lent respectability to the group's scientific assertions. Three central features of Jellinek's classic disease model articulated in *The Disease Concept of Alcoholism* were *tolerance*, or the need for increasing doses to produce the same effect over time; *physical dependence*, or the occurrence of withdrawal symptoms once alcohol consumption is stopped; and *loss of control*, or the inability to cease drinking. Although Jellinek's model of alcoholism came under attack, his work provided disease concept advocates with a scientific base for their educational and treatment directives.

This was especially true for a new organization that began as part of the Yale Center of Alcohol Studies, the National Committee for Education on Alcoholism (NCEA). The NCEA was founded in 1944 by a public relations mastermind and reformed alcoholic, a woman named Marty Mann. Under Mann's direction, the NCEA became independent of Yale in 1949 and continued its mission to educate the general public about alcoholism as a disease. Marty Mann proved an energetic and effective advocate for alcoholism and served thirty-five years as NCEA director, carrying the message that alcoholism was a disease throughout the United States and abroad. This marriage of scientific research and public relations outreach resembled in some respects the work of the WCTU and their campaign for scientific temperance education. However, this time the "propaganda machine" was promoting the disease concept rather than abstinence and advancing the reform of the individual problem drinker through institutional care rather than through the nationwide prohibition of alcohol.

As William White has observed, there were other factors that contributed to the rise of the modern alcoholism paradigm. Public health leaders such as Lawrence Kolb, the head of addiction research at the United States Public Health Service between World War I and World War II, strongly advocated the disease concept of alcoholism and the replacement of the term "drunkenness" with "alcoholism," even if he believed, with most mid-twentieth-century psychiatrists, that addiction was a symptom of underlying psychopathy. Indeed, Kolb encouraged the construction of public hospitals to treat people with alcohol problems. There also arose within the mainstream medical community a growing interest in addiction medicine. The formation of the New York City Medical Society on Alcoholism in 1954, which eventually evolved into what is known today as the American Society of Addiction Medicine, signaled this development.

Likewise, in the 1950s, important medical groups such as the World Health Organization, the American Medical Association, and the American Hospital Association all issued statements about alcoholism's status as a disease or condition that was best served by general hospital treatment (an important assertion, for many hospitals refused to admit alcoholics). The nature of institutional treatment was changing. "Isolation and segregation of patients into specialized hospitals, except for contagious disease, should be associated with the past and not with the future," declared E. M. Bluestone in 1944. "There is a wholesome tendency these days to draw the functions of all special hospitals, sanatoriums and asylums into the orbit of the general hospital. Another tendency," Bluestone observed, was "the replacement of the term 'cure' by the newer term 'rehabilitation,' since we are committed to the total cure of all patients by total means, and their restoration to full, or at least partial, service in the community."[19] The general hospital was to supervise the triage of, and provide some medical treatment for, alcoholics within the new medical paradigm. It would take decades for the general hospital to embrace such a role, however, and this happened only as insurance companies recognized addiction treatment services as reimbursable expenses in the 1970s and 1980s.

Employee assistance programs (EAPs) for alcoholics, too, made significant strides, playing an essential part in spreading the word about alcoholism to a larger audience. Pioneered in the 1940s by firms such as Eastman Kodak and the E. I. du Pont de Nemours Company, these programs grew from a few in the 1940s to more than six hundred by 1975. This trend continued into the 1980s. Frequently utilizing recovered alcoholic employees as spokespersons, EAPs capitalized on the expense to industry caused by alcoholic men and women in the workforce via absenteeism, inefficiency, and on-the-job injuries. Once again the Yale Center of Alcohol Studies figured importantly, for its staff developed the Yale Plan for Business and Industry—a program that attempted to persuade employers that they would save money by rehabilitating their alcoholic employees rather than firing them and hiring new personnel. EAP advocates stressed that employees with alcohol problems would come forward if they knew that they would receive a sympathetic helping hand.[20] Here again physicians played a role, but corporations interested in "the bottom line" pushed forward medical care.

The church formally reentered the alcoholism arena in 1949 with the formation of the National Clergy Council on Alcoholism and Related Drug Problems (NCCA), an organization established to guide the American Catholic Church in its ministry to alcoholics. Another important organization, the North Conway (New Hampshire) Institute (NCI), was established in 1951 by a reformed alcoholic minister, David Works, who had attended the Yale Summer School of Alcohol Studies. The NCI encouraged church participation in promoting the disease concept of alcoholism and addressing the needs of alcoholic Christians throughout the United States. By 1958, the National Council of Churches issued a statement on alcohol problems that urged ministers to assist problem drinkers within their parishes, recommended that parishes extend their community outreach efforts to education on alcohol and alcoholism, and supported the heavy regulation of alcohol distribution sales. Spirituality entered the treatment arena not just through AA but more directly through the sanctuary doors. Faith continued to be a pivotal part of alcoholism treatment, much as it had been in the early alcoholism movement.

Finally, as we saw in the Zottoli Report, state and municipal authorities began to address alcoholism as a medical problem, establishing outpatient and inpatient programs for chronic heavy drinkers, and launching initiatives to educate the public about alcoholism. In 1943 Oregon and Utah became the first states since Prohibition to initiate a range of programs related to alcoholism. States spent an aggregate total of $30,572 on alcoholism treatment programs in 1944; by 1955 the total had risen to $3 million, a figure that doubled by 1960. Connecticut was the first state to establish a division of government for alcohol problems. Chartered in 1945 but not in full operation until 1947, the Connecticut Commission on Alcoholism promoted both treatment and education, capitalizing on the innovative work of the Yale Center of Alcohol Studies. Indeed, the majority of alcoholics who received care through state facilities in the 1950s did so through outpatient programs affiliated with hospitals, programs that were usually modeled after the Yale Plan clinics. Between 1945 and 1955, 75 percent of state legislatures passed alcoholism initiatives. Even though these initiatives met with mixed success in terms of promoting institutional treatment, they established state laws that recognized alcoholism as a medical condition, facilitating the expansion of community treatment facilities in the 1960s and 1970s.[21]

State care was not what it had been at the dawn of the twentieth century. Long-term confinement in a state asylum was not the desired norm for either the alcoholic or the mentally ill patient.[22] The level of public disillusionment with state asylums and hospitals intensified by the 1940s and 1950s. Initiated by Mike Gorman's 1948 *Reader's Digest* exposé of the grievous condition of the state hospital system in Oklahoma and Albert Deutsch's *The Shame of the States*, which widened the scope of complaint from one state to dozens, awareness of the overcrowding, understaffing, and general neglect of the mentally ill in large state facilities prompted a critical review of state institutional care— a review that culminated in the "deinstitutionalization" movement of the late 1960s and 1970s.

Indeed, the uneasy relationship between state hospitals for the insane and their inebriate patients that had existed in the late nineteenth and early twentieth centuries remained during Prohibition and after Repeal. According to a 1947 issue of the *AA Grapevine*, "the average alcoholic, unless permanently insane, was a 'headache' to the staff of the psychiatric institution."[23] Alcoholics complained of abuse within the state hospital. According to White, the situation had changed little by the mid-1960s, when a survey of state hospitals for the insane found that many of the institutions refused alcoholics admission because they regarded these patients as troublesome and difficult cases.[24]

The difficulties in treating chronic alcoholics were acutely felt by general hospitals as well, observed Giorgio Lolli to an audience at the New York Academy of Medicine in 1955: "For a variety of reasons, including the fundamental one that it is very difficult for the therapist to withstand the impact of alcoholic patients only, I favor clinics which are not restricted entirely to alcoholics. Clinics for treatment and research on eating and drinking habits seem very much in order to me." Lolli wished to bring together treatment of obesity "as a problem of uncontrolled urges to eat solid food and alcoholism as a problem of uncontrolled urges to drink fluids"—a desire that seems prescient in the light of early-twenty-first-century concerns.[25]

The public asylum, however, had lost its privileged position as a preferred mode of alcoholism treatment. In the 1950s most alcoholic persons who received institutional care for their condition did so at "private sanitaria, small 'drying-out' facilities, small residential alcoholism treatment programs, and a growing number of outpatient alcoholism counseling clinics."[26] As ideas had passed between the state and private treatment communities in the early alcoholism movement, so too were they exchanged in the modern era. In the 1950s, for example, the Willmar (Minnesota) State Hospital for alcoholics established a regimen of short-term residential care followed by outpatient participation in AA meetings and "recovery-home" living—a regimen that became known as the "Minnesota Model" and was adopted by the Hazelden Foundation, a private treatment facility in the state.[27] It came to dominate treatment across the country in the 1960s and 1970s.

Indeed, midcentury psychodynamic, community, and social psychiatry all pointed away from the asylum to the community-situated clinic. This approach regarded alcoholism as a condition firmly rooted in the individual's inability to adapt to environmental circumstances. As Lawrence Wooley, M.D., associate professor of psychiatry at Johns Hopkins, asserted in 1946, "Alcoholism could not exist without alcohol, but alcoholics are people and it seems

ridiculous to blame that substance for what the human organism does with it.
... Alcoholism today is an outgrowth of our culture—of what our culture
does to some personalities."[28] Wooley's perspective matched well the ideas of
Yale physiologists Haggard and Jellinek in *Alcoholism Explored,* for they
asserted not only that people drank out of anxiety but also that society-wide
historical change such as "rapid industrialization" had resulted in waves
of drunkenness. The two Yale scientists argued that because cycles of industrialization had hit their high points at the beginning of the nineteenth
century and later in the 1870s, thousands of men and women who had
earned a good living at their trades previously were suddenly deprived of their
livelihoods":

> They became impoverished, and many began to drink excessively. The fact that
> impoverishment and addiction occurred simultaneously has led to the designa-
> tion "poverty drinkers." These artisans and craftsmen did not become addicted to
> alcohol because they were poor, but because they were frustrated, because they
> had been humiliated in the pride they had taken in their skills, and because they
> saw their future as utterly helpless. These men were dominated by anxieties and
> they drowned their anxieties in alcoholic intoxication.[29]

Haggard and Jellinek pinpointed wartime anxieties in promoting similar results in the early 1940s. Their interpretation highlighted the environmental-
ism pervading the alcoholism field at midcentury, and it harked back to the
explanations of their nineteenth-century predecessors—men such as Joseph
Parrish and George Miller Beard, who believed that the stress of modern civilization played an important etiological role in inebriety. This broad focus on
societal change also implied that state action not only could, but should, play
a role in mitigating alcohol problems, including alcoholism, for the state possessed the power to address problems so global in scope.

All of these developments suggest that there was a sea change in the way alcoholism was regarded throughout American society during the three decades
following Repeal. Indeed, the 1970 passage of the Hughes Act, which authorized the creation of the NIAAA, is widely regarded as the culmination of the
modern alcoholism movement. The Hughes Act—promoted by Iowa Senator
Harold Hughes, a reformed alcoholic—defined alcoholism as a disorder in its
own right, not as a symptom of some other psychiatric illness; it set nationally
funded alcoholism research on its own foundation, distinct from the National
Institute for Mental Health (NIMH); and it established federal grant programs

administered through the NIAAA that supported public and private institutions on a competitive grant basis, while it subsidized states according to their "per capita income, population, and demonstrated need." [30] In the 1970s and 80s, the NIAAA and the National Council on Alcoholism (NCA) worked alone and in tandem with states to push insurance companies to underwrite treatment for alcoholism at inpatient and outpatient facilities. This was crucial to the expansion of the treatment industry in these years, supporting unprecedented growth. From a few dozen alcoholism treatment programs in the 1950s, the number of treatment facilities reached 2,400 in 1977; 6,800 in 1987; and 9,057 by 1991.[31]

## Post-Repeal Resistance to the Disease Concept

Support for the medical approach to treating alcohol problems expanded steadily post-Repeal, but not without some resistance. The resistance was aimed more at the disease concept than at the medical approach to alcohol abuse, however. Indeed, in many respects, alcoholism science raised more questions than it answered about the nature of the condition. For all of his pioneering work, E. M. Jellinek was not without his critics. And in fairness to the founder of the "classic disease concept" model, Jellinek himself voiced genuine reservations about the utility of his multistage model as a designation for the wide variety of problem drinking habits he observed. Jellinek and Haggard, after all, did not endorse a unitary disease model, but instead proposed a typology of inebrieties. Researchers at the Yale Center of Alcohol Studies also rejected the allergy model in 1944, and they later failed to find a particular personality profile for "the" alcoholic, as posited by many psychiatrists in the 1940s and 1950s. Two studies cast doubt on alcoholism's status as a specific disease: University of Washington sociologist Joan Jackson, known primarily for her study of the dynamics of the alcoholic family, found the notion of a specific "alcoholism syndrome" extremely problematic in 1958; and the 1972 RAND Corporation Report found that a significant proportion of individuals diagnosed as alcoholics could "recover" and drink normally without losing control. In retrospect, the disease concept appears to have been more a public relations success than a scientific one.[32] Psychiatrist Harry Tiebout, who had concluded in 1951 that alcoholism should be treated as a disease rather than a symptom of an underlying or primary disorder, tellingly observed that "To change the metaphor [from vice to disease], we have stuck our necks out and not one of us knows if he will be stepped on individually or collectively. I sometimes tremble to think of how little we have to back up our claims."[33]

In the 1960s and 1970s, anthropologists and sociologists studying drinking practices demonstrated conclusively that the user's mind-set and the setting of use were instrumental factors in shaping the consumption patterns of alcohol and other psychoactive substances. Indeed, studies throughout the 1970s and 1980s continued to reveal the importance of culture in mediating behaviors linked to alcohol and to other drugs. Finally, and most important within today's medical arena, no specific causal mechanism—other than the ingestion of alcohol—was ever identified for the disease of alcoholism. The elusive quest has continued.

Outside the world of alcohol science, there was and continues to be controversy too. The Supreme Court cast doubt on the disease concept in the 1968 case of *Powell v. Texas* by narrowly upholding (a 5–4 vote) a lower court's decision that a chronic alcoholic was responsible for his conduct while under the influence, citing among other issues the lack of medical consensus on the disease concept of alcoholism. For its part, the American Psychiatric Association, which had recognized alcoholism as a personality disorder in its first *Diagnostic and Statistical Manual* (*DSM-I*) in 1952, reclassified alcohol problems in its *DSM-III* (1980), distinguishing between "alcohol dependency" and "alcohol problems." Only the former was regarded as a disease, but both could benefit from medical attention. This distinction mirrored the 1977 findings of the World Health Organization. That year, a WHO Expert Committee on alcohol-related disabilities drew attention to the incapacitating consequences of problem drinking that were distinct from alcoholism. This observation had significant political implications, for it suggested that reducing the *overall* alcohol consumption of the general public could yield tremendous benefits—more than might accrue from focusing on alcoholism alone. It was and is unlikely that Americans would ever again pass a prohibition amendment, but public health officials hoped that more headway might be made in addressing the diversity of alcohol problems that affected the United States.

Indeed, the more encompassing "alcohol problems" approach that arose in the late 1970s named alcoholism as but one of many negative consequences of alcohol consumption, and it has achieved tremendous currency within medical and public health circles in the past few decades. In 1990, for example, the Institute of Medicine of the National Academy of Sciences published *Broadening the Base of Treatment for Alcohol Problems*, reasoning that the "alcohol problems" moniker reflected the Institute's belief "that the focus of treatment needs to be expanded . . . 'Alcohol problems' is felt to be a more inclusive definition of the object of treatment than such current alternatives as 'alcoholism' or 'alcohol dependence syndrome,' but it is nevertheless compatible with these widely used conceptual frameworks."[34] Medical and social treatment were still in order, but for a wider range of drinkers. Included in the "alcohol problems" approach are fetal alcohol syndrome (FAS), industrial accidents, unemployment, drunken driving, violence (including rape), binge drinking, and alcohol consumption among minors.

Alcohol sociologists have emphasized the importance of interpreting these developments within alcohol studies in the context of late-twentieth-century politics and policies. A focus on individual consumption rates across society (and their manifold negative social and medical consequences) may suggest, for example, that the government, as the recipient of taxes from the sale of ethanol, has an obligation to assist individuals disabled by its consumption. Some alcohol policy experts have interpreted this focus as an effort on the part of social and medical scientists to redress the "end of big government" mission of the Reagan years and his administration's general preference for supply-side policies—two trends that were intellectually and programmatically at cross purposes with the expansion of treatment.[35] The rapid growth of addiction treatment facilities in the 1980s suggests, however, that the effect was offset by the rise of private care facilities, employee assistance programs, and insurance coverage of addiction treatment.

That the transition from alcoholism to alcohol problems began with the Reagan administration, however, may indicate that it was part of a rising tide of cultural conservatism, part of a larger "neo-temperance" movement that has attempted to rein in a host of behaviors perceived to be morally questionable and potentially damaging both to individuals and to society at large.[36] Ironically, it is also possible to see the "alcohol problems" approach as a return to the sociomedical perspective of the early alcoholism movement and the holistic orientation expressed by Jellinek and Haggard in *Alcohol Explored*. This broad-based approach may constitute a counterbalance to the growing focus within alcoholism research on genetic and neuropathic mechanisms. Regardless, one wonders what the consequences of the alcohol problems perspective will be for the alcoholic person. Will he or she be regarded as just one of many problem drinkers who deserve medical and social attention, or will association with the image of the drunken driver and the mother who drinks through her pregnancy demonize the alcoholic once again? Few public awareness campaigns have been more successful in terms of their visibility than Mothers Against Drunk Driving (MADD). In the interests of creating safer highways, has society become hardened to the plight of the alcoholic and less sympathetic to the disease concept of alcoholism? Only time will tell.

Nor is public health the only area in which a broader approach is gaining momentum. Within the fields of neurophysiology and neurochemistry, a more comprehensive perspective on alcoholism and other drug addictions appears to be winning supporters. Roughly one century after the unified theory of "inebriety" was abandoned for "alcoholism," "opiate addiction," "cocaine inebriety," and a host of other specific intoxications, pharmacological researchers are proposing a unitary theory of addiction in which a variety of psychoactive substances are believed to trigger similar neurochemical responses in their users. The policy and treatment sides of alcohol and drug use preceded this development by a few decades, of course, with their championing of the terms "chemical dependency" and "substance abuse," but the new "dopamine hypothesis" appears to be garnering attention.[37] This is one of the latest chapters in the elusive quest for a disease mechanism. Yet even within the world of neurochemistry, culture, social behavior, and learning remain key elements in mediating drug response. To quote a cover article from *Time* devoted to the dopamine hypothesis: "Realistically, no one believes better medications alone will solve the drug problem. In fact, one of the most hopeful messages coming out of current research is that the biochemical abnormalities associated with addiction can be reversed through learning."[38]

As we enter the new millennium, the elusive quest for mechanism and the essential tension between the physiological and social dimensions of alcoholism seem to have kept in lock step with the trajectory of alcohol research. Indeed, in the most recent *Special Report to the U.S. Congress on Alcohol and Health,* Enoch Gordis, the director of the NIAAA, observes: "Perhaps the single greatest influence on the scope and direction of alcohol research has been the finding that a portion of the vulnerability to alcoholism is genetic. This finding, more than any other, helped to establish the biological basis of alcoholism."[39] At the same time, Gordis is quick to add that this is not the end of the story, but rather the beginning, observing that "the whole human animal, not just its genetic or neural parts," will continue to be an important focus of research because so many within the field reject the "'reductionist' view, which seeks to define humankind in terms of its genes," and believe that "genes are not (or even mostly) destiny, just as humankind is not just the sum of its neurons and circuits."[40]

Gordis's message reminds us that all disease, inasmuch as it affects human beings, is social disease. Even after we have discovered the elusive mechanisms that drive the pathologies of the biological world, we will still have to contend with the social conditions that frame, and sometimes activate, pathological mechanisms. As Herbert Yahraes commented nearly sixty years ago, when asked if alcoholism could be prevented, the medical researcher usually answered "No" because he or she had not found out why certain people become immoderate drinkers; the temperance advocate generally answered "Yes" because he or she believed society would be best served by cutting out alcohol altogether. The wisest students of alcoholism also answered "Yes," claimed Yahraes, but for different reasons. They reasoned that the key to preventing alcoholism lay "in the building of a society in which the individual is better fed and housed, has better medical care, has better facilities for mental hygiene, has fewer money worries, and has the facilities and the encouragement to engage in recreational activities besides drinking."[41] Whether or not America has the wisdom or will to create such a society is an open question. However, if we can learn to appreciate the considerable impact that social values, political priorities, and economic circumstances have on the definition and prevalence of pathology, we will have made an important first step.

# Notes

## Preface

1. Sarah Whitney Tracy, "The Foxborough Experiment: Medicalizing Inebriety at the Massachusetts Hospital for Dipsomaniacs and Inebriates, 1833–1919" (Ph.D. diss., University of Pennsylvania, 1992).

2. For a historical discussion of the incomplete medicalization of and ongoing debate over alcoholism and several other behavioral disorders, see Charles Rosenberg, "Contested Boundaries: Disease, Deviance, and Diagnosis," unpublished manuscript. A useful multistage model for medicalization is that of sociologists Peter Conrad and Joseph Schneider, whose focus is on the medicalization of deviance from a historical perspective. They see these stages as: (1) the definition of certain behaviors as deviant; (2) the promotion of new medical diagnoses for these behaviors within specialized professional literature; (3) claims-making on the part of medical and nonmedical parties regarding the validity of the new medical framework for deviance; (4) the acquisition of medical turf via legislative and judicial bodies, essentially a successful appeal to state authority; and (5) The institutionalization of the new medical framework through codification and bureaucratization, that is, the acceptance of new laws as part of the legal canon and the creation of new organizations dedicated to furthering the medical approach (*Deviance and Medicalization: From Badness to Sickness* [Philadelphia: Temple University Press, 1992], 266–71). On the problematic status of alcoholism's medicalization in the twentieth century, see Ron Roizen, "How Does the Nation's 'Alcohol Problem' Change from Era to Era? Stalking the Social Logic of Problem-Definition Transformations since Repeal," in *Altering American Consciousness: The History of Alcohol and Drug Use in the United States, 1800–2000,* ed. Sarah W. Tracy and Caroline Jean Acker (Amherst: University of Massachusetts Press, 2004), 61–87; and Lynn M. Appleton, "Rethinking Medicalization: Alcoholism and Anomalies," in *Images of Issues: Typifying Contemporary Social Problems,* 2nd ed., ed. Joel Best (New York: Aldine De Gruyter, 1995), 59–80.

3. For an examination of the centrality of diagnosis in modern medicine, see Charles Rosenberg, "The Tyranny of Diagnosis: Specific Entities and Individual Experience," *Milbank Quarterly* 80, no. 2 (2002): 37–60. See also Caroline J. Acker, "Stigma or Legitimation? A Historical Examination of the Social Potentials of Addiction Disease Models," *Journal of Psychoactive Drugs* 25, no. 3 (July-September 1993): 193–205.

4. For a sampling of the rich literature on alcohol's many cultural functions, especially in the United States, see Tracy and Acker, eds., *Altering American Consciousness;*

Jack S. Blocker Jr., David M. Fahey, and Ian Tyrell, eds., *Alcohol and Temperance in Modern History: An International Encyclopedia*, 2 vols. (Santa Barbara, CA: ABC CLIO, 2003); Elaine Frantz Parsons, *Manhood Lost: Fallen Drunkards and Redeeming Women in the Nineteenth-Century United States* (Baltimore: Johns Hopkins University Press, 2003); Lori Rotskoff, *Love on the Rocks: Men, Women, and Alcohol in Post-World War II America* (Chapel Hill: University of North Carolina Press, 2002); David T. Courtwright, *Forces of Habit: Drugs and the Making of the Modern World* (Cambridge, MA: Harvard University Press, 2001); Richard Stivers, *Hair of the Dog: Irish Drinking and Its American Stereotype*, new rev. ed. (New York: Continuum International Publishing Group, 2000); Andrew Barr, *Drink: A Social History of America* (New York: Carroll and Graf, 1999); Madelon Powers, *Faces Along the Bar: Lore and Order in the Workingman's Saloon, 1870–1920* (Chicago: University of Chicago Press, 1998); Mariana Valverde, *Diseases of the Will: Alcohol and the Dilemmas of Freedom* (Cambridge: Cambridge University Press, 1998); Catherine Gilbert Murdock, *Domesticating Drink: Women, Men, and Alcohol in America, 1870–1940* (Baltimore: Johns Hopkins University Press, 1998); Wolfgang Schivelbusch, *Tastes of Paradise: A Social History of Spices, Stimulants, and Intoxicants* (New York: Vintage, 1993); Susanna Barrows and Robin Room, eds., *Drinking: Behavior and Belief in Modern History* (Berkeley: University of California Press, 1991); Mary Douglas, ed., *Constructive Drinking: Perspectives on Drink from Anthropology* (Cambridge: Cambridge University Press, 1987); Joseph Gusfield, *Symbolic Crusade: Status Politics and the American Temperance Movement* (Urbana: University of Illinois Press, 1963).

5. "Section on Public Hygiene and State Medicine," *Minutes of the Twenty-Seventh Annual Meeting of the American Medical Association, Held in the City of Philadelphia, June 6th, 7th, 8th, and 9th, 1876*, 27, 324.

6. Charles A. Rosenwasser, "A Plea for the Establishment of Hospitals for the Rational Treatment of Inebriates," *Medical Record*, 8 May 1909, 795.

7. Herbert Fingarette, *Heavy Drinking: The Myth of Alcoholism as a Disease* (Berkeley: University of California Press, 1988), esp. 48–69.

8. Jellinek was aware of the shortcomings of his multistage model of alcoholism, but others were less concerned with its qualifications; ibid., 19–21; Institute of Medicine, *Broadening the Base of Treatment for Alcohol Problems* (Washington, DC: National Academy Press, 1990).

9. Fingarette, *Heavy Drinking*, 99–113; Julie Bruns and Fred Hanna, "Abstinence versus Controlled Use: A Fresh Perspective on a Stale Debate," *Journal of Addictions and Offender Counseling* 16, no. 1 (October 1995): 14–30.

10. See also Linda and Mark Sobell, "Controlled Drinking after 25 Years: How Important was the Great Debate?" *Addiction* 90, no. 9 (September 1995): 1149.

11. For more on the importance of context and mind-set in shaping the experience of the psychoactive drug user, see Norman Zinberg, *Drug, Set, and Setting: The Basis for Controlled Intoxicant Use* (New Haven, CT: Yale University Press, 1984). See also the rich literature on the cross-cultural consumption of alcohol, including: Douglas, ed., *Constructive Drinking;* Dwight Heath, *Drinking Occasions: Comparative Perspectives on Alcohol and Culture* (Philadelphia: Brunner/Mazel, 2000); Dwight Heath, "Drinking Patterns: A Fresh Perspective," *DATA: The Brown University Digest of Addiction Theory and Application* 20 (October 2001): 8; Robin Room, "Intoxication and Bad Behaviour: Understanding Cultural Differences in the Link," *Social Science and Medicine* 53 (July 2001): 153–64.

12. For more on the "alcohol problems" perspective in public health, see Roizen, "How Does the Nation's 'Alcohol Problem' Change from Era to Era?" See also Institute of Medicine, *Broadening the Base of Treatment.*

13. George Vaillant, L. Gale, and E. S. Milofsky, "Natural History of Male Alcoholism. II. The Relationship between Different Diagnostic Dimensions," *Journal of Studies on Alcohol* 43 (1982): 216–32.

14. Berton Rouche, *The Neutral Spirit: A Portrait of Alcohol* (Boston: Little, Brown, 1960).

15. AMA, "House of Delegates Policies on Alcoholism as a Disease: H-95.983 Drug Dependencies as Diseases" (passed in 1987).

16. William A. White, "Alcoholic and Drug Intoxication," in *Reference Handbook of the Medical Sciences,* ed. A. H. Buck and T. L. Stedman (New York: W. Wood, 1900), 5:81.

## Introduction

1. "Minutes" in Proceedings of the First Meeting of the American Association for the Cure of Inebriates, held in New York, November 29th and 30th, 1870, in *Proceedings, 1870–1875* (New York: Arno Press, 1981), 8.

2. Ibid., 3.

3. Ibid., 8.

4. Benjamin Rush, *"An Enquiry into the Effects of Spiritous Liquors upon the Human Body, and their Influence upon the Happiness of Society,* 3rd ed. (Philadelphia: John McCulloch, 1791). Note that the original date of publication is said to be 1784 and the title that is commonly used is *An Inquiry into the Effects of Ardent Spirits upon the Human Body and Mind with an Account of the Means of Preventing and of the Remedies for Curing Them.* This last title dates from later editions of the text.

5. Benjamin Rush, "Plan for an Asylum for Drunkards to be called the Sober House," reprinted in *The Autobiography of Benjamin Rush,* ed. George Washington Corner (Princeton, NJ: Princeton University Press, 1948), 354–55.

6. "Minutes" in Proceedings of the First Meeting, 2; italics are in the original.

7. My use of the term "early alcoholism movement" distinguishes this attempt to promote the disease concept of inebriety from what alcohol historians and sociologists have labeled "the modern alcoholism movement," that is, the post-Repeal campaign to advance a medical understanding of alcoholism that was spearheaded by research, public relations groups, and treatment organizations such as the Yale Center of Alcohol Studies, the Research Council on Problems of Alcohol, and Alcoholics Anonymous during the 1940s and 50s.

8. There is evidence that initially, in the early years of Prohibition, rates of alcohol-related illness were significantly reduced. See John Burnham, "New Perspectives on the Prohibition Experiment of the 1920s," *Journal of Social History* 2 (Fall 1968): 51–68.

9. Mark Twain, *Pudd'nhead Wilson—A Tale* (London: Chatto and Windus, 1894), 197.

10. See Paul Boyer, *Urban Masses and Moral Order in America, 1820–1920* (Cambridge, MA: Harvard University Press, 1995), 196–97; *Women in Industry: Decision of the United States Supreme Court in Curt Muller vs. State of Oregon . . . and Brief for the State of Oregon by Louis D. Brandeis assisted by Josephine Goldmark* (New York: Arno Press, 1969).

11. For an excellent overview of the temperance movement in America, see Jack Blocker, *American Temperance Movements: Cycles of Reform* (Boston: Twayne Publishers,

1989); for an overview of drinking practices in America, see Mark Lender and James Kirby Martin, *Drinking in America, a History,* rev. ed. (New York: Free Press, 1987).

12. Roy Rosenzweig, *Eight Hours for What We Will* (New York: Cambridge University Press, 1983).

13. Merton M. Hyman, Marilyn A. Zimmerman, Carol Gurioli, and Alice Helrich, *Drinkers, Drinking, and Alcohol-Related Mortality and Hospitalizations: A Statistical Compendium* (New Brunswick, NJ: Journal of Alcohol Studies, 1980).

14. For more on the WCTU, see Janet Zollinger Giele, *Two Paths to Women's Equality: Temperance, Suffrage, and the Origins of Modern Feminism* (New York: Twayne Publishers, 1995); Ruth Bordin, *Women and Temperance: The Quest for Power and Liberty, 1873–1900* (New Brunswick, NJ: Rutgers University Press, 1990); Jack Blocker Jr., *"Give to the Winds Thy Fears": The Women's Temperance Crusade, 1873–74* (Westport, CT: Greenwood Press, 1985); Barbara Leslie Epstein, *The Politics of Domesticity: Women, Evangelism, and Temperance in Nineteenth-Century America* (Middletown, CT: Wesleyan University Press, 1981).

15. Philip Pauly, "The Struggle for Ignorance about Alcohol: American Physiologists, Wilbur Olin Atwater, and the Woman's Christian Temperance Union," *Bulletin of the History of Medicine* 64 (1990): 366–92; see also Jonathan Zimmerman, *Distilling Democracy: Alcohol Education in America's Public Schools, 1880–1925* (Lawrence: University Press of Kansas, 1999).

16. For overviews of public health policy in America, see John Harley Warner and Janet Tighe, eds., *Major Problems in the History of American Medicine and Public Health* (Boston: Houghton Mifflin, 2001); Judith Walzer Leavitt and Ronald L. Numbers, eds., *Sickness and Health in America—Readings in the History of Medicine and Public Health,* 3rd ed., rev. (Madison: University of Wisconsin Press, 1997); George Rosen, *A History of Public Health,* exp. ed. (Baltimore: Johns Hopkins University Press, 1993); John Duffy, *The Sanitarians: A History of American Public Health* (Urbana: University of Illinois Press, 1992).

17. It also hints at the questionable accuracy of statistics on alcohol consumption, which must be taken *cum grano salis,* as any "alcohologist" will tell you. Several efforts have been made to correct the accuracy of the statistics I quote here, but they have failed for various reasons, leaving the current figures as those that have served historians and sociologists for the past two and a half decades. Of course, the relatively stable level of consumption between 1870 and 1920, hovering around 2.0 gallons of absolute alcohol per person, may reflect stabilized drinking tendencies or a significant reduction on the part of some segment of society with a coincident increase in the drinking of another. Unfortunately, we do not have a level of statistical knowledge to provide a more refined analysis of the figures. The much-cited statistical source is the previously cited Hyman et al., *Drinkers, Drinking, and Alcohol-Related Mortality and Hospitalizations.*

18. Some may object to my use of the term "moral panic," which usually denotes an acute period of perceived social crisis. The fifty-year period covered in this book would seem to deviate from this usage. Yet I do see the concerns voiced over alcohol and its consumption as a sociological phenomenon that includes all of the criteria of a moral panic:

1. Something or someone is defined as a threat to values or interests.
2. This threat is depicted in an easily recognizable form by the media.

3. There is a rapid build-up of public concern.
4. There is a response from authorities or opinion-makers.
5. The panic recedes or results in social changes.

I take this understanding from Kenneth Thompson, *Moral Panics* (New York: Routledge, 1998), esp. 1–30. Perhaps the perceived magnitude of the problem within modern American society, when combined with such a deeply embedded and significant feature of American life as drinking alcohol, made for a protracted social reform campaign.

19. Morton Keller, *Regulating a New Society: Public Policy and Social Change in America, 1900–1933* (Cambridge, MA: Harvard University Press, 1994), 109–10.

20. Raymond Calkins, "A Summary of Investigations Concerning Substitutes for the Saloon," in Committee of Fifty, *The Liquor Problem: A Summary of Investigations Conducted by the Committee of Fifty, 1893–1903* (Boston: Houghton Mifflin, 1905), 150.

21. Perry Duis, *The Saloon: Public Drinking in Chicago and Boston, 1880–1920* (Urbana: University of Illinois Press, 1983), 205. For contemporary perspectives on the prostitution-saloon connection, see Calkins, *Substitutes for the Saloon* (Boston: Houghton Mifflin, 1901); for historians' perspectives, see Powers, *Faces Along the Bar;* Timothy Gilfoyle, *City of Eros: New York City, Prostitution, and the Commercialization of Sex, 1790–1920* (New York: Norton, 1992); Thomas C. Mackey, *Red Lights Out: A Legal History of Prostitution, Disorderly Houses and Vice Districts, 1870–1917* (New York: Garland, 1987); Ruth Rosen, *The Lost Sisterhood: Prostitution in America, 1900–1918* (Baltimore: Johns Hopkins University Press, 1982).

22. Jacob Ries, *How the Other Half Lives: Studies Among the Tenements of New York* (New York: Dover Publications, 1971 [orig. New York: Scribner's, 1890]), 165. For more on the gambling and saloon connection, see Robert Archey Woods, ed., *Americans in Process: A Settlement Study* (Boston: Houghton Mifflin, 1902); Calkins, *Substitutes for the Saloon;* Duis, *The Saloon.*

23. For more on working-class saloons, see Powers, *Faces Along the Bar.*

24. Roy Rosenzweig, "The Rise of the Saloon," in *Rethinking Popular Culture: Contemporary Perspectives in Cultural Studies,* ed. Chandra Mukerji and Michael Schudson (Berkeley: University of California Press, 1991).

25. Lillian Brandt, "Alcoholism and Social Problems," *The Survey* 25 (1 October 1910): 18.

26. See Rosenzweig, *Eight Hours for What We Will;* Kathy Peiss, *Cheap Amusements: Working Women and Leisure in Turn-of-the-Century New York* (Philadelphia: Temple University Press, 1986); Joseph Gusfield, "Passage to Play: Rituals of Drinking Time in American Society," in *Constructive Drinking,* ed. Mary Douglas; idem, *Contested Meanings: The Construction of Alcohol Problems* (Madison: University of Wisconsin Press, 1996).

27. See Thomas H. O'Connor, *The Boston Irish: A Political History* (Boston: Little, Brown, 1995), 156–59; Barbara Miller Solomon, *Ancestors and Immigrants: A Changing New England Tradition* (Boston: Northeastern University Press, 1989), 111–21.

28. O'Connor, *The Boston Irish,* 158.

29. Ibid., 157.

30. Robert A. Woods, *The City Wilderness: A Settlement Study* (Boston: Houghton Mifflin, 1898). Woods was a leader in the American settlement house movement. He played

an important role in alcohol policy making in Boston, and he served as chairman of the board of trustees of the state hospital for inebriates in Massachusetts.

31. Homer Folks, "Social Aspects of Alcoholism," *The Survey* 25 (1 October 1910): 14.

32. For more on the history of social work, see Roy Lubove, *Professional Altruist: The Emergence of Social Work as a Career, 1880–1930* (Cambridge, MA: Harvard University Press, 1968); Allen Davis, *Spearheads for Reform: The Social Settlements and the Progressive Movement, 1890–1914* (New York: Oxford University Press, 1976); Robyn Muncy, *Creating a Female Dominion in American Reform, 1890–1935* (New York: Oxford University Press, 1994); Eric Schneider, *In the Web of Class: Delinquents and Reformers in Boston, 1810s–1930s* (New York: New York University Press, 1992); Michael Katz, *In the Shadow of the Poorhouse: A Social History of Welfare in America* (New York: Basic Books, 1986); Camilla Stivers, *Bureau Men, Settlement Women: Constructing Public Administration in the Progressive Era* (Lawrence: University Press of Kansas, 2002).

33. Mary Richmond, *Social Diagnosis* (New York: Russell Sage Foundation, 1917), 430.

34. Folks, "Social Aspects of Alcoholism," 14–15.

35. See Zimmerman, *Distilling Democracy;* Pauly, "The Struggle for Ignorance over Alcohol."

36. Sam Bass Warner, *Crime and Criminal Statistics in Boston* (Cambridge, MA: Harvard University Press, 1934), 37–49, 136–38.

37. "Due to Drunks—Number of Prisoners on the Increase—Commitments, However, Are on the Decrease—Such Arrests in Boston Go Up 8029," *Boston Daily Globe,* 24 January 1893, 2.

38. Standing Committee on Hospitals of the State Charities Aid Association of New York, *Treatment of Public Intoxication and Inebriety, March 20, 1909* (New York: State Charities Aid Association of New York, 1909), 6.

39. Committee of Fifty, *The Liquor Problem: A Summary of Investigations,* 122–23.

40. For more on the history of professions, see Paul Starr, *The Social Transformation of American Medicine* (New York: Basic Books, 1982); Eliot Freidson, *Professionalism: The Third Logic* (Chicago: University of Chicago Press, 2001); idem, *Professional Dominance* (Chicago: Atherton Press, 1970); Andrew Abbott, *The System of Professions: An Essay on the Division of Expert Labor* (Chicago: University of Chicago Press, 1988); Nathan O. Hatch, *The Professions in American History* (Notre Dame, IN: University of Notre Dame Press, 1988); Thomas Haskell, *The Authority of Experts: Studies in History and Theory* (Bloomington: Indiana University Press, 1984); Burton Bledstein, *The Culture of Professionalism: The Middle Class and the Development of Higher Education in America* (New York: Norton, 1976), 81–101; Bruce Kimball, *The "True Professional Ideal" in America: A History* (Cambridge, MA: Blackwell, 1992); and Samuel Haber, *The Quest for Authority and Honor in the American Professions, 1750–1900* (Chicago: University of Chicago Press, 1991).

41. *Annual Report of the President of Harvard College, 1870–1871,* 20.

42. Elisha Harris, "Practical Points Relating to the Criminality, Repression and Cure of Drunkenness and Dypsomania," in Proceedings of the Third Meeting of the American Association for the Cure of Inebriates, held in New York, October 8th, 9th, and 10th, 1872, in *Proceedings, 1870–1875* (New York: Arno Press, 1981), 121.

43. J. W. Grosvenor, "What Shall We Do with Our Alcoholic Inebriates?" *Bulletin of the American Academy of Medicine* 2 (June 1895): 128.

44. J. W. Grosvenor, "The Relation of Alcohol to Preventive Medicine," *Bulletin of the American Academy of Medicine* 3 (August 1897): 132.

45. For more on the public health mission of the inebriety physicians, see State of Connecticut, *Report of Committee on State Board of Health and Vital Statistics Made to the General Assembly, May Session, 1875* (Hartford, CT: Case, Lockwood, and Brainard, 1875), 13.

46. Here I borrow the terms "social" and "cultural" authority from Paul Starr, who uses them in *The Social Transformation of American Medicine*: "Social authority involves the control of action through the giving of commands, while cultural authority entails the construction of reality through definitions of fact and value." Physicians may possess both types of authority, for they may direct other medical personnel, such as physician assistants or nurses, as well as patients, to follow their orders. When patients go to a physician to learn what's "wrong" with them, they are relying on the doctor's "authority to interpret signs and symptoms, to diagnose health or illness, to name disease, and to offer prognoses. . . . By shaping the patient's understanding of their own experience, physicians create the conditions under which their advice seems appropriate" (*Social Transformation of American Medicine*, 13–14).

47. Ibid., 15.

48. See Nancy Tomes, *A Generous Confidence: Thomas Story Kirkbride and the Art of Asylum-keeping, 1840–1883* (Cambridge: Cambridge University Press, 1984), 252–53, 262.

49. Charles Nichols in "Proceedings of the Association of Medical Superintendents," *American Journal of Insanity* 32, no. 3 (1876): 365.

50. Ibid.

51. Haven Emerson, "The Duty of the Health Departments in the Alcohol Question," *Boston Medical and Surgical Journal*, 18 January 1917, 79.

52. Elizabeth Lunbeck, *The Psychiatric Persuasion: Knowledge, Gender, and Power in Modern America* (Princeton, NJ: Princeton University Press, 1994), 21.

53. Of course demonstrating "success" was a contentious issue. Would reformers measure success in terms of rates of "cure," or abstinence? Or would "material improvement" suffice, that is, the return of the inebriate to the workforce and his support of himself and his family, even if he had occasional lapses or drank in moderation? If other hospitals for the insane and jails were "freed" of this recidivist clientele by the existence of the inebriate institution, would these other institutions regard the inebriate hospital as successful simply by dint of its tending a troublesome patient population for whom they had little regard? All of these questions were on the table in both Iowa and Massachusetts.

54. Katz, *In the Shadow of the Poorhouse;* Gerald Grob, *Mental Illness in American Society, 1875–1940* (Princeton, NJ: Princeton University Press, 1983); Ellen Dwyer, *Homes for the Mad: Life Inside Two Nineteenth-Century Asylums* (New Brunswick, NJ: Rutgers University Press, 1987).

55. David Rothman, *Conscience and Convenience—The Asylum and Its Alternatives in Progressive America,* rev. ed. (New York: Aldine de Gruyter, 2002), 5.

56. Ibid. The text was originally published by Little, Brown in 1980.

57. Erwin Ackerknecht, "A Plea for a 'Behaviorist' Approach in Writing the History of Medicine," *Journal of the History of Medicine and Allied Sciences* 22 (1967): 211–14. The comparison of ideology and actions was also an essential motivation behind the "new social history" of the 1970s and 80s, which imported insights and forms of analysis from

the social sciences into historical research. The field also endeavored to explore history from the ground up, inserting the everyday experiences of ordinary people into our picture of the past, and to move beyond the rhetoric of reform to the arenas (often institutions) in which reform was practiced.

58. Note that I use the male pronoun "he" to refer to "the inebriate." Although women were inebriates during the period covered by this book, habitual drunkenness was portrayed as a male disease in the United States. The vast majority of arrests for public drunkenness were men, and most institutions—medical and penal—served far more men than women. For this reason, I have chosen to use "he" instead of "she" or "he/she" or "they."

59. Jellinek, *The Disease Concept of Alcoholism*, 4–7.

ONE: Disease Concept(s) of Inebriety

1. "Proceedings of the Association of Medical Superintendents," *American Journal of Insanity* 32 (January 1876): 362–86.

2. "Minutes of the Section on Public Hygiene and State Medicine," Minutes of the Twenty-Seventh Annual Meeting of the American Medical Association, *Transactions of the American Medical Association* 27 (1876): 324.

3. Ibid., 325–26.

4. See Thomas Vancourt, "Brief Historical Sketch of the 'Chicago Washingtonian Home'; with Some General Remarks," in *Proceedings of the Second Meeting of the American Association for the Cure of Inebriates,* November 1871, 103–15.

5. "Minutes" in Proceedings of the First Meeting of the American Association for the Cure of Inebriates, 8.

6. Parrish, "Philosophy of Intemperance," in Proceedings of the First Meeting of the American Association for the Cure of Inebriates, 31.

7. Ibid., 29.

8. John Willett, "The Relation of the Church to Inebriates," in Proceedings of the First Meeting of the American Association for the Cure of Inebriates, 57.

9. William Sweetser, A Dissertation on Intemperance to which was awarded the Premium offered by the Massachusetts Medical Society (Boston: Hilliard, Gray, and Company, 1829), reprinted in *Nineteenth-Century Medical Attitudes Toward Alcoholic Addiction—Six Studies, 1814–1867,* ed. Gerald Grob (New York: Arno Press, 1981), 97–98.

10. Nathan Crosby, "Inebriate Asylums: Remarks in Opposition to Them before the Committee on Charitable Institutions" (Boston: Nation Press, 1871), 4; emphasis in the original.

11. Joseph Parrish, "Opening Address: Intemperance a Disease," in Proceedings of the Third Meeting of the American Association for the Cure of Inebriates, 3.

12. Joseph Parrish, "Dr. Parrish's Remarks," in Proceedings of the [Fourth and] Fifth Meetings of the American Association for the Cure of Inebriates, 76.

13. Theodore L. Mason, Willard Parker, and Robert Harris, "Report of the Special Committee on Principles," in Proceedings of the Third Meeting of the American Association for the Cure of Inebriates, 78. The Association's principles were as follows: (1) Intemperance is a disease. (2) It is curable in the same sense that other diseases are. (3) Its primary cause is a constitutional susceptibility to the alcoholic impression. (4) This constitutional tendency may be inherited or acquired. (5) Alcohol has its true place in the

arts and sciences. It is valuable as a remedy, and like other remedies, may be abused. In excessive quantity it is a poison, and always acts as such when it produces inebriety. (6) All methods hitherto employed having proved insufficient for the cure of inebriates, the establishment of asylums for such a purpose is the great demand of the age. (7) Every large city should have its local or temporary home for inebriates, and every state, one or more asylums for the treatment and care of such persons. (8) The law should recognize intemperance as a disease and provide other means for its management than fines, stationhouses, and jails.

14. Ibid., 80.

15. *Report of the Massachusetts Hospital for Dipsomaniacs and Inebriates,* January 1893 (Boston: Wright and Potter, 1893), 25.

16. William James, "The Explosive Will," in *The Principles of Psychology* (New York: Henry Holt and Company, 1890), 2:543.

17. Valverde, *Diseases of the Will,* 48–49.

18. See Elisha Harris, "Practical Points Relating to the Criminality, Repression and Cure of Drunkenness and Dypsomania," in Proceedings of the Third Meeting of the American Association for the Cure of Inebriates, 121. Harris was a prominent New York City physician and outspoken public health advocate who edited the 1865 *Report of the Council of Hygiene and Public Health of the Citizens' Association of New York Upon the Sanitary Condition of the City,* the first full-scale survey of New York City's health problems.

19. Of course, the meaning of *every* disease evolves over time. Consider the changing and competing images of tuberculosis before and after the discovery of the bacillus that caused the disease. The white plague vs. romanticized consumption vs. a constitutionally selective disease. Think of AIDS, early in its twentieth-century history known as GRID (gay related immune disorder), a disease associated with two minority populations: the homosexual community and intravenous drug users. Some social commentators focused on the modern technology of transportation in considering the disease's spread across the continents. AIDS was a disease of the late twentieth century, in this regard. Others focused on its deadly effects as well as on the individuals affected and regarded the condition as divine punishment. Some within the African-American community saw AIDS as the latest stage in a genocidal campaign against black people. Consider its "innocent"—transfusion cases—vs. "guilty"—sexually reckless or drug-addicted victims. All diseases, as well as their perception by the medical and lay communities, change over time. See Charles Rosenberg, "Introduction. Framing Disease: Illness, Society and History," and Steven Peitzman, "From Bright's Disease to End Stage Renal Disorder," in *Framing Disease: Studies in Cultural History,* ed. Charles Rosenberg and Janet Golden (New Brunswick, NJ: Rutgers University Press, 1992).

20. *Cherokee Democrat,* 22 July 1902, 2; *Cherokee Democrat,* 25 July 1902, 1; *Cherokee Democrat,* 21 October 1902, 2; *Cherokee Democrat,* 12 June 1903, 1; *Cherokee Democrat,* 27 November 1905, 1; *Cherokee Democrat,* 29 January 1906, 1.

21. "Poe in Pathology: His Faults were Not His Own, They were a Disease that he Inherited," *Boston Evening Transcript,* 24 August 1907, 11. The quotation within a quotation belongs to Charles H. Goudiss, M.D.

22. "Confessions of a Dipsomaniac," *St. Paul Medical Journal* 6, no. 9 (September 1904): 703–5. See also the original article, "The Confessions of a Dipsomaniac—edited by William Lee Howard, M.D.," *The Arena* 32, no. 176 (July 1904): 1–10. Howard was a Baltimore physician and acquaintance of the dipsomaniac in question. The various

conditions that the anonymous author described as dipsomania may be a result of the physician editing the article. An editorial aside in the piece notes: "The 'Confessions' were intrusted to his supervision and the psychologic facts are from his pen."

23. "Confessions of a Dipsomaniac," *St. Paul Medical Journal*, 703–5.

24. See Charles Rosenberg and Carroll Smith-Rosenberg, "The Female Animal: Medical and Biological Views of Women," in Rosenberg, *No Other Gods: On Science and American Social Thought* (Baltimore: Johns Hopkins University Press, 1976), 54–70; Carroll Smith-Rosenberg, *Disorderly Conduct: Visions of Gender in Victorian America* (New York: Oxford University Press, 1985); Elaine S. Abelson, *When Ladies Go A-Thieving—Middle-Class Shoplifters in the Victorian Department Store* (New York: Oxford University Press, 1989); and Cheryl Krasnick Warsh, "'Oh, Lord, Pour a Cordial in Her Wounded Heart': The Drinking Woman in Victorian and Edwardian Canada," in *Drink in Canada—Historical Essays*, ed. Warsh (Montreal: McGill-Queen's University Press, 1993), 70–91.

25. *Trial of Charles B. Huntington for Forgery; Principal Defense: Insanity*, Prepared for publication by the Defendant's Counsel, from Full Stenographic Notes Taken by Messrs. Roberts and Warburton, Law Reporters (New York: John S. Voorhies, 1857), 248. See Norman Dain, *Concepts of Insanity in the United States, 1789–1865* (New Brunswick, NJ: Rutgers University Press, 1964), 73–82, for an extended discussion of monomania and moral insanity. See also Charles Rosenberg, *The Trial of the Assassin Guiteau: Psychiatry and the Law in the Gilded Age* (Chicago: University of Chicago Press, 1995).

26. Edward C. Mann, "Intemperance and Dipsomania as Related to Insanity," in Proceedings of the Sixth Meeting of the American Association for the Cure of Inebriates, 66.

27. Jan Goldstein has argued persuasively that Georget, Esquirol's favorite student, saw monomania, or *manie sans delire*, as a useful tool to advance the psychiatric profession's status in post-revolutionary France. Medical jurisprudence offered an arena for the expansion of medical authority. Physicians in the courtroom would act as expert witnesses possessing the knowledge and skill to distinguish moral insanity from vicious and criminal behavior. The prestige of the legal profession might rub off on the coattails of the physicians while they offered these difficult cases of partial insanity. Georget was only partially successful, as were the advocates of moral insanity in the United States and Great Britain. American disease concept advocates, promoting dipsomania, attempted a similar feat. See Jan Goldstein, *Console and Classify: The French Psychiatric Profession in the Nineteenth Century* (Cambridge: Cambridge University Press, 1990), 152–96.

28. John O'Dea, "Methomania," *Medico-Legal Journal*, 1868: 67.

29. For more on Morel and degeneration theory, see Charles Rosenberg, "The Bitter Fruit: Heredity, Disease and Social Thought," in *No Other Gods;* Robert A. Nye, *Crime, Madness and Politics in Modern France: The Medical Concept of National Decline* (Princeton, NJ: Princeton University Press, 1984); Daniel Pick, *Faces of Degeneration: A European Disorder, c. 1848–c. 1918* (Cambridge: Cambridge University Press, 1989); Ian Dowbiggin, *Inheriting Madness: Professionalization and Psychiatric Knowledge in Nineteenth-Century France* (Berkeley: University of California Press, 1991); idem, *Keeping America Sane: Psychiatry and Eugenics in the United States and Canada, 1880–1940* (Ithaca, NY: Cornell University Press, 1997).

30. L. W. Baker, "The Alcohol Habit," reprinted from *Alienist and Neurologist*, April 1888. Baker's article was originally a speech given to the New England Psychological Society on 10 January 1888.

31. Charles Nichols, "Report of the Government Hospital for the Insane, for 1866

and 1867," quoted in Stephen Rogers, "Hereditary Diseases of the Nervous System Unattended by Mental Aberration," *Medico-Legal Journal*, 1869: 88.

32. George M. Beard, "Causes of the Recent Increase of Inebriety in America," *Quarterly Journal of Inebriety* 1 (1876): 25–48. For a biographical sketch of Beard and his contributions to American psychology, see Charles Rosenberg, "George M. Beard and American Nervousness," in *No Other Gods*, 98–108.

33. George Cutten, *The Psychology of Alcoholism* (New York: Scribner's, 1907), 257

34. By the "New Psychology" I refer to the late-nineteenth- and early-twentieth-century brand of psychology that replaced the philosophically oriented faculty psychology of the mid-nineteenth century with an evolutionary, physiological, and empirically testable understanding of mind. This new approach to the study of mind assumed that mental life was rooted in human experience and viewed the individual's mental development not only in relation to his physiological status but also in relation to his social experience. By psychodynamic psychiatry, I refer to the study of mental health and illness as it was believed to relate to various conscious and unconscious mental and emotional processes as they develop in the course of an individual's lifetime.

35. The term *inebriate* continued to be used, particularly in state bureaucracies and state hospitals throughout Prohibition and thereafter, although the term *alcoholism* had become much more popular by the 1920s. See Austin Barney, "Report of the Public Welfare Council on a Study of the State Farm for Inebriates," 1938, Archives of the Connecticut State Law Library, State Capitol, Hartford, Connecticut. See also Joseph Zottoli, "Report of the Special Commission to Investigate the Problem of Drunkenness in Massachusetts, Under Chapter 62 of the Resolves of 1943" (Boston: Wright and Potter, 1945). Both documents provide examples of the persistence of the term within state governments.

36. See L. W. Baker, "The Alcohol Habit," *Alienist and Neurologist* (April 1888), reprint, 2.

37. C. W. Earle, "The Responsibilities and Duties of the Medical Profession Regarding Alcoholic and Opium Inebriety," Thirty-Ninth Annual Meeting of the Illinois State Medical Society, May 21, 1889, pamphlet, 5–6.

38. Norman Kerr, *Inebriety: or Narcomania—Its Etiology, Pathology, Treatment, and Jurisprudence* (London: H. K. Lewis, 1894), 19–68.

39. Kerr served as president of the British Society for the Study of Inebriety between 1884 and 1899. For more on his tenure, see Virginia Berridge, *British Journal of Addiction: Special Issue—The Society for the Study of Addiction, 1884–1988* 85, no. 8 (August 1990), esp. 991–1003, where Berridge discusses the exchange of ideas between the AACI and the BSSI.

40. These terms are taken from patient records, annual and biennial reports of Iowa's inebriate hospitals, and published accounts of the treatment of Iowa's inebriates between 1902 and 1920.

41. Lewis D. Mason, "Inebriety A Disease," Speech delivered at a special meeting of the American Association for the Study and Cure of Inebriety, held at Harlem Lodge, Catonsville, MD, 3 June 1903, reprinted in the *Quarterly Journal of Inebriety* 25 (1903): 211–22.

42. Ibid., 213.

43. See George Miller Beard, *Stimulants and Narcotics; Medically, Philosophically, and Morally Considered* (New York: G. P. Putnam and Sons, 1871), esp. 5–25.

44. Thomas D. Crothers, ed., *The Disease of Inebriety from Alcohol, Opium and Other Narcotic Drugs—Its Etiology, Pathology, Treatment, and Medico-Legal Relations, arranged and compiled by the American Association for the Study and Cure of Inebriety* (New York: E. B. Treat, 1893).

45. Austin O'Malley, *The Cure of Alcoholism* (St. Louis, MO: B. Herder, 1913), ii.

46. Caroline J. Acker, "Stigma or Legitimation? A Historical Examination of the Social Potentials of Addiction Disease Models," *Journal of Psychoactive Drugs* 25 (July–September 1993): 198.

47. For excellent analyses of the physiological and temperance communities, see Philip Pauly, "The Struggle for Ignorance about Alcohol: American Physiologists, Wilbur Olin Atwater, and the Woman's Christian Temperance Union," *Bulletin of the History of Medicine* 64 (1990): 366–92; idem, "How Did the Effects of Alcohol on Reproduction Become Scientifically Uninteresting?" *Journal of the History of Medicine and Allied Sciences* 48 (1993): 171–97; and Jonathan Zimmerman, *Distilling Democracy: Alcohol Education in America's Public Schools, 1880–1925* (Lawrence: University of Kansas Press, 1999). Virtually any article that addressed the disease of intemperance, dipsomania, inebriety, or alcoholism paid copious reference to the demonstrated effects of alcohol on the body and mind, focusing especially on the cellular lesions occasioned within the nervous system, reproductive system, liver, and stomach.

48. See Magnus Huss, *Alcoholismus chronicus, eller chronisk alcoholsjukdom; ett bidrag till dyskrasiernas kännedom, enlight egen och andras erfarenhet* (Stockholm, 1849). The German translation followed a year later. There remains no English translation. For more on Huss's career and interests, see Jean-Charles Sournia, *A History of Alcoholism* (Oxford: Basil Blackwell, 1990), esp. 43–50.

49. See Ron Roizen, "How Does the Nation's 'Alcohol Problem' Change from Era to Era?" and Lender and Martin, *Drinking in America*, on the changing placement of responsibility for alcoholism—individual vs. beverage. See also G. Craig Reinarman, "The Social Construction of an Alcohol Problem: The Case of Mothers Against Drunk Drivers and Social Control in the 1980s," *Theory and Society* 17 (1988): 91–120.

50. Thomas D. Crothers, "The Physical Character of Crimes of the Alcoholic," *Bulletin of the American Academy of Medicine* 15 (February 1914): 33–40.

51. Thomas D. Crothers, "Criminality from Alcoholism," *Journal of Criminal Law and Criminology* 6, no. 4 (November 1915): 859–66.

52. [Samuel E. Gill], *Alcohol and Alcoholism—Alcoholism a Preventable Disease* (Philadelphia: Pennsylvania State Sabbath School Association, 1913), 8.

53. See Caroline Jean Acker, *Creating the American Junkie: Addiction Research and the Classic Era of Narcotic Control* (Baltimore: Johns Hopkins University Press, 2002); and David Courtwright, *Dark Paradise: A History of Opiate Addiction in America before 1940* (Cambridge, MA: Harvard University Press, 1982).

54. The iatrogenic addiction of women who had been prescribed opiates for pain had contributed significantly to the mid- and late-nineteenth-century addict population. However, once the medical profession began to curtail its prescription of opiates (by the dawn of the twentieth century), many of these women weaned themselves of opiates, and the remaining addicts were usually young, urban, working-class or underemployed men.

55. L. Pierce Clark, "A Psychological Study of Some Alcoholics," *Psychoanalytic Review* 4 (1919): 270–71, 274.

56. The term "New Woman" caught on in the late nineteenth century to describe a single woman with job and usually a college education. Economically and intellectually independent, she frequently engaged in sports such as bicycling, golf, and tennis, and she was usually a supporter of women's suffrage.

57. Clark, "A Psychological Study," 271–75.

58. Smith Ely Jelliffe and William A. White, *Diseases of the Nervous System: A Textbook of Neurology and Psychiatry* (Philadelphia: Lea and Febiger, 1919), 925–29.

59. Branches of the movement were established in Brooklyn, Buffalo, Detroit, Philadelphia, Baltimore, and Seattle. Although originally based in Boston's Episcopal Church, the movement extended to the Baptist, Presbyterian, Congregational, Unitarian, and Universalist churches in these cities. See Eric Caplan, *Mind Games: American Culture and the Birth of Psychotherapy* (Berkeley: University of California Press, 1998), 117–48.

60. Ibid. See also John Gardner Greene, "The Emmanuel Movement, 1906–1929," *New England Quarterly* 7, no. 3 (September 1934): 494–532. James's quotation is taken from his letter to Emmanuel movement leader and Harvard medical school professor of neurology, James Jackson Putnam. See James to Putnam, 19 August 1908, in *James Jackson Putnam and Psychoanalysis: Letters between Putnam and Sigmund Freud, Ernest Jones, William James, Sandor Ferenczi, and Morton Prince, 1877–1917* (Cambridge, MA: Harvard University Press, 1971), 74.

61. Greene, "The Emmanuel Movement," 526.

62. "Cures Already; Mental Healing Class Largely Attended; Moral Character Needed to Resist Disease; Cheerfulness and Smiles Strong Factors," *Boston Globe*, 3 January 1906, 9.

63. Mark Lender, "A Special Stigma: Women and Alcoholism in the Late 19th and Early 20th Centuries," in *Alcohol Interventions*, ed. David L. Strug, Merton M. Hyman, and S. Priyadarsini (Binghamton, NY: Haworth Press, 1986); Cheryl Krasnick Warsh, "'Oh, Lord, Pour a Cordial in Her Wounded Heart,'" 70–91.

64. Mark Lender, "Women Alcoholics: Prevalence Estimates and their Problems as Reflected in Turn-of-the-Century Institutional Data," *International Journal of Addiction* 16, no. 3 (1981): 443–48. Lender notes that the ratio of men to women was highest at the private hospitals and lowest at the public institutions. This difference may be attributed in part to the fact that so few private institutions admitted women.

65. See Rosenzweig, "The Rise of the Saloon," and Robin Room, "Cultural Contingencies of Alcoholism: Variations Between and Within Nineteenth-Century Urban Ethnic Groups in Alcohol-Related Death Rates," *Journal of Health and Social Behavior* 9 (March 1968): 99–113. Twice as many Irish men died of alcohol-related problems as Irish women, but the number of Irish men who died of liver diseases (except jaundice) was 80 percent that of Irish women.

66. Horatio R. Storer, "Appendix," in Albert Day, *Methomania: A Treatise on Alcoholic Poisoning* (Boston: James Campbell, 1867). These words are incorrectly credited to Day. See Lender and Martin, *Drinking in America*, 118.

67. M. E. J. Kelley, "Women and the Drink Problem," *Catholic World* 69 (1899): 679.

68. V. V. Anderson and C. M. Leonard, "Drunkenness as Seen among Women in Court," *Mental Hygiene* 3 (1919): 271.

69. For more on the ideology of womanhood that evolved in the first half of the nineteenth century, see Barbara Welter, "The Cult of True Womanhood, 1800–1860," *Ameri-*

*can Quarterly* 18 (Summer 1966): 151–74; Smith-Rosenberg, *Disorderly Conduct;* Nancy Cott, *The Bonds of Womanhood: "Woman's Sphere" in New England, 1780–1835* (New Haven, CT: Yale University Press, 1977); Rosalind Rosenberg, *Beyond Separate Spheres: Intellectual Roots of Modern Feminism* (New Haven, CT: Yale University Press, 1982).

70. Thomas D. Crothers, "Is Alcohol Increasing Among American Women?" *North American Review* 155 (1892): 733.

71. Rosenberg and Smith-Rosenberg, "The Female Animal."

72. E. Decaisne, "Dipsomania in Women," *Quarterly Journal of Inebriety* 7 (1884): 248.

73. Crothers, *Disease of Inebriety,* 384. Here the influence of George Miller Beard is unmistakable. As Beard had argued for neurasthenia, so the inebriety physicians claimed for the habitually drunk woman: the overly sensitive nervous systems of society's top ranks, easily exhausted, predisposed them to seek stimulants, while the lower tiers of society were simply sliding down the degenerative slope to extinction. Yet Beard appeared relatively consistent in his belief that women, by dint of their more refined nature and sensitive nervous systems, were not vulnerable to alcohol. See Beard, *Stimulants and Narcotics,* 116–17, 138–40, and "Chronic Alcoholism," in Proceedings of the Fifth Meeting of the American Association for the Cure of Inebriates, 56–57.

74. Lady Henry Somerset, "The Duxhurst Industrial Farm Colony for Female Inebriates," *British Journal of Inebriety* 10 (1912–13): 82.

75. Heywood Smith, "Alcohol in Relation to Women," *Quarterly Journal of Inebriety* 23 (1901): 191.

76. Crothers, *Disease of Inebriety,* 384; the original article appeared in the *Quarterly Journal of Inebriety* 11 (1889): 272.

77. M. E. J. Kelley, "Women and the Drink Problem," 680.

78. Lucy M. Hall, "Inebriety in Women—Its Causes and Results," *Quarterly Journal of Inebriety* 5 (October 1883): 222.

79. M. E. J. Kelley, "Women and the Drink Problem," 682–83.

80. Smith, "Alcohol in Relation to Women,": 190–91.

81. Crothers, "Is Alcoholism Increasing Among American Women?" 735.

82. See Crothers, *Disease of Inebriety;* Parrish, *Alcoholic Inebriety;* The Editor, "On the Use and Abuse of Alcohol by Women," *The Practitioner* 6 (1871): 87–96; Lucy M. Hall, "Inebriety in Women: Its Causes and Results"; idem, "Some Aspects of the Inebriety Question," *Quarterly Journal of Inebriety* 6, no. 4 (1884): 233–38; Agnes Sparks, "Alcoholism in Women—Its Cause, Consequence, and Cure," *Medico-Legal Journal* 15 (1897): 213–20; Crothers, "Is Alcoholism Increasing Among American Women?"; M. E. J. Kelley, "Women and the Drink Problem"; William C. Sullivan, "The Children of the Female Inebriate," *Quarterly Journal of Inebriety* 22, no. 2 (1900): 129–38; Lena Beach, "Inebriety and the Use of Drugs Among Women," *Women's Medical Journal* 16 (1906): 106–8; Alfred Lawrence, "Alcoholism in Women, In Some of Its Medico-Legal Aspects," *The Post-Graduate* 26, no. 1 (1911): 53–63; Victor Anderson, "The Alcoholic as Seen in Court," *Boston Medical and Surgical Journal* 174 (1916): 492–515; idem, "The Immoral Woman as Seen in Court: A Preliminary Report," *Boston Medical and Surgical Journal* 177, no. 26 (1917): 899–907; idem, "Drunkenness as Seen Among Women in Court," *Mental Hygiene* 3 (1919): 266–74.

83. Agnes Sparks, "Alcoholism in Women," 215.

84. British concern for the female inebriate was tied to fears of racial degeneration as well. For more on the female inebriate in Britain, see Peter Carpenter, "Missionaries with

the Hopeless? Inebriety, Mental Deficiency and the Burdens," *British Journal of Learning Disabilities* 28 (2000): 60–64; G. Hunt, J. Mellor, and J. Turner, "Wretched, Hatless, and Miserably Clad; Women and the Inebriate Reformatories from 1900," *British Journal of Sociology* 40 (1989): 244–70; David Gutzke, "'The Cry of the Children: The Edwardian Medical Campaign Against Maternal Drinking," *British Journal of Addiction* 79 (1984): 71–84; and Roy MacLeod, "The Edge of Hope: Social Policy and Chronic Alcoholism, 1870–1900," *Journal of the History of Medicine,* July 1967: 215–45.

85. Smith, "Alcohol in Relation to Women," 191.

86. C. W. Earle, "The Responsibilities and Duties of the Medical Profession Regarding Alcoholic and Opium Inebriety," *Address of the President, Delivered at the 39th Annual Meeting of the Illinois State Medical Society, May 21st, 1889,* reprint, 3.

87. "Alcohol," in *Standard Encyclopedia of the Alcohol Problem,* ed. Ernest Cherrington, (Westerville, OH: American Issue Publishing Company, 1925), 128.

88. See *Nostrums and Quackery: Articles on the Nostrum Evil and Quackery Reprinted, with Additions and Modifications, from the Journal of the American Medical Association* (Chicago: American Medical Association, 1912); William L. White, *Slaying the Dragon: The History of Addiction Treatment and Recovery in America* (Bloomington, IL: Chestnut Health Systems/Lighthouse Institute, 1998); and William H. Helfand, *Quack, Quack, Quack: The Sellers of Nostrums in Prints, Posters, Ephemera and Books* (New York: Grolier Club, 2002).

89. Martha Meir Allen, *Alcohol: A Dangerous and Unnecessary Medicine* (Marcellus, NY: Department of Medical Temperance of the National Woman's Christian Temperance Union, 1900), 417–18; "Is Liquor Drinking Among Women on the Increase?" in *Report of the National Woman's Christian Temperance Union—Thirtieth Convention, November 13–18, 1903,* 113.

90. Allen, "The Pure Food Law," and "Medical Temperance," in *National Woman's Christian Temperance Union Report, Thirty-third Annual Convention, October 26–31, 1906,* 114, 118–20.

91. Belle Lindner Israels, "The Way of the Girl," *The Survey* 22 (3 July 1909): 495; M. E. J. Kelley, "Women and the Drink Problem," 678–87; William C. Sullivan, "The Children of the Female Inebriate," *Quarterly Journal of Inebriety* 22, no. 2 (1900): 129–38; Hall, "Inebriety in Women," 203–24.

92. W. P. Howle, "Inebriety," in Correspondence, *JAMA* 24, 2 March 1895, 321–22.

93. For historical works on diphtheria and "Typhoid Mary" Mallon, see Evelynn Hammonds, *Childhood's Deadly Scourge: The Campaign to Control Diphtheria in New York City, 1880–1930* (Baltimore: Johns Hopkins University Press, 2002), and Judith Walzer Leavitt, *Typhoid Mary—Captive to the Public's Health* (Boston: Beacon Press, 1997).

94. Parrish, "The Philosophy of Intemperance," 25.

95. William Hammond, Thomas D. Crothers, Elon Carpenter, Cyrus Edson, "Is Drunkenness Curable?" *North American Review,* 1891, 346–74.

96. Ibid.

97. See Joseph W. Warren, "Alcohol Again: A Consideration of Recent Misstatements of its Physiological Action," *Boston Medical and Surgical Journal* 117, no. 1 (1887), reprint; Editorial, "An Appeal to Truth," *The Outlook* 64 (1900): 390–91; Clark Bell, "Alcohol and Its Use as a Diet or for Medical Purposes," *Medico-Legal Journal* 25 (1908): 53–60, 175–81; John Shaw Billings, William H. Welch et al., *Physiological Aspects of the Liquor Problem* (Boston: Houghton Mifflin, 1903); Henry Smith Williams, "Alcohol

and the Individual," *McClure's Magazine,* October 1908, 704–12; M. A. Rosanoff and A. J. Rosanoff, "Evidence Against Alcohol," *McClure's Magazine,* March 1909, 557–66; and Eugene Lyman Fisk, "Alcohol and Physiology," *Atlantic,* January 1917, 43–50.

98. Thomas D. Crothers, "The Disease of Inebriety and Its Social Science Relations," paper read before the American Social Science Association, at Saratoga, New York, September 5, 1883, *Journal of Social Science* (1883), reprint.

99. "The Analogy of Slavery and Intemperance before the Law," *New Englander,* May 1880, identifies the similarities and differences between these types of enslavement; see also "The Story of an Alcohol Slave as Told By Himself," *McClure's Magazine,* August 1909, 426–30; "Struggling Out of the Toils," *Literary Digest,* 23 September 1916, 777–78; and "Slaves of an Invisible Command! The Soul and the Drunkard," *McClure's Magazine,* November 1918, 21–22, 34.

100. Crothers, "The Disease of Inebriety and Its Social Science Relations."

101. J. W. Grosvenor, "The Relations of Alcohol to Preventive Medicine," *American Academy of Medicine* 3 (August 1897): 143.

102. Frederick Peterson, "Alcohol and Public Health," *Ninth New York State Conference of Charities and Corrections Proceedings,* November 16–20, 1908 (Albany, NY: J. B. Lyon Company, State Printers, 1909), 28–29.

103. Barron Lerner, *Contagion and Confinement: Controlling Tuberculosis Along the Skid Row* (Baltimore: Johns Hopkins University Press, 1998). See also Thomas Mays, "Alcohol as a Factor in the Causation of Pulmonary Tuberculosis," *JAMA* 68, 2 February 1907, 398–99; and Haven Emerson, "The Duty of the Health Departments in the Alcohol Question," *Boston Medical and Surgical Journal* 176 (18 July 1917) 77–79.

104. Henry Smith Williams, "Alcohol and the Individual," 710–11.

105. Mays, "Alcohol as a Factor," 399.

106. See Pauly, "Struggle for Ignorance over Alcohol." Medical temperance advocates, led by the WCTU, attempted to rein in physicians' prescription of alcohol to their patients but were not entirely unsuccessful. The dizzying array of contradictory scientific evidence levied between "Wet" and "Dry" physicians and physicians kept the public and the profession confused (See Editorial, "An Appeal to Truth," *The Outlook* 64 [17 February 1900]: 390–92). The defining truth for many if not most practitioners was what worked in the clinic, not what was determined in the laboratory, as was noted by one physician, who was conversant in the latest literature on alcohol's depressive effects yet prescribed it as a stimulant in cases of snake-bite, fainting, and pneumonia (see William Brady, "What a Glass of Whiskey Does to You," *Illustrated World* [December 1916]: 542–45, 620). With national Prohibition imminent and wartime prohibition in effect, the AMA finally passed a resolution in 1917 opposing the use of alcohol as a beverage and as a medicine.

107. Homer Folks, "Social Aspects of Alcoholism," *The Survey* 25 (October 1910): 15.

108. J. W. Grosvenor, "Is the Habitual Use of Alcoholic Intoxicants in the Home Consistent with Its Thorough Sanitation?" *Bulletin of the American Academy of Medicine* 5 (October 1900): 80.

109. Mary Richmond, *Social Diagnosis* (New York: Russell Sage Foundation, 1917), 431. Having consulted with inebriety specialist Irwin Neff, the superintendent of the Massachusetts state hospital for inebriates, Richmond drew up a questionnaire to assist social workers managing cases of inebriety.

110. Maudsley quoted in Joseph Parrish, "The Philosophy of Intemperance," 28.

111. Stephen Rogers, "Hereditary Diseases of the Nervous System, Unattended by Mental Aberration," 77.

112. J. M. Toner, W. W. Godding, and N. S. Lincoln, "Meeting of December 22: Report of Committee on Dr. Godding's Paper," *JAMA* 8, no. 2, 8 January 1887, 50.

113. Ibid., 49.

114. L. W. Baker, "The Alcohol Habit," 8.

115. Gonzalva Smythe, "Influence of Heredity in Producing Disease and Degeneracy. The Remedy. Insanity, Inebriety, Crime and Pauperism Considered Diseases. Acquired Physical Conditions Transmitted to Offspring. An Unprejudiced Discussion of the Alcohol Question. All Sin Reduced to a Physical Basis," *Transactions of the Indiana State Medical Society, 42nd Annual Session held in Indianapolis, Indiana, June 10 & 11, 1891*, 21–22.

116. Editorial, "The Restriction of Procreation," *St. Paul Medical Journal* 10 (April 1908): 214–15. This editorial endorsed the restriction of procreation through vasectomy for "the large class embraced under the comprehensive term, 'degenerates,' which includes the chronic insane, the imbecile, the epileptic, the confirmed inebriate, the sexual pervert, and the habitual criminal," 214. The editorial also discussed Indiana as a model state, leading the way for eugenic sterilization efforts.

117. Arthur H. Estabrook and Charles B. Davenport, *The Nam Family—A Study in Cacogenics*, Eugenics Record Office Memoir No. 2 (Cold Spring Harbor, Long Island, NY: Eugenics Record Office, 1912). "Nam" was a pseudonym used by investigators to protect the identity of the people studied.

118. Ibid., 2.

119. Ibid., 84.

120. E. E. Southard, "Notes on the Geographical Distribution of Insanity in Massachusetts, 1901–1910," *American Breeders' Magazine* 3 (1912): 11–20. For more on Southard's eugenical views and the relations between psychiatry and eugenics, see Lunbeck, *The Psychiatric Persuasion*, and Ian Robert Dowbiggin, *Keeping America Sane: Psychiatry and Eugenics in the United States and Canada, 1880–1940* (Ithaca, N.Y.: Cornell University Press, 1997).

121. Edward C. Mann, "The Disease of Inebriety," *JAMA*, 1 December 1894, 822; Mann quoted Willard Parker's speech from the 1871 meeting of the AACI, "What Science and the Inebriate Asylums have Taught Us."

122. Irwin H. Neff, "Some of the Medical Problems of Alcoholism," *Boston Medical and Surgical Journal* 164, no. 4 (26 January 1911), 112.

TWO: Cultural Framing of Inebriety

1. Joseph Parrish, "The Philosophy of Intemperance," 29.

2. Albert Day, "Inebriety and Its Cure—A Paper Read before the Suffolk District Medical Society, December 30, 1876" (Boston: C. W. Calkins & Co., Printers, 1877), 8.

3. George M. Beard, "Causes of the Recent Increases of Inebriety in America," 30.

4. Ibid., 31. Note that similar concerns motivated Benjamin Rush to write his "Inquiry into the Effects of Ardent Spirits" and launch the nineteenth-century temperance movement. While Beard was more concerned with the pace of life and the impact of modern technology on it, Rush was more concerned with life in a democratic republic, for he regarded the republic's health as only as good as its citizens'.

For Rush, sobriety was essential if all individuals were to have a say in the republic's governance.

5. C. H. Hughes, "The Successful Management of Inebriety without Secrecy in Therapeutics," paper read before the Section on Diseases of the Mind and Nervous System, Pan-American Medical Congress, Washington, DC, September 7, 1893, reprint, 4.

6. Lillian Brandt, "Alcoholism and Social Problems," *The Survey* 25 (1910): 21.

7. For more on the relationship between industrial workers and alcohol, see Roy Rosenzweig, "The Rise of the Saloon," and *Eight Hours for What We Will*. See also W. Mattieu Williams, "Stimulants and Work," *Knowledge*, 18 August 1882, 201; "Industrial Alcoholism," *American Journal of Sociology* 11, no. 5 (March 1906): 715; Lewis Edwin Theiss, "Industry versus Alcohol," *The Outlook* (8 August 1914): 856–61; Theiss, "The People versus Alcohol," *The Outlook* (9 June 1915): 315–16, 325–27; "No Booze for Big Business," *Literary Digest*, March 1916, 569–70; and Eugene Lyman Fisk, "Alcohol and Human Efficiency," *Atlantic*, February 1917: 203–10.

8. Brandt, "Alcoholism and Social Problems," 19.

9. The pace of society may have proven too great even for Frederick Taylor, the great champion of scientific management, who along with many other prominent Americans—Jane Addams, William James, Woodrow Wilson—experienced a nervous breakdown at the turn of the century. See Nathan G. Hale, *Freud and the Americans: The Beginnings of Psychoanalysis in the United States, 1876–1917* (New York: Oxford University Press, 1995), 233–34.

10. Elizabeth Tilton, "The Rising Anti-Alcohol Movement," *The Survey* 32 (8 August 1914): 488.

11. P. A. Lovering, "The Alcohol Question in the Navy," *JAMA* 47, 16 February 1907, 591.

12. Rosenzweig, "The Rise of the Saloon," 124.

13. U.S. Commissioner of Labor, *Twelfth Annual Report: Economic Aspects of the Liquor Problem* (Washington, DC: Government Printing Office, 1898), 69–72.

14. Day, "Inebriety and Its Cure," 14.

15. Crothers, *Disease of Inebriety;* Norman Kerr, *Inebriety or Narcomania—Its Etiology, Pathology, Treatment and Jurisprudence* (London: H. K. Lewis, 1894), 654–77.

16. For more on the rise of consumer society, see Richard W. Fox and T. Jackson Lears, eds., *The Culture of Consumption: Critical Essays in American History* (New York: Pantheon Books, 1983); Daniel Horwitz, *The Morality of Spending: Attitudes Toward the Consumer Society in America, 1875–1940* (New York: Ivan Dee, 1992); Richard Bushman, *The Refinement of America: Persons, Houses, Cities* (New York: Alfred Knopf, 1992); William Leach, *Land of Desire: Merchants, Power, and the Rise of a New American Culture* (New York: Pantheon Books, 1993); Lawrence Glickman, *A Living Wage: American Workers and the Making of Consumer Society* (Ithaca, NY: Cornell University Press, 1997); Lawrence Glickman, ed., *Consumer Society in American History: A Reader* (Ithaca, NY: Cornell University Press, 1999). For more on the changing definition of freedom in American culture and the changing relations of individual and state, see Eric Foner, *The Story of American Freedom* (New York: W. W. Norton, 1998), and Jeffrey Sklansky, *The Soul's Economy: Market Society and Selfhood in American Thought, 1820–1920* (Chapel Hill: University of North Carolina Press, 2002).

17. Lewis D. Mason, "The Relation of the Pauper Inebriate to the Municipality and

the State from the Economic Point of View," *Quarterly Journal of Inebriety* 26 (October 1904): 327.

18. For more on the "worthy" vs. "unworthy" poor, see Michael B. Katz, *In the Shadow of the Poorhouse: A Social History of Welfare in America* (New York: Basic Books, 1986), 19.

19. Brandt, "Alcoholism and Social Problems," 19.

20. Day, "Inebriety and Its Cure," 9.

21. Ernest Emerson, "Alcoholism," *Boston Medical and Surgical Journal* 173 (15 July 1915): 114–15.

22. Stewart Paton, *Psychiatry: A Textbook for Students and Physicians* (Philadelphia: J. B. Lippincott, 1905), 310.

23. Henry George, *Progress and Poverty: An Inquiry into the Cause of Industrial Depressions, and of Increase of Want with Increase of Wealth—The Remedy* (San Francisco: W. M. Hinton & Co., Printers, 1879).

24. For one of the most strident anti-immigrant statements from a medical author, see Gonzalva Smythe, "Influence of Heredity," esp. 10–11. Smythe argued in favor of immigration restrictions, believing that "it is the duty of our authorities to inform these people that they can not be admitted to the United States unless they furnish a consular certificate, that neither tuberculosis, scrofula, cancer, insanity, inebriety, crime or pauperism is hereditary in the families from which they sprung.... No one should be admitted who can not be perfectly digested, absorbed, assimilated and Americanized."

25. Henry William Blair, *The Temperance Movement or the Conflict between Man and Alcohol* (Boston: William E. Smythe Company, 1888).

26. See Roy Lubove, *The Professional Altruist*; Michael Katz, *In the Shadow of the Poorhouse*; Allan J. Davis, *Spearheads for Social Reform*; Camilla Stivers, *Bureau Men, Settlement Women*.

27. As late as 1900, 50 percent of the nation's biggest cities offered almost no poor relief other than that afforded at the poorhouse (Eric Foner, *The Story of American Freedom* (New York: W. W. Norton, 1998), 121. See also Katz, *In the Shadow of the Poorhouse.*

28. Parrish, "The Philosophy of Intemperance," 31.

29. Orpheus Everts, *What Shall We do for the Drunkard? A Rational View of the Use of Brain Stimulants* (Cincinnati, OH: Robert Clarke & Co., 1883), 34–35.

30. For more on the necessity of regarding the inebriate as a state ward, see W. W. Godding, "The Problem of the Inebriate," *JAMA* 8, 8 January 1887, 29–32; and William Hammond, Thomas D. Crothers, Elon Carpenter, and Cyrus Edson, "Is Drunkenness Curable?" *North American Review* 153 (1891): 346–74.

31. Willard quoted in Gerald Grob, *Mental Institutions in America: Social Policy to 1875* (New York: Free Press, 1973), 309–10.

32. Jack D. Pressman, *Last Resort: Psychosurgery and the Limits of Medicine* (Cambridge: Cambridge University Press, 1998), 20.

33. Paton, *Psychiatry,* 10, 310.

34. Joseph Gusfield, "Passage to Play: Rituals of Drinking Time in American Society," in *Constructive Drinking,* ed. Mary Douglas (Cambridge: Cambridge University Press, 1987).

35. Jack London, *John Barleycorn* (New York: Century, 1913), 339.

36. For more on treating and clubbing and their symbolic significance, see Powers, *Faces along the Bar*; Marianna Alder, "From Symbolic Exchange to Commodity Con-

sumption: Anthropological Notes on Drinking as a Symbolic Practice," in *Drinking: Behavior and Belief*, ed. Barrows and Room, 376–98; and Lewis Hyde, *The Gift: Imagination and the Erotic Life of Property* (New York: Vintage, 1979).

37. The crisis of masculinity received a great deal of attention from historians in the 1980s and 1990s. See Jeffrey P. Hantover, "Progressivism and the Masculinity Crisis," in *The American Man*, ed. Elizabeth Pleck and Joseph Pleck (Englewood Cliffs, NJ: Prentice Hall, 1980); Mark C. Carnes and Clyde Griffen, *Meanings for Manhood—Constructions of Masculinity in Victorian America* (Chicago: University of Chicago Press, 1990); Gail Bederman, *Manliness and Civilization: A Cultural History of Gender and Race in the United States, 1880–1917* (Chicago: University of Chicago Press, 1995); Michael Kimmel, *Manhood in America: A Cultural History* (New York: Free Press, 1996); E. Anthony Rotundo, *American Manhood: Transformations in Masculinity from the Revolution to the Modern Era* (New York: Basic Books, 1994).

For excellent discussions of the relationship of gender, alcohol, and alcoholism, see Michelle McClellan, "'Lady Tipplers': Gendering the Modern Alcoholism Paradigm, 1933–1960," in Tracy and Acker, *Altering American Consciousness*; Lori Rotskoff, "Sober Husbands and Supportive Wives: Marital Dramas of Alcoholism in Post–World War II America," in Tracy and Acker, *Altering American Consciousness*; Murdock, *Domesticating Drink*; and Lori Rotskoff, *Love on the Rocks—Men, Women, and Alcohol in Post–World War II America* (Chapel Hill: University of North Carolina Press, 2002).

38. Lunbeck, *The Psychiatric Persuasion*, 245.

39. For more on the significance of the saloon in working-class life, see Rosenzweig, *Eight Hours for What We Will*; Powers, *Faces along the Bar*; Alder, "From Symbolic Exchange to Commodity Consumption"; Duis, *The Saloon*; Jon M. Kingsdale, "The Poor Man's Club: Social Functions of the Urban Working-Class Saloon," in Pleck and Pleck, *The American Man*; Randy D. McBee, *Dance Hall Days: Intimacy and Leisure among Working-Class Immigrants in the United States* (New York: New York University Press, 2000).

40. Robert Archey Woods, "Drunkenness: How the Local Community Can Be Brought to Do Its Part," Address at the Probation Conference, Norfolk State Hospital, Pondville, Massachusetts, 28 October 1915.

41. For a splendid statement in favor of middle-class sobriety and the necessity of self-control for success in life, see Francis Landry Patton, *The Duty of Self Control, An Address to Students of Princeton University, in Marquand Chapel, Sunday afternoon, January 30, 1898* (Princeton, NJ: Princeton University Press, 1898). For a statement of the necessity of drinking to the working classes, see Samuel Gompers, "Labor and Beer—When You Invade a Man's Habits, What Happens?" *McClure's Magazine*, June 1919, 30–31.

42. See Carroll Smith-Rosenberg, "The New Woman as Androgyne: Social Disorder and Gender Crisis, 1870–1936," in *Disorderly Conduct: Visions of Gender in Victorian America* (New York: Oxford University Press, 1985), 245–96.

43. Karl Abraham, "The Psychological Relations between Sexuality and Alcoholism," (originally published in *Zeitschrift für Sexualwissenschaft*, 1908) *Selected Papers of Karl Abraham, M.D.*, ed. Douglas Bryan and Alix Strachey (London: Hogarth Press and the Institute of Psychoanalysis, 1949), 83.

44. Ibid., 86–87.

45. Daniel Dorchester, *The Liquor Problem in All Ages* (New York: Phillips and Hunt, 1884), 464.

46. Hammond, Crothers, Carpenter, and Edison, "Is Drunkenness Curable?" 354.

47. See Pauly, "The Struggle for Ignorance about Alcohol."

48. Frances Willard, "The Summing Up of the Whole Matter," speech to the Fourth Biennial Meeting of the World's WCTU, Toronto, 23 October 1897, reprinted in Anna A. Gordon, *The Beautiful Life of Frances E. Willard* (Chicago: Woman's Temperance Publishing Association, 1898), 175.

49. Pauly, "The Struggle for Ignorance about Alcohol." See also Zimmerman, *Distilling Democracy.*

50. Ibid.

51. Irwin Neff, "Inebriety from a Medical Viewpoint," *Boston Medical and Surgical Journal* 176 (8 February 1917): 205.

52. As Jonathan Zimmerman has shown, the medical critique of scientific temperance education was not alone. Educators resented the intrusion of temperance agitators in their professional realm, decrying and often disregarding the Scientific Temperance Instruction Bureau's officially recommended texts for physiology instruction. See Zimmerman, *Distilling Democracy,* esp. 59–80, 81–100.

53. Frances Willard, "The Summing Up of the Whole Matter," 175.

54. Ibid., 173–74.

55. Willard, "Summing Up," 175.

56. Murdock, *Domesticating Drink,* 47.

57. "Matron's Report," *Second Report of the New England Home for Intemperate Women* (Boston: Franklin Press, 1881), 7.

58. Frances Willard, "President's Annual Address," *Minutes of the National Woman's Christian Temperance Union at the Twentieth Annual Meeting, October 18–21, 1893* (Chicago: Woman's Temperance Publishing Association, 1893), 138.

59. "Resolutions," *Minutes of the National Woman's Christian Temperance Union, November 16–21, 1894* (Chicago: Woman's Temperance Publishing Association, 1894), 48.

60. Zimmerman, *Distilling Democracy,* 51.

61. Editorial, "The Care of the Inebriate," *JAMA* 24, no. 8, 23 February 1895, 288–89; editorial, "State Care of Inebriates," *JAMA* 48, no. 9, 2 March 1907, 803.

62. For more on the interesting tale of Leslie E. Keeley, see White, *Slaying the Dragon,* 50–63; Cheryl Warsh, "Adventures in Maritime Quackery: The Leslie E. Keeley Gold Cure Institute of Frederickton, NB," *Acadiensis* 17 (Spring 1988): 109–30; H. Wayne Morgan, "No. Thank you. I've been to Dwight: Reflections on the Keeley Cure for Alcoholism," *Illinois Historical Journal* 82, no. 3 (1989): 147–66; Lender and Martin, *Drinking in America,* 122–24; "Keeley, Leslie Enraught," "Keeley Cure," and "Keeley League," in *Standard Encyclopedia of the Alcohol Problem,* ed. Ernest Cherrington (Westerville, OH: American Issue Publishing Company, 1928); "Keeley, Leslie Enraught," in *Dictionary of American Temperance Biography—From Temperance Reform to Alcohol Research, the 1600s to the 1980s,* ed. Mark Lender (Westport, CT: Greenwood Press, 1984), 270–72; C. S. Clark, *The Perfect Keeley Cure: Incidents at Dwight . . .* (Milwaukee, WI: n.p., 1893); Fred B. Hargreaves, *Gold—A Cure for Drunkenness, Being an Account of the Double Chloride of Gold Discovery Recently Made by Leslie E. Keeley,* pamphlet (Dwight, IL: Keeley Institute, 1880); Henry Wood, "Does Bi-Chloride of Gold Cure Inebriety?" *The Arena* 7 (January 1893): 145–52; Leslie E. Keeley, *Inebriety is a Disease,* pamphlet issued by Keeley Bureau of Information, Boston, MA, 1894; and "Editor Medill and the Keeley Cure," *Golden Magazine,* 1895, 315–22; "Treatment for Oklahoma Inebriates," *JAMA* 25, 26 October 1895, 730.

63. For more on Keeley's own views about inebriety and its treatment, see Leslie E. Keeley, "Drunkenness, A Curable Disease," *American Journal of Politics* (July 1892): 27–43; idem, "Does Bi-Chloride of Gold Cure Inebriety?" *The Arena* 7 (March 1893): 450–60; idem, "Inebriety and Insanity," *The Arena* 8 (August 1893); idem, "Drunkenness and Its Cure," *Golden Magazine*, January 1895, 307–13; idem, *The Non-Heredity of Inebriety* (Chicago: S. C. Griggs and Company, 1896).

64. These are John Medill's words as reported in "Editor Medill and the Keeley Cure," *Golden Magazine*, January 1895, 321.

65. Warsh, "Adventures in Maritime Quackery," 121–22.

66. Chauncey Chapman, "The Bichloride of Gold Treatment of Dipsomania," *Chicago Medical Recorder* 4 (1893): 105–9.

67. Warsh, "Adventures in Maritime Quackery," 119; Chapman, "The Bichloride of Gold Treatment," 106, 109.

68. Warsh, "Adventures in Maritime Quackery," 119.

69. Leslie E. Keeley, *Opium—Its Use, Abuse, and Cure; or From Bondage to Freedom* (Chicago: Banner of Gold Co., Publishers, 1897), 89. This text is fascinating also for its discussion of the general surfeit of drugs that people took at the turn of the century, or what Keeley called "The Age of Drugs" (115–18).

70. J. M. French, "The Treatment of Alcoholism," *Boston Medical and Surgical Reporter*, 12 September 1897, reprint, 1–2.

71. Lewis D. Mason, "Inebriety a Disease," Speech delivered at a special meeting of the American Association for the Study and Cure of Inebriety, held at Harlem Lodge, Catonsville, MD, 3 June 1903, reprinted in the *Quarterly Journal of Inebriety* 25 (1903): 211–22.

72. H. R. Chamberlain, "Modern Methods of Treating Inebriety," *The Chautaquan*, July 1891, 497.

73. Keeley, *The Non-Heredity of Inebriety*, 281.

74. Ibid., 286–89.

75. Hammond, Crothers, Carpenter, and Edson, "Is Drunkenness Curable?" 366.

76. Ibid., 351.

77. Frederick Peterson, "The Treatment of Alcoholic Inebriety," *JAMA* 20, 15 April 1893, 408–9.

78. Irwin Neff, "Some of the Medical Problems of Alcoholism," *JAMA* 164, no. 4, 26 January 1911, 114.

79. Charles Dana, "Inebriety: A Study of Its Causes, Duration, Prophylaxis, and Management," *Medical Record* 60, no. 4, 27 July 1901, 123.

80. Editorial, "The Care of the Inebriate," *JAMA* 24, no. 8, 23 January 1895, 288.

THREE: Institutional Solutions for Inebriety

1. Thomas Rand to Irwin Neff, September 1914, Inpatient Case Files of the Norfolk State Hospital, Agency Code HS7.17, Series 260X, Department of Mental Health (held by agreement with the Massachusetts Archives at Columbia Point, Boston, MA). Hereafter, I will refer to these files with the Patient Pseudonym, date of document, and any other identifying information, such as the individual to whom a letter was written or the patient history. I have changed the names of patients to conceal their identities. The only

exceptions to this practice are individuals whose names were published in newspapers, which was often the custom in states such as Massachusetts and Iowa.

2. Irwin Neff to Thomas Rand, 21 September 1914, Inpatient Case Files of NSH.

3. Jim Baumohl, "Inebriate Institutions in North America, 1840–1920," in *Drink in Canada: Historical Essays,* ed. Cheryl Krasnick Warsh (Montreal: McGill-Queen's University Press, 1993), 111. Baumohl was one of the earliest and most thorough historians of the inebriate asylum movement, and most of his initial assessments of inebriate institutions have stood the test of time. He subsequently revised his assessment of the state industrial hospital when presented with the case of Massachusetts, see Jim Baumohl and Sarah W. Tracy, "Building Systems to Manage Inebriates: The Divergent Paths of California and Massachusetts, 1891–1920, *Contemporary Drug Problems* 21 (Winter 1994): 557–95.

4. See White, *Slaying the Dragon,* 21–127, for an excellent discussion of the wide range of treatment solutions for inebriates of all varieties between 1850 and 1930.

5. See *Annual Report of the Washingtonian Home, Boston for the Year 1871* (Boston: Wright and Potter, 1871); White, *Slaying the Dragon,* 47–48; *Annual Reports of the Norfolk State Hospital, 1914–1919* (Boston: Wright and Potter, 1914–20); *Annual Reports of the Foxborough State Hospital, 1905–1914* (Boston: Wright and Potter, 1905–15); "Report of the Commissioners on Inebriate Asylums," Senate Document No. 9, 9 January 1871; "Report of the Joint Special Committee on Establishment of an Inebriate Asylum in the Commonwealth," Senate Document No. 235, 28 April 1864; Parton, *Smoking and Drinking.*

6. John Shaw Billings in "Minutes of the Section on Public Hygiene and State Medicine," 325; Lyman Abbott, "Letter to the Editor—Reply," *The Outlook* 98 (12 August 1911): 816.

7. Benjamin Rush, "Plan for an Asylum for Drunkards to be called the Sober House," 354–55.

8. Elihu Todd, Samuel B. Woodward et al., *Report of a Committee of the Connecticut Medical Society Respecting an Asylum for Inebriates,* 5. For an excellent overview of Pinel's work at Bicêtre and Saltpetriere and Tuke's career at the York Retreat, see Gerald Grob, *The Mad Among Us: A History of the Care of America's Mentally Ill* (New York: Free Press, 1994), 25–29. For more detailed treatments of their moral therapy, see Dora B. Weiner, "Philippe Pinel's 'Memoir on Madness' of December 11, 1794: A Fundamental Text of Modern Psychiatry," *American Journal of Psychiatry* 149 (1992): 725–32; and Anne Digby, *Madness, Morality and Medicine: A Study of the York Retreat, 1796–1914* (Cambridge: Cambridge University Press, 1985). For an excellent analysis of the early drive to establish an inebriate asylum in Connecticut, see Mark Luoto, "The Public Inebriate in Connecticut," undergraduate thesis, Yale University, 1974.

9. Samuel B. Woodward, *Essays on Asylums for Inebriates* (Worcester, MA, 1838), reprinted in *Nineteenth-Century Medical Attitudes toward Alcoholic Addiction, Six Studies, 1814–1867,* ed. Gerald N. Grob (New York: Arno Press, 1981), 1.

10. Tomes, *A Generous Confidence,* 252–53, 262.

11. For more on the experience of Elizabeth Packard, see Barbara Sapinsley, *The Private War of Mrs. Packard* (New York: Paragon House, 1991).

12. Charles Hughes in "Proceedings of the Association of Medical Superintendents, 21 May 1875," *American Journal of Insanity* 32 (January 1876): 353.

13. Todd and Woodward, *Report of a Committee*, 2. See also Mark Luoto, "The Public Inebriate in Connecticut."

14. "Disabilities of Inebriates—A Communication from the Inmates of the Pennsylvania Sanitarium," in Proceedings of the First Meeting of the American Association for the Cure of Inebriates, 41.

15. See Katherine Chavigny, "Reforming Drunkards in Nineteenth-Century America: Religion, Medicine, Therapy," in Tracy and Acker, *Altering American Consciousness*, 108–23.

16. The pledge is quoted in several authoritative secondary texts on the Washingtonian movement. See White, *Slaying the Dragon*, 8; Cherrington, ed., *Standard Encyclopedia of the Alcohol Problem*, 2807; Leonard U. Blumberg and William L. Pittman, *Beware the First Drink! The Washington Temperance Movement and Alcoholics Anonymous* (Seattle: Glen Abbey Books, 1991), 59.

17. White, *Slaying the Dragon*, 10.

18. For more on John Augustus, see *A Report of the Labors of John Augustus, for the Last Ten Years in Aid of the Unfortunate: Containing a Description of his Method of Operations; Striking Incidents, and Observations upon the Improvement of Some of our City Institutions, with a View to the Benefit of the Prisoner and of Society* (Boston: Wright and Hasty, 1852).

19. Edgar K. Whitaker, Sarah Holland et al. "Memorial to the Honorable Legislature of the State of Massachusetts," House Document No. 38, Commonwealth of Massachusetts, 1845, 2–7.

20. Why the Home for the Fallen was renamed the Washingtonian Home remains a mystery. However, it is important to emphasize that it had no ties to the earlier Washingtonian temperance movement. I will refer to the Washingtonian Homes—all of which had no affiliation with the temperance drive of the same name—as neo-Washingtonian homes, following the convention of other historians.

21. Quotations from the *Annual Reports of the Washingtonian Home* for the years 1859 through 1863, quoted in House Document No. 2000, Joseph Zottoli et al., *Report of the Special Commission to Investigate the Problem of Drunkenness in Massachusetts*, March 7, 1945 (Boston: Wright and Potter, 1945), 140.

22. Day's biography is recounted in Parton, *Smoking and Drinking*, 117–22. Day was instrumental in lobbying the Massachusetts Legislature for the incorporation of the Boston Washingtonian Home and securing the state's financial support.

23. Day, *Annual Report of the Washingtonian Home for 1866*, quoted in House Document No. 2000, Zottoli et al., *Report of the Special Commission*, 145.

24. Day, *Annual Report of the Washingtonian Home for 1873*, quoted in House Document No. 2000, Zottoli et al., *Report of the Special Commission*, 148.

25. Day served two terms as superintendent at the Boston Washingtonian Home. The first was between 1857 and 1867. He left Boston to take on the troubled New York State Inebriate Asylum in Binghamton, where he served from 1867 to 1870, returning to Massachusetts to run the private Greenwood Institute for Inebriates between 1870 and 1873. He returned to manage the Boston Washingtonian Home between 1873 and 1893.

26. Thomas Vancourt, "Brief Historical Sketch of the Chicago Washingtonian Home; with Some General Remarks," in Proceedings of the Second Meeting of the American Association for the Cure of Inebriety, 103.

27. Ibid., 114.

28. P. J. Wardner, "The Moral and Social Treatment of Inebriates," in Proceedings of the First Meeting of the American Association for the Cure of Inebriates, 71.

29. Bederman, *Manliness and Civilization,* 172.

30. Vancourt, "Brief Historical Sketch," 115.

31. Although both homes were closed with Prohibition, their patient censuses dwindling to dozens rather than hundreds of inebriates, the managers used the facilities' collective assets to erect a new hospital for Chicagoans, the Martha Washington Hospital, which served general as well as addicted patients. It functioned until 1991, when a number of Chicago's hospitals were consolidated under managed care.

32. Robert Harris, "Report of the Franklin Reformatory Home for Inebriates, 913–915 Locust Street, Philadelphia," in Proceedings of the Fifth Meeting of the American Association for the Cure of Inebriates, 80–81.

33. Ibid., 85–86.

34. "Dr. Parrish's Remarks," in Proceedings of the Fifth Meeting of the American Association for the Cure of Inebriates, 75, 78.

35. *Second Report of the New England Home for Intemperate Women* (Boston: Franklin Press, 1881), 7–8.

36. For more on the pernicious effects of the city, see Kimmel, *Manhood in America,* 81–155. Of course, concerns about the unhealthy nature of urban living were nothing new. The early sanitarians had highlighted the city's assault on physical health, promoting better public water supplies and sewage disposal, among other municipal services for the benefit of the public's health. See George Rosen, *A History of Public Health* (Baltimore: Johns Hopkins University Press, 1993); Charles E. Rosenberg, *The Cholera Years: The United States in 1832, 1849, 1866* (Chicago: University of Chicago Press, 1962). George Miller Beard, William Alexander Hammond, and Silas Weir Mitchell, however, were most concerned with the ravaging effects of the urban world on humankind's nervous system and psyche. For these physicians it was the fast-paced, technologically driven nature of the modern city that bedeviled humans at the end of the nineteenth century. See Beard, *American Nervousness, Its Causes and Consequences; a Supplement to Nervous Exhaustion (Neurasthenia)* (New York: G. P. Putnam's Sons, 1881); Mitchell, *Fat and Blood: An Essay on the Treatment of Certain Forms of Neurasthenia and Hysteria* (Philadelphia: J. B. Lippincott and Co., 1884); and Hammond, *A Treatise on Diseases of the Nervous System* (New York: D. Appleton and Co., 1871).

37. Charles Rosenberg, "Community and Communities: The Evolution of the American Hospital," in *The American General Hospital: Communities and Social Context,* ed. Diana Long and Janet Golden (Ithaca, NY: Cornell University Press, 1989), 6; see also Rosenberg, *The Care of Strangers: The Rise of America's Hospital System* (New York: Basic Books, 1987).

38. Samuel Godwin, "How our Home is Conducted," in Proceedings of the Third Meeting of the American Association for the Cure of Inebriates, 49.

39. D. Banks McKenzie, "The Inebriate and His Treatment," in Proceedings of the Fourth Meeting of the American Association for the Cure of Inebriates, 45.

40. Albert Day, *Annual Report of the Washingtonian Home for 1866,* quoted in House Document No. 2000, Joseph Zottoli et al., *Report of the Special Commission,* 144. Zottoli adds that Day is quoting James Hutchinson of Glasgow.

41. *Second Report of the New England Home for Intemperate Women* (Boston: Franklin Press, 1881), 8.

42. Joseph Parrish, *First Annual Report of the President of the Citizens Association of Pennsylvania, 1867* (Philadelphia: Henry B. Ashmead, 1867), 18. The Citizens Association was a voluntary philanthropic association drawing Philadelphia luminaries together to establish an inebriate asylum. Parrish was its leader, as well as the first director of the Pennsylvania Sanitarium, the institution the association erected for treating inebriates.

43. *Second Report of the New England Home for Intemperate Women,* 7.

44. Lady Henry Somerset, "The Duxhurst Industrial Farm Colony for Female Inebriates," *British Journal of Inebriety* 10 (1912–13). Although this report of the Duxhurst Colony is later than the work of the New England Home for Intemperate Women, it highlights the essence of the gendered approach to treating women in the inebriate home of the late nineteenth century.

45. Jim Baumohl, "Inebriate Institutions in North America," 101.

46. J. Edward Turner, *The History of the First Inebriate Asylum in the World by Its Founder; An Account of His Indictment, also a Sketch of the Woman's National Hospital* (New York: J. Edward Turner, 1888), 62–82. For an insightful treatment of the New York asylum, see John W. Crowley and William L. White, *Drunkard's Refuge: Lessons of the New York State Inebriate Asylum* (Amherst: University of Massachusetts Press, 2004).

47. Turner, *History of the First Inebriate Asylum,* 55.

48. Ibid., 56.

49. Parton, *Smoking and Drinking,* 115.

50. Ibid., 193–221. For an analysis of later demographics, see Jim Baumohl and Robin Room, "Inebriate Institutions and the State Before 1940," in *Recent Developments in Alcoholism,* ed. Marc Galanter (New York: Plenum, 1987), 5:135–74.

51. Though the charges were eventually dropped, this spelled the beginning of the end for the embittered Turner. In the early 1880s, he attempted to establish a Woman's National Hospital for Inebriates in Wilton, Connecticut, but was unsuccessful because the scandal from Binghamton undermined his credibility. Exhausted after years of difficulties managing the hospital, Willard Parker deeded the asylum to the State of New York for one dollar in 1867, and twelve years later, in 1879, the state converted it to a public hospital for the insane.

52. Parton, *Smoking and Drinking,* 125–26.

53. See Anonymous "Our Inebriates, Classified and Clarified by an Inmate of the New York State Inebriate Asylum," *Atlantic Monthly,* April 1869, 477–83; and Anonymous, "Our Inebriates, Harbored and Helped, by an Inmate of the New York Inebriate Asylum," *Atlantic Monthly,* July 1869, 109–19.

54. Parton, *Smoking and Drinking,* 134.

55. Anonymous, "Our Inebriates, Harbored and Helped," 111.

56. Ibid., 116.

57. Ibid., 114–15.

58. Zimmerman, *Distilling Democracy.* See also Thomas D. Crothers, "New School of Psychology," *Quarterly Journal of Inebriety* 19 (1897): 209–10.

59. *First Annual Report of the President of the Citizen's Association of Pennsylvania,* 5.

60. Ibid., 18–19.

61. For more on the cottage plan, see Ellen Dwyer, *Homes for the Mad,* 137; see Tomes,

*A Generous Confidence,* 267–70, for a description of the far more common "Kirkbride plan" of linear, monolithic institutional construction; see also *Report of the Trustees of the Massachusetts Hospital for Dipsomaniacs and Inebriates (at Foxborough), January 1893* (Boston: Wright and Potter, 1893), esp. 6–7; and "Trustees' Report," *Twenty-first Annual Report of the Foxborough State Hospital for the Year Ending 30 November 1912* (Boston: Wright and Potter, 1912), 8–9. When a new hospital was built by Foxborough patients on new grounds in Norfolk, Massachusetts, the "cottage system" was chosen once again.

62. *First Annual Report of the President of the Citizen's Association,* 19. Parrish's typology of alcoholism was as follows: *Casual drinkers* were the least diseased, for they only occasionally drank, and only occasionally when they drank did they drink to excess; *impulsive and unbalanced youthful drinkers* gave way to their appetites and temptation, posing significant dangers to themselves and others; *periodical drinkers* were usually model citizens in their sober periods, but were afflicted by irregularly spaced compulsions to drink; the *habitually moderate drinker,* like the casual drinker, only occasionally drank to excess, but he drank every day; *habitual drunkards* drank themselves into oblivion every day; the *vagrant inebriate* lived in the pub and though capable of work, had chosen an idle, inebriated life; and the *minor drinkers* were underage drinkers of all varieties (their numbers, Parrish believed, were increasing to a horrifying extent). Women fell into each of these categories as well (ibid., 6–11).

63. Ibid., 19.

64. *Pennsylvania Sanitarium Address to the People by the Directors* (Philadelphia: Henry B. Ashmead, Printer, 1871), 9.

65. Ibid., 15.

66. Edward C. Mann, *The Nature and Treatment of Inebriety also the Opium Habit and Its Treatment* (New York: Charles A. Coffin, 1878), 43.

67. Ibid.

68. Thomas D. Crothers, "Private Asylums and their Difficulties," *Quarterly Journal of Inebriety* 16 (1894): 325.

69. Ibid., 325–27.

70. White, *Slaying the Dragon,* 50, 60.

71. Thomas D. Crothers, *The Walnut Lodge Hospital,* pamphlet, n.d., publisher unknown.

72. White, *Slaying the Dragon,* 54–55.

73. Ibid., 54.

74. Ibid., 55.

75. Ibid., 50–63. Warsh, "Adventures in Maritime Quackery."

76. William Hammond in William Hammond, Thomas D. Crothers, Elon Carpenter, Cyrus Edson, "Is Drunkenness Curable?" *North American Review* (1891): 346–74.

77. By labeling their cures with the "gold" moniker, Keeley and other specific cure proprietors were paying explicit homage to the gold standard championed by Republican presidential candidate William McKinley as a means of stabilizing the nation's banking and credit practices and facilitating trade with other nations. Gold became a measure of authenticity, stability, and security.

78. At the dawn of the twentieth century, proprietary medicine companies ran huge ad campaigns in local newspapers across the country. Revenues from these ads ac-

counted for a substantial portion of the budgets of local papers. The claims made by patent medicine manufacturers were hyperbolic, and they were not required to label the ingredients in their products, many of which contained addictive compounds such as alcohol, opiates, and cocaine. Muckraking journalist Samuel Hopkins Adams brought the quackery and profit-making of the patent medicine trade to the attention of the general public through his publications in *Collier's*. Adams's expose of the charlatanism of the industry and as its distribution of dangerous drugs helped facilitate the passage of the Pure Food and Drug Act of 1906. For more on muckraking, see Louis Filler, *Appointment at Armageddon: Muckraking and Progressivism in the American Tradition* (New Brunswick, NJ: Transaction Publishers, 1996). For more on the patent medicine trade, see James Harvey Young, *Toadstool Millionaires: A Social History of Patent Medicines in America Before Federal Regulation* (Princeton, NJ: Princeton University Press, 1961).

79. White, *Slaying the Dragon*, 85.

80. Charles B. Towns, "The Sociological Aspect of the Treatment of Alcoholism," *Modern Hospital* 8 (February 1917), reprint, 27.

81. Charles B. Towns, "Successful Medical Treatment in Alcoholism," *Modern Hospital* 8 (January 1917): 18.

82. See Alexander Lambert, "Some Statistics and Studies from the Alcoholic Wards of Bellevue Hospital," *Medical and Surgical Report of Bellevue and Allied Hospitals in the City of New York, Volume One* (New York: Martin Brown Press, 1904), 113–54. This report was issued the same year that Lambert observed Towns's work and became favorably impressed by it. Lambert's personal interest in the problem of the alcoholic stemmed from his work as a general practitioner of medicine: "The clinical picture of the disease in Bellevue Hospital is so complicated by the morbid processes produced by alcohol that it seemed necessary to study carefully the alcoholic patients in order to clearly appreciate the medical service in the general medical wards of the hospital" (113). Lambert asserted that one quarter of all admissions to the Bellevue Hospital passed through the alcoholic wards.

83. Alexander Lambert, "Care and Control of the Alcoholic," *Boston Medical and Surgical Journal* 166 (1912): 615–21.

84. Towns, "The Sociological Aspect of the Treatment of Alcoholism," reprint, 27–28.

85. The Charles B. Towns Hospital did successfully treat one very important case, however. In 1934, Bill Wilson, a stock analyst and confirmed alcoholic, checked himself into the Manhattan hospital and had a conversion experience that not only helped him achieve sobriety but led to his eventual founding of Alcoholics Anonymous with physician Robert Smith. Of course, one could argue that Towns's treatment had relatively little to do with Wilson's "conversion experience" or subsequent decision to pursue sobriety. It is also possible to explain what Wilson called his "Hot Flash" as the result of delirium or hallucinations induced by the belladonna treatment patients received at Towns. Regardless, Wilson's experience is undoubtedly the most famous, and perhaps the most important, of any of Towns's patients, for he went on to establish the world's most successful treatment for alcoholism, Alcoholics Anonymous. For more on Wilson's experience at Towns, see White, *Slaying the Dragon*, 87, 128–30. See also Ernest Kurtz, *Not God: A History of Alcoholics Anonymous* (Minneapolis, MN: Hazelden Information Education, 1998).

86. Towns, "The Sociological Aspect of the Treatment of Alcoholism." reprint, 35.

87. Charles B. Towns, "Care of Alcoholics in the Modern Hospital," *Modern Hospital* 7 (December 1916): 5–7.

FOUR: Public Inebriate Hospitals and Farm Colonies

1. For more on the history of inebriate care in California and Massachusetts, see Jim Baumohl and Sarah W. Tracy, "Building Systems to Manage Inebriates: The Divergent Paths of California and Massachusetts, 1891–1920," *Contemporary Drug Problems* (Winter 1994): 557–96. The trustees of the Southern California State Asylum for the Insane successfully petitioned the California legislature to alter the institution's charter in 1891.

2. *Report of the Committee on the Necessity and Expediency of an Inebriate Asylum,* 26 May 1870 (publisher unknown), 13.

3. State of Connecticut, *Report of Committee on Penal Treatment of Inebriates and Inebriate Asylum* [sic] (New Haven: Stafford Printing Office, 1874), 3–4.

4. Ibid., 22.

5. The Depression of 1873–79 led to a great increase in the number of homeless and jobless individuals across the country. This problem was felt acutely in urban areas, where these indigents, commonly called "tramps" and "vagrants," were most visible. The Connecticut General Assembly was quite concerned with the drain these people placed upon the state's finances. Thus, the state legislature asked the Committee on Penal Treatment of Inebriates to consider the predicament of the state in relation to its burgeoning vagrant population. The inebriate's situation was similar to the vagrant's, for they were both frequently recidivists at the jailhouse and almshouse; both groups siphoned off the state's financial resources; and often inebriates were vagrants and vice versa.

6. See State of Connecticut, *Report of Committee on Penal Treatment of Inebriates and Vagrants, Made to the General Assembly* (Hartford, CT: Press of the Case, Lockwood and Brainard Co., 1875), 20–21. The inebriate who ended up in jail for petty crimes committed while under the influence of alcohol was not to be treated as a criminal, but as a drunkard with a "besetting vice." Vagrants who were forced out of neighboring states and flocked to Connecticut, and recidivists who repeatedly committed mild offences demanded the same work-centered care, claimed the committee. This was also true for the dipsomaniac who did not respond to asylum treatment. At the very least, they hoped that they might make incorrigible cases self-supporting.

7. *Report of Committee on State Board of Health and Vital Statistics Made to the General Assembly, May Session 1875* (Hartford, CT: Press of the Case, Lockwood and Brainard Co., 1875), 6.

8. Ibid., 16.

9. *First Annual Report of the Executive Committee of the Asylum at Walnut Hill to the Corporators at their Annual Meeting* (Hartford, CT: Press of the Case, Lockwood and Brainard Co., 1875).

10. *Fourth Annual Report of the Officers and Superintendent of the Asylum at Walnut Hill, Hartford, Conn., at their Annual Meeting, October 9, 1878, also Petition to the Legislature* (Hartford, CT: Press of the Case, Lockwood and Brainard Co., 1878), 5.

11. "Superintendent's Report of State Farm for Inebriates," *Report of the Norwich State Hospital for the Insane,* Hartford, CT, 1916, 18–19.

12. "Dr. T. D. Crothers," in *Hartford in 1912, Story of the Capitol City Present and Prospective, Its Resources, Achievement, Opportunities and Ambitions* (Hartford, CT: The Hartford Post, 1912), 257.

13. Ibid. Doubtless, this was something of an exaggeration.

14. *Report of the Norwich State Hospital for the Insane to the Governor for the Two Years Ended September 30, 1916,* Hartford, CT, 1916, 20.

15. *Report of the Norwich Hospital for the Insane to the Governor for the Two Years Ended September 30, 1910,* Hartford, CT, 1911, 14.

16. *Report of the Norwich Hospital for the Insane to the Governor for the Two Years Ended September 30, 1908,* Hartford, CT, 1908, 10. Pollock drew the connection between the accumulating number of insane custodial cases and alcohol in his 1908 biennial report to the board of trustees and governor, listing "intemperance" among the "extrinsic sociological factors" that were potent causes of insanity.

17. *Report of the Norwich Hospital for the Insane to the Governor for the Two Years Ended September 30, 1910,* 14.

18. *Report of the Norwich Hospital for the Insane to the Governor for the Two Years Ended September 30, 1912,* Hartford, CT, 1912, 11.

19. *Report of the Norwich State Hospital for the Insane to the Governor for the Two Years Ended September 30, 1916,* 21.

20. *Report of the Norwich State Hospital to the Governor for the Twenty-One Months Ended June 30, 1920,* Hartford, CT, 1920, 7.

21. See John Burnham, "New Perspectives on the Prohibition Experiment of the 1920s," *Journal of Social History* 2 (Fall 1968): 51–68.

22. *Report of the Norwich State Hospital to the Governor, for the Twenty-Four Months Ending June 30, 1928,* 11.

23. See "Report of the Public Welfare Council on a Study of the State Farm for Inebriates, 1938," manuscript, Archives of the Connecticut State Library, Hartford.

24. The central tenets or principles of the Minnesota Model were heavily influenced by Alcoholics Anonymous. They included: (1) acknowledgment of alcoholism as a disease; (2) a holistic approach toward addressing the physical, psychological, social, and spiritual dimensions of the condition; (3) a belief that vulnerability to alcohol is one expression of a more generalized vulnerability to psychotropic drugs, or the potential for chemical dependency; (4) endorsement of a professional, multidisciplinary context for treatment; and (5) endorsement of Alcoholics Anonymous principles as a means of securing and maintaining sobriety. The model itself was developed in the 1940s and 50s, when AA became a large presence within the state and three separate treatment facilities—Pioneer House, Willmar State Hospital, and Hazelden pioneered new methods of treatment that included and complimented AA's "12-step" program. See Forrest Richeson, *"Courage to Change": Beginnings, Growth, and Influence of Alcoholics Anonymous in Minnesota* (Minnesota: M & M Printing, 1978); Jerry Spicer, *The Minnesota Model: The Evolution of the Interdisciplinary Approach to Addiction Recovery* (Center City, MN: Hazelden Educational Materials, 1993); and White, *Slaying the Dragon,* 199–212.

25. Richeson, *"Courage to Change,"* 2.

26. R. M. Phelps, "Inebriety as a Disease, Analytically Studied," *Medical News,* 23 June 1894 (reprint), 2–3.

27. R. M. Phelps, "Inebriety," *Northwestern Lancet,* 15 October 1893 (reprint), 7.

28. Ibid., 6, 8.

29. Frederick Peterson, *The Influence of Alcohol* (New York: St. Mark's Healing Mission, 1909); idem, "Alcohol and Public Health," in *Ninth New York State Conference of Charities and Correction* (Albany, NY: J. B. Lyon State Printers, 1909), 26–32.

30. "Concerning the Commitment of Inebriates," *St. Paul Medical Journal* 5 (September 1903): 691.

31. Ibid., 692.

32. Ibid.

33. "What Shall We Do with the Chronic Drunkard?" *St. Paul Medical Journal* 8 (February 1906): 97.

34. Ibid., 100.

35. Haldor Sneve, "Treatment of Inebriety and Narcotism," paper read before the Quarterly Conference of Executive Officers of State Institutions with the State Board of Control, St. Paul, Minnesota, 6 August 1907, 3.

36. Silas W. Leavitt and Others vs. City of Morris, Nos. 15, 774–(241), Supreme Court of Minnesota, 1909, *105 Minn. 170,* 175–76.

37. "The Minnesota Hospital Farm for Inebriates," *St. Paul Medical Journal* 13 (July 1911): 317.

38. "Need of Provision for Treating Inebriates," *Charities and the Commons* 20 (8 August 1908): 573.

39. Standing Committee on Hospitals of the State Charities Aid Association of New York, *Treatment of Public Intoxication and Inebriety,* pamphlet no. 108, 20 March 1909, 5–6.

40. Homer Folks, "Social Aspects of Alcoholism," *The Survey* 25 (1 October 1910): 14–15.

41. Bailey Burritt to Burdette G. Lewis, 24 August 1917, in Correspondence of Mayor Mitchell, Departmental Correspondence Received, Board of Inebriety, 1917, Box 043, Folder 444, Microfilm, New York City Municipal Archives.

42. *Treatment of Public Intoxication and Inebriety,* 20.

43. Standing Committee on Hospitals of the State Charities Aid Association, *The Alcoholic 'Repeater' or Chronic Drunkard,* pamphlet no. 113, 15 February 1910, 24.

44. Ibid., 25. This passage appeared in bold face in the text. There is clearly a typographical error in which "useful" is printed in place of "useless."

45. For more on the prickly interagency relations among the Municipal Board of Estimate and Apportionment, the New York City Comptroller's Office, and the Board of Inebriety, see William Morrison to William J. Gaynor, Mayor and Head of the Board of Estimate and Apportionment, 30 January 1912, in Correspondence of Mayor Gaynor, Departmental Correspondence Received, Board of Inebriety, 1912, Box 036, Folder 303, Microfilm, New York City Municipal Archives; William Morrison to William J. Gaynor, Mayor and Head of the Board of Estimate and Apportionment, 26 February 1912, in Correspondence of Mayor Gaynor, Departmental Correspondence Received, Board of Inebriety, 1912, Box 036, Folder 303, Microfilm, New York City Municipal Archives; Very Reverend John J. Hughes to Gaynor, 18 June 1913, in Correspondence of Mayor Gaynor,

Departmental Correspondence Received, Box 036, Folder 304, Microfilm, New York City Municipal Archives.

46. John Kingsbury, President of the Board of Inebriety, to John Purroy Mitchell, Mayor of the City of New York, 29 June 1914, Correspondence of Mayor Mitchell, Departmental Correspondence Received, Board of Inebriety, 1917, Box 043, Folder 444, Microfilm, New York City Municipal Archives.

47. See Charles Samson, Executive Secretary of the Board of Inebriety, to Judge John J. Eagan, 7 May 1915, Correspondence of Mayor Mitchell, Departmental Correspondence Received, Board of Inebriety, 1915, Box 043, Folder 442, Microfilm, New York City Municipal Archives; Charles Samson, "The Care and Treatment of Inebriates in New York," *British Journal of Inebriety* 11 (1913): 27–31.

48. Ibid., 29.

49. Samson to Eagan, 7 May 1915.

50. R. A. Tighe to John Purroy Mitchell, Mayor of the City of New York, 5 November 1917, Correspondence of Mayor Mitchell, Departmental Correspondence Received, Board of Inebriety, 1917, Box 043, Folder 444, Microfilm, New York City Municipal Archives.

51. New York Academy of Medicine, "The Modern Conception of Inebriety," Discussion at the Meeting Held February 1, 1917, *New York Medical Record*, 24 March 1917, 522–23.

52. Charles F. Stokes, "A Preliminary Report on the Pathology and Treatment of Narcotic Addiction," *New York Medical Record*, 9 June 1917, 969–71; "The Cure of Narcotic Addiction," *New York Medical Record*, 9 June 1917, 993.

53. Charles F. Stokes to John Purroy Mitchell, Mayor of the City of New York, 18 June 1917, Correspondence of Mayor Mitchell, Departmental Correspondence Received, Board of Inebriety, 1917, Box 043, Folder 444, Microfilm, New York City Municipal Archives; idem, "A Preliminary Report on the Pathology and Treatment of Narcotic Addiction," *New York Medical Record*, 9 June 1917.

54. Burdette Lewis to Bailey Burritt, 22 August 1917, Correspondence of Mayor Mitchell, Departmental Correspondence Received, Board of Inebriety, 1917, Box 043, Folder 444, Microfilm, New York City Municipal Archives. Lewis's choice of the term "cosmopolitan" is odd. It is likely that he meant "metropolitan."

55. Bailey Burritt to Burdette Lewis, 24 August 1917, Correspondence of Mayor Mitchell, Departmental Correspondence Received, Board of Inebriety, 1917, Box 043, Folder 444, Microfilm, New York City Municipal Archives.

56. Ibid.

57. M. L. Fleming quoted in "Failure of Prohibition—Institutions for Alcoholics Have About as Many Patients as Before Law Became Effective," *New York Times*, 7 September 1919, 10. The title of the *Times* article is a bit misleading since "about as many patients as before law became effective" masks the fact that the number of patients had been declining prior to Prohibition. See also, "Big Drop in Alcoholism: Statistics Show 70 to 90 Percent Cut in Hospital Cases," *New York Times*, 4 April 1920, 16.

58. James A. Hamilton, "Drug Addicts on Riker's Island," *New York Medical Journal and Medical Record*, 20 December 1920, 715–17; Hamilton, "Classification and Segregation of Male Inmates in New York City Prisons," *New York Medical Journal*, 18 May 1921, 757–59.

59. Of course, it is possible to argue that the response of public institutions, with their focus on "work as therapy," constituted a regressive punitive action predicated on a concept of inebriety as a vice or crime, but in so doing we would miss an important part of the picture. Such work "therapy" also should be seen in Meyerian mental hygiene terms as an effort to help the inebriate "adapt" to his social role and the societal expectation that one was to support oneself, as well as one's family—an expectation that coincided nicely with the economic interests of the state. Gilded Age and Progressive Era psychiatrists and mental hygienists were intent on expanding their professional reach outside the insane asylum. Addressing inebriety as a medical problem was one way to do this. Focusing therapy on activities that would help inebriates readjust to the working world outside the asylum was yet another piece of this project. Regardless of the interpretation, though, all of the forms of therapy reviewed in this chapter placed emphasis on work or employment in reforming the inebriate. Institutional motives may have varied—from rebuilding the inebriate's self-esteem to strengthening his constitution to lessening the inebriate's drain on taxpayers—but work was a constant element of care and cure.

60. Similarly, patients at the Keeley Institute in Dwight, Illinois, could co-mingle with that town's citizens, who, according to William White, acted as their informal guardians; all that was required was their mandatory presence for shots and lectures. See White, *Slaying the Dragon*, 53–54.

61. See Jim Baumohl, "Inebriate Institutions in North America," 104–12.

FIVE: The "Foxborough Experiment"

1. On inebriate institutions, see Crowley and White, *Drunkard's Refuge;* White, *Slaying the Dragon*, 20–87; Baumohl, "Inebriate Institutions in North America"; Baumohl and Tracy, "Building Systems to Manage Inebriates"; and Sarah W. Tracy, "Contesting Habitual Drunkenness: State Medical Reform for Iowa's Inebriates, 1902–1920," *Annals of Iowa* (Summer 2002): 241–86. On other types of institutions, see Lawrence Goodheart, *Mad Yankees: A Study of the Changing Treatment of Madness in Nineteenth-Century America* (Amherst: University of Massachusetts Press, 2003); Tomes, *A Generous Confidence;* Dwyer, *Homes for the Mad;* Barbara Brenzel, *Daughters of the State: A Social Portrait of the First Reform School for Girls in North America, 1856–1905* (Cambridge: MIT Press, 1983); Lunbeck, *The Psychiatric Persuasion;* Grob, *The State and the Mentally Ill;* Katz, *In the Shadow of the Poor House;* Barbara Bates, *Bargaining for Life: A Social History of Tuberculosis, 1876–1938* (Philadelphia: University of Pennsylvania Press, 1992); Rothman, *The Discovery of the Asylum;* Leila Zenderland, *Measuring Minds: Henry Herbert Goddard and the Origins of American Intelligence Testing* (Cambridge: Cambridge University Press, 1998).

2. Charles A. Rosenwasser, "A Plea for the Establishment of Hospitals for the Rational Treatment of Inebriates," *Medical Record*, 8 May 1909, 795–98.

3. Although the New York State Inebriate Asylum preceded the Massachusetts Hospital for Dipsomaniacs and Inebriates by about thirty years, it was a private facility until it was sold to the state for a dollar.

4. *Report of the Trustees of the Massachusetts Hospital for Dipsomaniacs and Inebriates* (Boston: Wright and Potter, March 1893), 11.

5. These figures come from Richard W. Wilkie and Jack Tager, *Historical Atlas of Mass-*

*achusetts* (Amherst: University of Massachusetts Press, 1991). For more data on immigration in the Bay State, see Oscar Handlin, *Boston's Immigrants, 1790–1880: A Study of Acculturation,* rev. and enl. ed. (Cambridge, MA: Harvard University Press, 1991); Stephan Thernstrom, *The Other Bostonians: Poverty and Progress in the American Metropolis, 1880–1970* (Boston: Harvard University Press, 1973); Thomas O'Connor, *The Boston Irish—A Political History* (Boston: Little, Brown, 1995); Kerby A. Miller, *Emigrants and Exiles—Ireland and the Irish Exodus to North America* (Oxford: Oxford University Press, 1985); and Barbara Miller Solomon, *Ancestors and Immigrants: A Changing New England Tradition* (Boston: Northeastern University Press, 1989).

6. House Document No. 38, Edgar Whitaker, Sarah Holland et al., "Memorial to the Honorable Legislature of the State of Massachusetts," 1845, 1–2.

7. Ibid., 7.

8. Senate Document No. 94, "Report of the Joint Committee on an Asylum for Inebriates," 1850, 1.

9. Ibid., 3–4.

10. Ibid., 28–29.

11. Senate Document No. 75, April 1851, 2.

12. Ibid. The Sons of Temperance were a large fraternal temperance society with divisions nationwide, committed to the reformation of individual inebriates. In many ways the Sons picked up where the Washingtonians left off, maintaining far more institutionalized and structured organizations than their predecessors. In Massachusetts, the Sons seemed younger, poorer, less religious, and more often of an artisanal, working-class background than their contemporary middle-class, nativist, and neo-republican colleagues, the prohibitionists. By the late 1840s, however, the Sons had joined their more affluent comrades to endorse prohibition for all. Nationwide, the Sons' membership levels coincided with the crest of anti-liquor legislation in the United States. For more on the Sons of Temperance and other temperance fellowships, see Jack Blocker, *American Temperance Movements: Cycles of Reform* (Boston: Twayne, 1989), 48–51.

13. According to Jack Blocker, "Prohibition does not prohibit!" was the rallying cry of liquor dealers and anti-prohibitionists in the late 1850s, usually accompanied by "arguments that prohibition destroyed 'personal property' along with private property and community property" (*American Temperance Movements,* 59).

14. William Lloyd Garrison, abolitionist and founding editor of Boston's *Liberator,* devoted considerable attention to the temperance cause (and women's suffrage) after the Civil War. Even Neal Dow, the mayor of Portland, Maine, and his state's most visible temperance leader in the 1840s and 50s, was equally committed to the anti-slavery crusade. Yet Andrew parted ways with his fellow Free-Soilers when it came to prohibition.

15. Indeed, Andrew went further. Following the Civil War, in 1867, he represented several prominent hotels and innkeepers against the prohibitionists, reinterpreting the biblical passages that foes of alcohol were quick to quote, and assembled a mass of expert testimony from Boston's medical and scientific communities testifying to alcohol's multiple therapeutic uses. The Maine Law was repealed in Massachusetts the following year. According to his biographer, only the exigencies of the Civil War and Andrew's commitment to "keep the State steady in the support of the national government" prevented the governor from taking a more aggressive stance against the Maine Law during

his five terms." See Henry Greenleaf Pearson, *The Life of John A. Andrew* (Cambridge, MA: Riverside Press, 1904), 2:308.

16. Governor Andrew quoted in *Report of a Joint Special Committee Appointed to Consider the Matter of Inebriation as a Disease, and the Expediency of Treating the Same at Rainsford Island* (Boston: Wright and Potter, 1868), 1

17. Senate Document No. 235, 28 April 1864, 2.

18. *Report of a Joint Special Committee*, 4.

19. Ibid., 11, 14–15.

20. Senate Document No. 389, *Report of the Joint Special Committee Appointed to Consider the 'necessity of treating inebriation as a disease, and of legislating for its cure as well as its punishment,'* 9 June 1869, 1.

21. Senate Document No. 9, *Report of the Commissioners on Inebriate Asylums,* January 1871, 10.

22. Barbara Rosenkrantz, *Public Health and the State: Changing Views in Massachusetts, 1842–1936* (Cambridge, MA: Harvard University Press, 1972), 66. For more on the Massachusetts Board of Health during this period, see pp. 74–96. The board was already beginning to take an interest in matters alcoholic, however, as Henry I. Bowditch conducted his 1871 survey by mail; see Senate Document No. 50, "Correspondence Concerning the Effects of the Use of Intoxicating Liquor," January 1871, 255–347.

23. Nathan Crosby, "Inebriate Asylums: Remarks in Opposition to Them," 4. For Crosby, the inebriate asylum project was one more sign of a declining criminal justice system that coddled prisoners, providing them with "parlors and hops, plum puddings and Thanksgiving turkeys" (3).

24. Henry I. Bowditch, "Inebriate Asylums or Hospitals," *Annual Report of the State Board of Health,* January 1875, 27.

25. Ibid., 35.

26. Ibid., 37.

27. Several factors worked against taking action on the Board of Health report. First, the state's depressed economy made any new expenditures less attractive to the legislature. Second, a growing tradition of private philanthropy continued to discourage state spending for the drunkards' reform. Indeed, the number of private asylums for the Bay State's inebriates was growing. By 1881, the Washingtonian Home had company. Both the Appleton Temporary Home, incorporated in 1876 by members of the Boston YMCA Temperance Society, and the New England Home for Intemperate Women, opened in 1879 by a host of Boston social and medical luminaries, were serving patients in the Commonwealth's capital city. Nearly a decade would pass before Massachusetts would take real action on behalf of the state's alcoholics.

28. *Report of the State Board of Health, Lunacy, and Charity on the Proposed Institution for Inebriates,* January 1885 (Boston: Wright and Potter, 1885), 5.

29. *Report of the Massachusetts Hospital for Dipsomaniacs and Inebriates,* January 1893 (Boston: Wright and Potter, 1893), 25.

30. Theodore Fisher, "Insane Drunkards," paper presented to the Massachusetts Medical Society, June 1879, 19.

31. Senate Document No. 23, *Report of the Proceedings of the Trustees of the Massachusetts Hospital for Dipsomaniacs and Inebriates* (Boston: Wright and Potter, 30 January 1891).

32. The other members of the board included secretary of the Massachusetts Prison

Association, Warren F. Spaulding; Boston charity organizer, Anna Phillips Williams; Springfield and Boston entrepreneur, Tilly Haynes; Lowell physician, Burnham Benner; and Bostonian Samuel Carr Jr.

33. See Crowley and White, *Drunkard's Refuge.*

34. House Document No. 155, *Report of the Trustees of the Massachusetts Hospital for Dipsomaniacs and Inebriates,* February 1890, 2.

35. Ibid.

36. Senate Document No. 23, *Report of the Trustees of the Massachusetts Hospital for Dipsomaniacs and Inebriates,* 6.

37. Noyes had also studied psychology with G. Stanley Hall at the Johns Hopkins University following his graduation from Harvard. From there he had entered clinical work at Bloomingdale Asylum in New York and then proceeded to McLean. According to Sylvia B. Sutton, by 1893 Noyes had tired of laboratory life at McLean and wished to return to clinical work with patients; this was the basis of his decision to take the assistant physician's post at MHDI. Certainly, Noyes's research on the sources of nervous energy and the methods of replenishing it in relation to nutrition, exercise, and rest would have found welcome application at the Massachusetts Hospital for Dipsomaniacs and Inebriates. See Sylvia Sutton, *Crossroads in Psychiatry: A History of the McLean Hospital* (Washington, DC: American Psychiatric Press, 1986).

38. *Report of the Trustees of the Massachusetts Hospital for Dipsomaniacs and Inebriates,* January 1893, 6.

39. See ch. 3, n. 61 for more on the cottage plan. The New York State Craig Colony for Epileptics utilized the cottage plan in 1896, a few years after the Massachusetts Hospital for Dipsomaniacs and Inebriates had opened, as did the Duxhurst Industrial Colony for Inebriate Women in England, which opened in 1901.

40. In one sense, the separate cottages, with their distinctive patient populations, reproduced Bowditch's vision of several different asylums catering to different social and clinical clienteles. In another sense, the cottage design replicated the successful residential style of the long-admired Washingtonian Home in Boston.

41. Senate Document No. 23, *Report of the Trustees of the Massachusetts Hospital for Dipsomaniacs and Inebriates,"* 7.

42. "Inebriates' New Home," *Boston Globe,* evening edition, 18 February 1893, p. 8.

43. "Superintendent's Report," *Annual Report of the Massachusetts Hospital for Dipsomaniacs and Inebriates* (Boston: Wright and Potter, January 1894), 11.

44. "Trustees' Report," *Annual Report of the Massachusetts Hospital for Dipsomaniacs and Inebriates* (Boston: Wright and Potter, January 1894), 5.

45. Ibid., 8.

46. Underlining is in the original handwritten text. Julia Osborne to Marcello Hutchinson, 21 May 1894, Michael Osborne, Inpatient case files of NSH.

47. "Trustees' Report," *Annual Report of the Massachusetts Hospital for Dipsomaniacs and Inebriates* (Boston: Wright and Potter, 1897), 10.

48. "Trustees' Report," *Annual Report of the Massachusetts Hospital for Dipsomaniacs and Inebriates* (Boston: Wright and Potter, 1896), 7.

49. "Trustees' Report," *Annual Report of the Massachusetts Hospital for Dipsomaniacs and Inebriates* (Boston: Wright and Potter, 1894), 6.

50. "Superintendent's Report," *Annual Report of the Massachusetts Hospital for Dipsomaniacs and Inebriates* (Boston: Wright and Potter, 1896), 10.

51. House Document No. 900, 13 June 1894, 2.

52. "Superintendent's Report," *Annual Report of the Massachusetts Hospital for Dipso-maniacs and Inebriates* (Boston: Wright and Potter, 1896), 13.

53. The annual report for 1897 went so far as to print the chest outlines and admission (before) and discharge (after) torso profiles of sixteen patients ranging in age between 25 and 63 years, in addition to the usual tables of improved strength, lung capacity, and weight for the entire patient population. See "Superintendent's Report," *Annual Report of the Massachusetts Hospital for Dipsomaniacs and Inebriates* (Boston: Wright and Potter, 1897), 17–36.

Critics of the hospital indicted its methods as encouraging laziness among its patients. The emphasis annual reports placed on physical exercise may have been, in some way, a response to such criticism. Certainly, Foxborough's critics would have a difficult time accusing the hospital's physical culture program of promoting lethargy.

54. The superintendent's use of the term "personal equation" is idiosyncratic. Usually this term denotes the source of different experimental results obtained by two scientists performing the same experiment, following the same protocol; a synonym would be "observer error."

55. "Editorial: The Charges Against Foxborough Asylum," *Quarterly Journal of Inebriety* 16 (1894): 191–93.

56. "Editorial," *Foxborough Reporter,* 8 February 1896, 2.

57. Certainly, the hospital had no trouble integrating work and exercise into its rehabilitative program. However, the hospital continued to receive commitments from the courts who were often unable or unwilling to participate in the hospital's regimen. The trustees eventually acquired legal authority to dismiss many of these many of these individuals, but the public may have seen these inebriates' departure from Foxborough (in the same state as they had arrived) as an indication of the hospital's failure to rehabilitate its charges.

58. "Superintendent's Report," *Annual Report of the Massachusetts Hospital for Dipso-maniacs and Inebriates* (Boston: Wright and Potter, 1898), 17–20.

59. "Superintendent's Report," *Annual Report for the Massachusetts Hospital for Dipso-maniacs and Inebriates* (Boston: Wright and Potter, 1897), 20.

60. Among Hutchinson's accomplishments were: reducing per capita expenses by 50 percent, increasing the hospital's census, cutting the number of elopements by a third, and securing the funding to build the hospital's new gymnasium and hydrotherapeutic facilities.

61. "Trustees' Report," *Annual Report of the Massachusetts Hospital for Dipsomaniacs and Inebriates* (Boston: Wright and Potter, 1900), 5.

62. *Report of the Advisory Committee on the Penal Aspects of Drunkenness* (Boston: Mayor's Office, 1899), 39.

63. "Trustees' Report," *Annual Report for the Massachusetts Hospital for Dipsomaniacs and Inebriates* (Boston: Wright and Potter, 1901).

64. "Trustees' Report," *Annual Report of the Massachusetts Hospital for Dipsomaniacs and Inebriates* (Boston: Wright and Potter, 1902), 7.

65. Foxborough was certainly not unique among Progressive Era institutions in this regard. Indeed, one could easily interpret Woodbury's concern with institutional order and finances as the "convenience" that undermined the "conscience" of Progressive reform.

66. "Trustees' Report," *Annual Report of the Foxborough State Hospital* (Boston: Wright and Potter, 1907), 9.

67. "Superintendent's Report," *Annual Report of the Foxborough State Hospital* (Boston: Wright and Potter, 1907), 19.

68. J. F. Kennedy, "Inebriety and Its Management," *Bulletin of the Iowa Institutions* 4, no. 2 (April 1902): 190–92.

69. "No Hospital Probe: Executive Council to Defer Investigation Because of the Pending Grand Jury Inquiry," *Boston Evening Transcript*, 13 July 1906, 3.

70. Commonwealth of Massachusetts, "Hearing Before the Lieutenant Governor Eben S. Draper and Committee of the Council on the State Hospital at Foxborough, 24 January to 21 February 1907," 1:28.

71. "Visit to Foxboro, Council Sent No Tip Ahead, Talked with Inmates," *Boston Globe*, morning edition, 11 January 1907, 1.

72. See "Thaw Crazed by Liquor—Drink Changed Him into a Mad Man, Says Former Valet," *Boston Globe*, 16 February 1907, 1.

73. "Light on Foxboro, Governor's Council Begins Its Investigation, Sensational Charges at the First Hearing," *Boston Evening Transcript*, 24 January 1907, 2.

74. "Told Terrible Tales of Cruelty, Former Inmates of Foxboro Said They Were Beaten, Put Through 'Pumping,' and 'Death Room' Treatments, Investigators Hear Startling Testimony," *Boston Herald*, 25 January 1907, 2.

75. See "Foxboro Probe In: Inmates as Accusers, Hicks and Brennan Tell of Brutality," *Boston Globe*, 25 January 1907, morning edition, p. 2, for a lengthy summary of the charges Long brought against the inebriate hospital.

76. Ibid., 1.

77. Commonwealth of Massachusetts, "Hearing on the State Hospital at Foxborough," 1:55.

78. Ibid., 164.

79. "No Special Appropriation: Lively Debate Over Foxboro Institution Results in Cutting Off $15,757 Item Recommended," *Boston Globe*, evening edition, 2.

80. Commonwealth of Massachusetts, "Hearing on the State Hospital at Foxborough," 2:806.

81. S. Weir Mitchell, "Address Before the Fiftieth Annual Meeting of the American Medico-Psychological Association," *American Journal of Insanity* 51 (1894): 171–81.

82. "The Foxboro Case," *Boston Evening Transcript*, 8 February 1907, 10.

83. "Foxboro Hospital May Be Abolished: Belief Growing at State House That It Will Be Made Into a Reformatory," *Boston Herald*, 10 February 1907, 12.

84. Commonwealth of Massachusetts, "Hearing on the State Hospital at Foxborough," 2:1561.

85. Ibid., 1559, 1575–76, 1587, 1617–18.

86. For a recent history of McLean Asylum, see Alex Beam, *Gracefully Insane: Life and Death Inside America's Premier Mental Hospital* (New York: Public Affairs, 2003); see also Sutton, *Crossroads in Psychiatry.*

87. House Document No. 984, 24 January 1907, 1.

88. Commonwealth of Massachusetts, "Hearing on the State Hospital at Foxborough," 2:1620.

89. Ibid., 1621–22.

90. Editorial, *Boston Evening Transcript*, 16 February 1907, 2.

91. Commonwealth of Massachusetts, "Hearing on the State Hospital at Foxborough," Appendix I, p. 29.

92. "Complete Overturn: Governor Guild Makes a Foxboro Reorganization," *Boston Evening Transcript,* 11 September 1907, 1.

93. Ibid. At the investigation's outset in January 1907, the *Boston Globe* had reported: "The governor has long been dissatisfied with existing conditions at the Foxboro hospital. . . . While lieutenant governor he had been brought into contact with the institution on numerous occasions and had observed closely its workings. . . . He wants an inquiry that will lay bare the institution "to the very bone," as he is said to have expressed it not long ago. . . . To him it is not a question as to the fitness of this superintendent or that of the board of trustees. Governor Guild is after more than this. He wants to discover whether "the theory" [that inebriety is a curable disease] of the Foxboro institution is right or wrong" (*Boston Globe,* 25 January 1907, 1).

94. The Committee on Drunkenness was in large part responsible for two important changes in the penal statutes concerning drunkenness: the empowerment of probation officers to release those arrested for drunkenness without appearing in court (first or second time offenders), and the conditional release of persons unable to pay their fines at the time of sentencing for drunkenness, provided they were supervised by a probation officer and paid the fines gradually over time.

95. Besides Woods, the trustees included Frank Locke, an MIT graduate who had resigned his position as superintendent of the Boston Rubber Shoe Company to assume the presidency of the Young Men's Christian Union in Boston; he was also director of the Malden branch of Boston Associated Charities. Selectman, assessor, and overseer of the poor in his hometown of Rockland, Edwin Mulready, was also secretary of the National Union of Temperance Societies of America; Mulready had served nine years as the probation officer for the Norfolk and Plymouth superior courts. Drs. T. J. Foley of Worcester and William H. Prescott of Boston gave the board an aura of medical competence. Foley was a Holy Cross and Yale College of Medicine graduate, while Prescott took his medical degree from Harvard. Foley had ties to four Worcester hospitals, while Prescott was the assistant superintendent at Boston City Hospital. James Perkins brought able business and financial experience to the board as vice president of the American Trust Company. Finally, W. Rodman Peabody, a well-known Boston lawyer, lent legal expertise to the group.

96. "Trustees' Report," *Annual Report of the Foxborough State Hospital for the Year Ending November 30, 1908* (Boston: Wright and Potter, 1909), 7. Neff came highly recommended, receiving the endorsement of Henry M. Hurd, M.D., superintendent of the Johns Hopkins Hospital; E. A. Christian, M.D., the superintendent at Eastern Michigan State Hospital in Pontiac, Michigan; and neurologist Albert M. Barrett, director of the University of Michigan's new psychopathic hospital and later president of the American Psychiatric Association.

97. See "A Reorganized Foxboro," editorial in the *Boston Evening Transcript,* 19 February 1908, 14.

98. See David Rothman, "Individual Justice: The Progressive Design," in *Conscience and Convenience,* 43–81.

99. "Trustees' Report," *Annual Report of the Foxborough State Hospital for the Year Ending November 30, 1908* (Boston: Wright and Potter, 1909), 9. If trustees drew the analogy

between the inebriate patient and the general hospital patient, they nevertheless drew one important distinction: they saw the inebriate forfeit his right to be discharged when he accepted his institutionalized sick role, whereas the sick patient's right to be discharged was usually not in question in the general hospital. This is a clear indication that in spite of changes in the hospital's administration, certain dimensions of the moral/penal "therapeutic" framework persisted.

100. William A. White, *Outlines of Psychiatry*, 12th ed. (Washington, DC: Journal of Nervous and Mental Disease Publishing Company, 1911), 40.

101. Irwin H. Neff, "Some of the Medical Problems of Alcoholism," *Boston Medical and Surgical Journal*, 26 January 1911, 114.

102. The proceedings of the conference were published in the *Boston Medical and Surgical Journal*. See "Some Medical and Social Aspects of Mental Disease Due to Alcohol: Notes of a Conference Held at the Psychopathic Hospital, Boston, Massachusetts, Before the Legislative Commission on Drunkenness," *Boston Medical and Surgical Journal*, 18 December 1913, 929–42.

103. "Superintendent's Report," *Annual Report of the Foxborough State Hospital for the Year Ending November 30, 1909* (Boston: Wright and Potter, 1910), 16.

104. "Trustees Report," *Annual Report of the Norfolk State Hospital for the Year Ending November 30, 1915* (Boston: Wright and Potter, 1916), 16.

105. Trustees, *Drunkenness in Massachusetts: Conditions and Remedies* (Boston: Wright and Potter, 1910), 24.

106. Ibid. See also House Document No. 2053, *Report of the Commission to Investigate Drunkenness in Massachusetts* (Boston: Wright and Potter, 1914).

107. Nevertheless, Foss was convinced the New York City Board of Inebriety, the central agency that reviewed each inebriate case and determined if it was fit for hospital farm colony treatment, was a better mechanism for screening patient applicants and tracking their progress (or lack of it) following their hospital stays. Foss believed the courts were too overtaxed and decentralized to manage inebriate cases effectively. His recommendation to the Massachusetts General Assembly noted Massachusetts' priority in establishing an inebriate colony, specifying that the Bay State's progress in this direction was directly responsible for the creation of the Board of Inebriety.

108. *Report of the Commission to Investigate Drunkenness in Massachusetts*, 16.

109. Ibid., 11.

110. "Superintendent's Report," *Annual Report of the Norfolk State Hospital for the Year Ending November 30, 1914* (Boston: Wright and Potter, 1915), 11.

111. "Trustee's Report," *Annual Report of the Norfolk State Hospital for the Year Ending November 30, 1914* (Boston: Wright and Potter, 1915), 7.

112. Ibid., 10.

113. Trustees' Report," *Annual Report of the Norfolk State Hospital for the Year Ending 30 November 1917* (Boston: Wright and Potter, 1918), 8–9.

114. "Superintendent's Report," *Annual Report of the Norfolk State Hospital for the Year Ending 30 November 1917* (Boston: Wright and Potter, 1918), 14.

115. *Report of the Commission to Investigate Drunkenness in Massachusetts*, 21.

116. "Trustees' Report," *Annual Report of the Norfolk State Hospital for the Year Ending 30 November 1919* (Boston: Wright and Potter, 1920), 9.

SIX: Building a Boozatorium

1. Ed Harris to Governor William L. Harding, 5 January 1917, "State Institutions: Inebriate Hospital, Knoxville" in Governor Harding Papers; General Correspondence of the State's Institutions, Box 29, Folder 1–3732, State Historical Society of Iowa, Des Moines, Iowa. Unless otherwise noted, all archival materials are held at the SHSI-DM. The patient case files and records are kept at each of the state hospitals where they were originally recorded. All of the names appearing, with the exception of names published in newspapers, are pseudonyms.

2. Little did Harris know that the superintendents of the hospitals treating inebriates (Mount Pleasant for women; Knoxville for men) thought the "common drunkard" had a better prognosis than the "the true inebriate." See George Donohoe, "The Inebriate," *Bulletin of the Iowa Board of Control* 16 (March 1914). Of course, Harris might have been making a distinction based on economic class rather than pathological status, though the two have never been unrelated in discussions of inebriety or alcoholism. Nevertheless, his use of the term "common drunkard" was in stark contrast to the views of the superintendents of the inebriate hospitals, highlighting two important issues: the difference between lay and medical understandings of inebriety, and the changing terms used to describe those with hopeful versus pessimistic prognoses.

3. Ed Harris to Governor William L. Harding, 13 February 1917, Governor Harding Papers, SHSI-DM.

4. William L. Harding to Ed Harris, 17 February 1917, Governor Harding Papers, SHSI-DM. As we shall see, the length of time required for successful treatment was a hotly contested issue. Harris's "favorable" patient status quite likely had to do with the fact that he was better off financially than most inebriates and had a job waiting for him upon his return—indeed, a position that was beckoning him during his stay at Knoxville.

5. Ed Harris to Governor William L. Harding, 28 February 1917, Governor Harding Papers, SHSI-DM. By 1917, the inebriate hospital at Knoxville was not eager to use the term "cure" in relation to inebriety. They preferred to refer to patients as improved sufficiently for parole, and offered as a comparative case the insane patient who was discharged as cured from the state hospital but who might easily return to the hospital at a later date suffering again from mental illness. In addition, the term "cure" smacked of the patent medicine trade that sold "specifics" or "cures" for inebriety. As we shall see, this was a business that virtually every superintendent of an Iowa institution found loathsome professionally and therapeutically.

6. M. C. Mackin to Governor William L. Harding, 6 March 1917.

7. Ed Harris to Governor William L. Harding, 11 April 1917.

8. The uneven nature of the archival correspondence and patient records from Iowa's turn-of-the-century inebriate experiment makes it hard to judge the representativeness of Harris's case, even if we can determine how frequently businessmen were admitted to the state inebriate hospital for their alcohol problems. The minutes of the Iowa State Medical Society's annual meetings, the quarterly reports of the Board of Control, the annual reports of the inebriate hospitals, and the daily reportage of local newspapers tell an interesting tale of Iowa's Progressive Era struggle to define the nature of habitual drunkenness and to devise an acceptable sociomedical solution to this vexing problem.

9. Dorothy Schwieder, *Iowa: The Middle Land,* (Ames: Iowa State University Press, 1996), 212. For a discussion of the nineteenth-century temperance debates, see also Richard Jensen, "Iowa, Wet or Dry? Prohibition and the Fall of the GOP," in *Iowa History Reader,* ed. Marvin Bergman (Ames: State Historical Society of Iowa and Iowa State University Press, 1996), 263–88.

10. For more on the history of the Iowa State Psychopathic Hospital, see Paul E. Huston, "The Iowa State Psychopathic Hospital (part one)," *The Palimpsest* 54, no. 6 (November/December 1973): 11–27; and Huston, "The Iowa State Psychopathic Hospital (part two)," *The Palimpsest* 55, no. 1 (January/February 1974): 18–30. For an earlier discussion of the value of such an institution to Iowa, see Max N. Voldeng, "The Present Status of Mental Hygiene and Mental Control," *Journal of the Iowa State Medical Society* 3, no. 6 (December 1913). Voldeng, one of the first superintendents to treat inebriates at Cherokee State Hospital for the Insane, noted: "All observation hospitals, all institutions with psychopathic departments are replete with instances where early and proper control resulted in speedy recovery of various mental diseases. The prompt response to immediate supervision and treatment of alcoholic cases is apparent to everyone" (381).

11. Bailey Burritt, "The Habitual Drunkard," *The Survey* 25 (1910): 25–41.

12. Josiah F. Kennedy, "Inebriety and Its Management," *Bulletin of Iowa Institutions* 4, no. 2 (April 1902): 184–95.

13. Ibid., 185.

14. Ibid., 186. Kennedy's recommendations that corporations enforce a temperate workplace and "homeplace" for their workers presaged Henry Ford's Sociological Department by a few years. The Ford Sociological Department inspected workers' homes to see if their home life qualified them for the famous $5 Day. Drinking was grounds for exemption from this level of pay.

15. Iowans' concerns about alcoholism and degeneracy were plain to see in the state's first eugenics law, enacted in 1911. Said to be among the country's strictest eugenics legislation, the law encouraged the sterilization of "habitual criminals, degenerates and other persons," which included "criminals, rapists, idiots, feeble-minded, imbeciles, lunatics, drunkards, drug fiends, epileptics, syphilitics, moral and sexual perverts, and diseased and degenerate persons" held within state institutions—in other words, any person who was believed to run the risk of producing "children with a tendency to disease, deformity, crime, insanity, feeble-mindedness, idiocy, imbecility, epilepsy, or alcoholism" (*Supplement to the Code of Iowa [1913],* sec. 2600). Likewise, in a less extreme vein, the WCTU, according to historian Hamilton Cravens, saw eugenics as a means of "socialization into the proper habits of health, diet, and sobriety for the young" (*Before Head Start: The Iowa Station and America's Children* [Chapel Hill: University of North Carolina Press, 1993], 36–37).

16. See Amy Vogel, "Regulating Degeneracy: Eugenic Sterilization in Iowa, 1911–17," *Annals of Iowa* 54 (Winter 1995): 119–43; and Lee Anderson, "'Headlights Upon Sanitary Medicine': Public Health and Medical Reform in Late Nineteenth-Century Iowa," *Journal of the History of Medicine and Allied Sciences* 46 (1991): 178–200. See also idem, "A Case of Thwarted Professionalization: Pharmacy and Temperance in Late-Nineteenth-Century Iowa," *Annals of Iowa* 50 (Winter 1991): 751–71; and Philip L. Frana, "Smallpox: Local Epidemics and the Iowa State Board of Health," 1880–1900," *Annals of Iowa* 54 (Spring 1995): 87–118. For more on alcoholism and eugenics, see

Zenderland, *Measuring Minds,* esp. 186–221, and Dowbiggin, *Keeping America Sane,* esp. 85–88.

17. "Minutes of the 1871 Annual Meeting," in *One Hundred Years of Iowa Medicine: Commemorating the Centenary of the Iowa State Medical Society, 1850–1950* (Iowa City: Athena Press, 1950), 37.

18. "Minutes of the 1880 Annual Meeting," in *One Hundred Years of Iowa Medicine,* 43.

19. "Minutes of the 1885 Annual Meeting," in *One Hundred Years of Iowa Medicine,* 46.

20. "Minutes of the 1887 Annual Meeting," in *One Hundred Years of Iowa Medicine,* 48. McClure, an advocate of "mental therapy," was for many years the president of the board of trustees of the Mt. Pleasant State Asylum for the Insane (later the Mt. Pleasant State Hospital), and had firsthand knowledge of alcohol's role in mental illness.

21. "Minutes of the 1892 Annual Meeting," in *One Hundred Years of Iowa Medicine,* 51.

22. See, for example, minutes of 1878, 1879, 1881, 1884, 1886, 1890, 1893, 1903, and 1906 in *One Hundred Years of Iowa Medicine,* 41–43, 45, 47, 49, 51, 61, and 65.

23. "Minutes of the 1906 Annual Meeting," in *One Hundred Years of Iowa Medicine,* 65.

24. For more on the Board of Health's acquisition of state administrative power, see Harold Martin Bowman, *The Administration of Iowa: A Study in Centralization* (New York: Columbia University Press, 1903), 129–58, esp. 139–41; and Frana, "Smallpox: Local Epidemics."

25. Anderson, "'Headlights Upon Sanitary Medicine," 196.

26. Hubert Wubben in his *Civil War Iowa and the Copperhead Movement* (Ames: Iowa State University Press, 1980), notes that Dubuque's Judge Hamilton thought the state needed a center to deal with problem drinkers, a state Inebriate Asylum. If denied liquor "for a sufficient time they might" he believed, "acquire strength to resist temptation." (186). Citations from *Davenport Democrat,* 12 January 1863; *Dubuque Herald,* 14, 18 October 1863; *Waverly Phoenix* in *Dubuque Times,* 12 December 1863. Hamilton's language, particularly his note about acquiring the strength to resist temptation, is revealing, for it suggests that in spite advocating for a medical treatment facility, Hamilton clung to a moralistic interpretation of the disease of inebriety.

27. In 1868 the briefly lived circuit judges (who replaced the county judges) and district attorneys were charged with these responsibilities (see Bowman, *Administration of Iowa,* 96–98). Between 1870 and 1898 these remained the only administrative agencies for charitable and correctional institutions. One exception to this supervisory trend was the 1872 establishment of a gubernatorially appointed visiting committee to monitor the treatment of patients at the state's institutions for the insane (see Bowman, *Administration of Iowa,* 106–7).

28. Bowman, *Administration of Iowa,* 110.

29. Ibid., 115.

30. Ibid., 21.

31. Kennedy, "Inebriety and Its Management." Kennedy, a prominent general practitioner and public health activist who had turned down an appeal from the reorganized state medical school to become its first professor of medical theory and practice in 1870, devoted his energies instead to the State Board of Health, where his role and influence

were legendary. See also Anderson, "'Headlights Upon Sanitary Medicine,'" 193; L. F. Andrews, "Iowa State Health Board's Grand Old Man," *Des Moines Register and Leader,* 23 February 1908, 2.

32. Dan Elbert Clark, "The History of Liquor Legislation in Iowa," *Iowa Journal of History and Politics* 6 (1908): 55–87, 339–74, 503–608; Jensen, "Iowa, Wet or Dry?" 263–88.

33. Dorothy Schwieder, *Iowa: The Middle Land* (Ames: Iowa State University Press, 1996), 216–17.

34. Dan E. Clark, "Recent Liquor Legislation in Iowa," *Iowa Journal of History and Politics* 15 (1917): 42–43.

35. The senate committee's vision of inebriety as a mental health problem rather than a penal problem is significant. Recall that the legislature in Massachusetts followed a similar course, establishing a state hospital for inebriates and dipsomaniacs in 1893 that fell within the jurisdiction of the State Board of Lunacy and Charity. When that institution was reorganized as the Norfolk State Hospital for Inebriates in 1911, the Commonwealth had already split its Board of Lunacy and Charity into a Board of Insanity and a Board of Charities. Norfolk was placed under the supervision of the Board of Charities, locating it outside the state's mental health system, even though its protocols for admission, treatment, and release were modeled on those for the insane. In the Massachusetts case, I believe the switch signaled the state's view of the inebriate as a drain on the Bay State's economy first, and as a person with mental disease second.

36. "Dipsomaniac Law Put into Effect," *Cherokee Democrat,* 22 July 1902, 2.

37. Charles Applegate, the superintendent of the Mount Pleasant State Hospital voiced his objections to the new inebriate law's penal aspects in his biennial report to the Board of Control: "I believe that the inebriate should be committed by the commissioners of insanity the same as in the case of an insane person, and not allowed to remain in jail awaiting trial when in need of treatment, and when the greatest amount of good could be accomplished. If inebriety is a disease and the inebriate is to be treated in a hospital, his commitment should not convey the penal aspect of a criminal until he has been found guilty of a criminal offense" (*Twenty-second Biennial Report of Iowa State Hospital, Mount Pleasant, to the Board of Control of State Institutions—for Biennial Period ending June 30, 1903* [Des Moines: State of Iowa], 65).

38. At the request of the state hospital superintendents, the General Assembly revised the inebriate commitment laws in 1904 and again in 1907, giving hospital physicians more governing power over their inebriate patients, taking the power to parole patients away from the governor and placing it in the hands of the hospital superintendents, and making it possible for inebriates to voluntarily commit themselves to the hospital without a court trial. By 1907 the courts and inebriate hospitals (by then Knoxville for men and Mt. Pleasant for women) further restricted admission to people "not of bad character or repute aside from the habit for which the commitment was made," and to individuals who stood a reasonable chance of being cured. See John Briggs, *History of Social Legislation in Iowa* (Iowa City: State Historical Society of Iowa, 1915), 185–95.

39. "Inebriates Go to Mt. Pleasant," *Des Moines Register and Leader,* 22 July 1902.

40. "Dislike Dipsomaniacs—Insane Patients Incensed at Being Confined with Bestial Drunks," *Cherokee Democrat,* 21 October 1902, 2.

41. "Are a Big Nuisance: Such is Dr. Voldeng's Opinion of the Inebriates. No Place for

Them in Hospitals—Won't Work—Have a Bad Influence," *Cherokee Democrat,* 19 November 1903, 1.

42. The governor, however, still could issue a patient's parole upon the recommendation of the board or the hospital superintendent.

43. "New Hospital Open—Superintendent Willhite now Ready to Extend Hearty Welcome to the "Dipsies"—State Has Spent Thousands of Dollars in Fitting Up a Hospital for Treatment of Inebriates—Can Care for About Two Hundred," *Knoxville Journal,* 26 January 1906, 2.

44. John E. Briggs, *History of Social Legislation in Iowa* (Iowa City: State Historical Society of Iowa, 1915), 185–95.

45. Ibid., 194–95.

46. Board of Control of State Institutions, *Laws of Iowa Relating to the Care of Inebriates in State Hospitals* (Anamosa: Penitentiary Press, 1910). Cited in this pamphlet is a note that on the 18th day of January, A.D. 1906, the Board of Control designated the Mt. Pleasant State Hospital as the one to which female inebriates would be sent.

47. Coincidentally, this is what happened to the Massachusetts Hospital for Dipsomaniacs and Inebriates, although Knoxville, unlike MHDI, remains a VA Hospital to this day. Westborough State Hospital in Massachusetts remained the state institution that treated women, just as Mount Pleasant continued to receive female inebriates. Iowa, however, designated Independence State Hospital as the facility for the treatment of the dwindling number of inebriate men, while Massachusetts sent inebriates to state hospitals throughout the state. See Tracy, "The Foxborough Experiment"; "Laws of the 39th General Assembly of the State of Iowa," Chapter 187, Board of Control, S.F. 790," *Acts and Joint Resolutions Passed at the Regular Session of the Thirty-Ninth General Assembly of the State of Iowa* (Des Moines: State of Iowa, 1921), 194; and *Twelfth Biennial Report of the Board of Control of State Institutions* (Des Moines: State of Iowa, 1920), 11–12.

48. Whereas Knoxville had customarily had patient censuses ranging from 200 to 300, Independence treated fewer than one hundred persons per year immediately following Knoxville's closing. See *Twelfth Biennial Report*; and *Thirteenth Biennial Report of the Board of Control of State Institutions for the Period Ending June 30, 1922* (Des Moines: State of Iowa).

49. For more on George Miller Beard, see ch. 1, n. 32.

50. Charles Applegate, "Inebriety, and the Care and Treatment of Inebriates in the Mt. Pleasant State Hospital, Mt. Pleasant, Iowa," *Bulletin of Iowa Institutions* 5 (1903): 155.

51. W. S. Osborn, "State Care and Treatment of Inebriates," *Bulletin of the Iowa Board of Control* 9 (1907): 3.

52. Of course, Applegate's remarks are also colored by his own desire to win the support of others in this cause and obtain a separate facility for their inebriate charges.

53. Applegate, "Inebriety, and the Care and Treatment of Inebriates," 165.

54. Ibid., 155.

55. L. G. Kinne, "Alcoholism," *Bulletin of Iowa Institutions* 6 (1904): 184.

56. Osborn, "State Care and Treatment of Inebriates," 9.

57. On the rise of consumer society in America, see Daniel Horowitz, *The Morality of Spending: Attitudes toward the Consumer Society in America* (Baltimore: Johns Hopkins University Press, 1985), and Lawrence B. Glickman, ed., *Consumer Society in American History: A Reader* (Ithaca, NY: Cornell University Press, 1999).

58. The biennial reports of the Board of Control contain a wealth of demographic data that suggest that the majority of male patients were engaged in three types of work: "domestic and personal" services (ranging from bartending to hotel clerking and egg candling); here the largest job category was "laborer," which occupied between 26 and 36 percent of the patients prior to their institutionalization; "manufacturing, mechanical, and building" trades (ranging from painters to bakers to watchmakers to miners), occupying 25 to 29 percent of those institutionalized; and "agriculture and rural trades" (ranging from farmers to nurserymen and horsemen). Farmers were consistently the second largest occupation listed, next to laborers, occupying 15 to 17 percent of patients.

The aggregate statistics of the male patient population during the early years of the inebriate treatment experiment further indicate that between 1902 and 1906 counties that were either close to or boasted one of Iowa's larger cities supplied just over half of the inebriate hospital patients. By far the largest group of inebriates hailed from Polk County, seat of the state's capital, Des Moines. Other counties in close proximity to Ames, Cedar Rapids, Clinton, Council Bluffs, Davenport, Fort Dodge, Marshalltown, and Sioux City supplied substantial numbers as well, but at approximately one-fourth the rate of Polk. The top ten counties contributed approximately 40 percent of the patients. Half of the men were married, while another third were single; the widowed and divorced made up a fifth of the male patients. Over 80 percent of the inebriates had received their common school certificates; another 15 percent had obtained their high school diplomas or college degrees. "Constant users" were most common, with "occasional" and "periodic" drinkers together comprising the other 50 percent of the population. The average age of the male patients at the time of their admission was about 40 years, with the largest ten-year cohort between the ages of 40 and 49 years. The average age at which most men began getting intoxicated was about 25 years, with approximately 50 percent beginning between the ages of 15 and 24. On average, 90 percent of the men smoked and/or chewed tobacco, in addition to drinking.

59. John Cownie in "Minutes of the Quarterly Meeting of Executive Officers of the Board of Control," *Bulletin of Iowa Institutions* 5 (1903): 246.

60. Ibid., 246.

61. Ibid., 247. "Treating" was the term used to refer to the practice of paying for another's drink. In the United States, during the first half of the nineteenth century, political candidates and political bosses would often hire the saloons and groceries who sold liquor to provide it free for voters several weeks before an election. Following the Civil War, however, treating was done on a more individual level, where saloon keepers would buy a round of drinks for their patrons, thus starting a custom that the drinkers themselves kept up. Treating was a sign of masculine solidarity, or camaraderie with the other patrons. It brought saloon keepers a small fortune. See Lender and Martin, *Drinking in America*, 10, 54–56, 60, 104.

62. "Minutes of the Quarterly Meeting," 248.

63. Ibid., 247.

64. *State Hospital for Inebriates, Knoxville, Iowa—First Biennial Report*, 6.

65. H. S. Miner, "For What Was Our Inebriate Hospital Established and What Should be Its Aim?" *Bulletin of the Iowa Board of Control* 10 (1908): 152.

66. For a philosophical examination of the definition of inebriety in the United States and Great Britain, see Mariana Valverde, *Diseases of the Will: Alcohol and the Dilemmas of Freedom* (Cambridge: Cambridge University Press, 1998).

67. The courts were often agents of misinformation in this regard. As Knoxville superintendent George Donohoe remarked in 1914, "It is discouraging day after day to see cases come into the institution drunk or half drunk, in the peculiarly fatuous mental condition produced by persistent intoxication, and have them tell you that they have come for the 'cure' and want to now how many days it will take. Upon questioning them when they are sober, you learn that the person himself has been led to believe, if not actually promised, that if he pleads guilty when charged with being an inebriate, he will be sent to Knoxville for a few days or a few weeks to be cured of a disease from which he is suffering. The drinker who is easily led while drinking, pleads guilty and is committed under the inebriate laws for a term until cured, not to exceed three years. Is it any wonder that he rebels and is discontented when he finds conditions not at all as pictured to his family?" ("The Inebriate," *Bulletin of the Iowa Board of Control* 16 [1914]): 108.

68. Applegate, "Inebriety, and the Care and Treatment of Inebriates," 164.

69. "Lo the Poor Drunkard," *Knoxville Express,* 7 December 1904, 2.

70. "New Hospital Open—Superintendent Willhite now Ready to Extend Hearty Welcome to the 'Dipsies'," *Knoxville Journal,* 26 January 1906, 2.

71. "Biennial Message of B. F. Carroll, Governor of Iowa to the Thirty-Fourth General Assembly," *Legislative Documents Submitted to the Thirty-Fourth General Assembly of the State of Iowa* (Des Moines: Emory H. English, State Printer, 1911), 29.

72. For more on Massachusetts' efforts to treat inebriates through state institutions, see Tracy, "The Foxborough Experiment," and Baumohl and Tracy, "Building Systems to Manage Inebriates."

73. Applegate, "Inebriety, and the Care and Treatment of Inebriates," 162.

74. For more on the treatment regimens for patients, see O. C. Willhite, "The Care and Treatment of the Inebriate at the Cherokee State Hospital," *Bulletin of the Iowa Institutions* 5 (1903); Applegate, "Inebriety, and the Care and Treatment of Inebriates"; Kinne, "Alcoholism"; Osborn, "State Care and Treatment of Inebriates"; H. S. Miner, "Cause, Prevention, and Cure of Inebriety," *Bulletin of the Iowa Board of Control* 2 (1909).

75. "The Wheelbarrow Cure—A New Way of Making Teetotalers of the Dipsos," *Knoxville Express,* 8 August 1906, 2.

76. "Terrors of Dipsomaniac Ward Now Increased by Probability of Wife's Divorce," *Cherokee Democrat,* 6 February 1903, 2. See also, "Quits Drink to Stop Divorce: Harvey Conner Agrees to Go to Inebriate Asylum for Cure in Order to Keep His Wife," *Des Moines Register,* 5 February 1906, 3.

77. "Terrors of Dipsomaniac Ward," 2.

78. "Quits Drink to Stop Divorce," 3.

79. Karl Pedersen, 30 October 1903 notes, Patient #93, Case Files of the Independence State Hospital for Inebriates, Independence Mental Health Institute, Independence, IA. Recall that the Independence State Hospital for Inebriates was really an inebriate ward at Independence State Hospital for the Insane.

80. At this time, only the governor could issue an official letter of parole.

81. Karl Pederson, 15 September 1904 "Mental Condition notes," Patient #93, Case Files of the Independence State Hospital for Inebriates, Independence Mental Health Institute, Independence, IA.

82. Petition from the Undersigned Citizens and Residents of Anamosa and Jones County, (no date), Jan Vickers, Patient #4, Case Files of the Independence State Hospital for Inebriates, Independence Mental Health Institute, Independence, IA.

83. Sheriff W. A. Hogan to W. P. Crumbacker, Superintendent, 17 January 1905, Jan Vickers, Patient #4, Case Files of the Independence State Hospital for Inebriates, Independence Mental Health Institute, Independence, IA.

84. Cooper, Clemens, and Lamb, Lawyers to Superintendent O. C. Willhite, 13 December 1902, Dennis Rowley, Patient #14, Case Files of the Independent State Hospital for Inebriates, Independence Mental Health Institute. Note that Rowley was initially committed to Cherokee State Hospital but was transferred to Independence to relieve the overcrowding at Cherokee once the new Independence facility opened in 1903.

85. For a sampling of the literature on turn-of-the-century and Progressive Iowa, see Thomas J. Morain, *Prairie Grass Roots: An Iowa Small Town in the Early Twentieth Century* (Ames: Iowa State University Press, 1988); Morain, "To Whom Much Is Given: The Social Identity of an Iowa Small Town in the Early Twentieth Century," John L. Larson, "Iowa's Struggle for State Railroad Control," and Richard Jensen, "Iowa, Wet or Dry? Prohibition and the Fall of the GOP," in *Iowa History Reader,* ed. Marvin Bergman (Ames: State Historical Society of Iowa and Iowa State University Press, 1996); Frana, "Smallpox: Local Epidemics"; Vogel, "Regulating Degeneracy"; Keach Johnson, "The Roots of Modernization: Educational Reform in Iowa at the Turn of the Century," *Annals of Iowa* 50, no. 8 (Spring 1991): 892–918; Joyce McKay, "Reforming Prisoners and Prisons: Iowa's State Prisons—The First Hundred Years," *Annals of Iowa* 60, no. 2 (Spring 2001): 139–73; Lee Anderson, "A Case of Thwarted Professionalization: Pharmacy and Temperance in Late Nineteenth-Century Iowa," *Annals of Iowa* 50, no. 7 (Winter 1991): 751–71; H. Roger Grant, "Railroaders and Reformers: The Chicago and North Western Encounters Grangers and Progressives," *Annals of Iowa* 50, no. 7 (winter 1991): 772–86.

86. See Vogel, "Regulating Degeneracy."

87. W. S. Osborn, "State Care and Treatment of Inebriates," 4.

SEVEN: On the Vice and Disease of Inebriety

1. Elaine Frantz Parsons, *Manhood Lost: Fallen Drunkards and Redeeming Women in the Nineteenth-Century United States* (Baltimore: Johns Hopkins University Press, 2003), 5.

2. Ibid., 11.

3. Jim Baumohl briefly addressed gospel temperance leaders such as Jerry McAuley and Samuel Hadley, his successor at the Water Street Mission in New York City, in conjunction with the creation of institutions to manage inebriates (see "Inebriate Institutions in North America," 110–12). The literature on temperance narratives is far more extensive, however, and includes: Parsons, *Manhood Lost;* Katherine Chavigny, "Reforming Drunkards in Nineteenth-Century America: Religion, Medicine, Therapy," in Tracy and Acker, *Altering American Consciousness;* Edmund B. O'Reilly, *Sobering Tales: Narratives of Alcoholism and Recovery* (Amherst: University of Massachusetts Press, 1997); John W. Crowley, ed., *Drunkard's Progress: Narratives of Addiction, Despair, and Recovery* (Baltimore: Johns Hopkins University Press, 1998); and Valverde, *Diseases of the Will.*

4. For examples, see "Slaves of an Invisible Command! The Soul and the Drunkard," *McClure's Magazine,* November 1918, 21–22, 34; "Is This Why You Drink? An Anonymous Confession: The First of a Brilliant New Series on Booze," *McClure's Magazine,* September 1917, 16–17, 35–36, 38; "Patient Number 24," *Everybody's Magazine,* September 1916, 321–34; "The Confessions of a Dipsomaniac," *The Arena* 32 (July 1904): 1–10.

5. Lucian Cary, "What Drives Men to Drink?" *Illustrated World,* September 1915, 97–102.

6. Inmates of the Pennsylvania Sanitarium, "Disabilities of Inebriates," in Proceedings of the First Meeting of the American Association for the Cure of Inebriates, 40–41.

7. Ibid.

8. Parrish hardly neglected the moral side of inebriety. In fact, he argued that "human sympathy" was a "blessed messenger to the needy" and a "perpetual benediction" when it governed an inebriate asylum ("Disabilities of Inebriates," 43). He hoped that his new specialty of inebriety medicine would encourage "science and Christianity . . . to join their kindred ministries" in the reformation of the habitual drunkard ("Philosophy of Intemperance," 40).

9. Journals such as *Illustrated World* (later absorbed by *Popular Science Monthly*) and *Atlantic Monthly* featured articles on inebriates' journeys to sobriety. See Walter V. Woehlke's transcription of a story told to him by an engineer and publisher, "Coming Back from 'Booze'," *Illustrated World,* May 1916, 315–20, 404, 406; "Our Inebriates, Classified and Clarified by an Inmate of the New York State Asylum," and "Our Inebriates, Harbored and Helped by an Inmate of the New York State Asylum," *Atlantic Monthly,* July 1869, 477–83 and 109–19.

10. "Our Inebriates, Harbored and Helped," 112.

11. "Our Inebriates, Classified and Clarified," 477.

12. The anonymous author of "Slaves of an Invisible Command!" also emphasized the superior qualities of the drunkard: "We agreed that there was no place like the saloon for meeting bright brains; the cleverest men always drank—the greatest lawyers, mechanics, writers, and even clergymen" (22).

13. Anonymous, "Is This Why You Drink?" 16.

14. Anonymous, "The Hardest Ride a Man Can Take: Showing How Very Easy it is to Swear off, but—As told to Maximilian Foster," *McClure's Magazine,* August 1915, 25.

15. Anonymous, "Patient Number 24," *Everybody's Magazine,* September 1916, 322–28.

16. Ibid., 334.

17. Ibid., 329–31.

18. Ibid., 332–34.

19. Ibid., 330–31, 334.

20. For more on White, see Nathan G. Hale Jr., *Freud and the Americans—The Beginnings of Psychoanalysis in the United States, 1876–1917* (New York: Oxford University Press, 1995), esp. 379–83.

21. Anonymous, "Is This Why You Drink?" 16.

22. Ibid., 17.

23. Ibid., 36.

24. Ibid.

25. Ibid.

26. For more on the mutual aid tradition, see Chavigny, "Reforming Drunkards"; Crowley, *Drunkard's Progress*"; Jonathan Zimmerman, "Dethroning King Alcohol: The Washingtonians in Baltimore, 1840–1854," *Maryland Historical Magazine* 87, no. 4 (1992): 374–98; Leonard U. Blumberg with William Pittman, *Beware the First Drink! The*

*Washingtonian Temperance Movement and Alcoholics Anonymous* (Seattle: Glen Abbey Books, 1991); Blocker, *American Temperance Movements;* Lender and Martin, *Drinking in America.*

27. The reference to Jonathan and David is biblical. In the book of Samuel, Jonathan, King Saul's son, and David, slayer of Goliath, are recognized as having a very special and loving relationship that some interpreters of the Bible have regarded as homosexual.

28. "The Confessions of a Dipsomaniac," *The Arena* 32, no. 176 (July 1904): 1–10. According to the anonymous author, a "well-known American novelist and essayist, whose name for obvious reasons is withheld," Puritanism made individuals see nothing but vice and sin in the case of the dipsomaniac, while "the religious ascetic and hysteric" was regarded as under the influence of a mysterious power. The "acts of a religious maniac being governed by God; those of the dipsomaniac by the Devil" (2).

29. "Notes and Comments," *The Arena* 32 (July 1904): 1. These words are taken from the introductory remarks of the editor for *The Arena.*

30. "Confessions of a Dipsomaniac," 1.

31. Ibid., 2, 4.

32. Ibid., 4, 10.

33. Ibid., 3.

34. Psychiatrists understood a *neurosis* to be a relatively minor impairment of the psychic constitution, less disabling than a psychosis and less damaging to one's personality. The neurotic's mental and social functioning was impaired but still present to a reasonable degree. *Psychosis,* by contrast, was a term reserved for severe cases of mental illness, often prolonged and incapacitating. A psychosis often obliterated one's personality and ability to function mentally and socially.

35. "Slaves of an Invisible Command: The Soul and the Drunkard," *McClure's Magazine,* November 1918, 21.

36. Ibid.

37. Ibid., 22, 34.

38. Ibid.

39. Ibid.

40. Ibid., 21.

41. Anonymous, "Twelve Years with Alcohol: The Story of a Man Who Spent $70,000 Before He Quit," *McClure's Magazine,* October 1915, 35.

42. Ibid., 41. Because his father's drinking tortured his mother so, he will not marry until he swears off alcohol for good.

43. "Slaves of an Invisible Command!" 21.

44. Ibid., 34.

45. Anonymous, "How I Broke Away from Alcohol," *Illustrated World* 26 (1917): 764.

46. Anonymous, "The Story of an Alcohol Slave as Told by Himself," *McClure's Magazine* 33, August 1909, 429.

47. "The Hardest Ride a Man Can Take," 26.

48. Walter V. Woehlke, "Coming Back from 'Booze'," *Illustrated World* 25 (May 1916): 315.

49. Ibid, 406.

50. Ibid.

51. Patient case file of Nicholas Felson, inpatient case files of the NSH.

52. Worcester Technology later became the Worcester Polytechnical Institute, its current institutional name.

53. Admission form, patient case file of Nicholas Felson, inpatient case files of the NSH, 25 February 1916.

54. Lawrence Felson to Irwin Neff, 5 April 1916, inpatient case files of the NSH.

55. This was an uncharacteristically tardy response for Neff, who almost always wrote back to patients and their families within a few days of receiving their letters or phone calls.

56. Irwin Neff to Lawrence Felson, 2 May 1916, inpatient case files of the NSH.

57. Nicholas Felson to Irwin Neff, 15 May 1916, inpatient case files of the NSH.

58. Nicholas Felson to Irwin Neff, 6 July 1916, inpatient case files of the NSH.

59. Nicholas Felson to Irwin Neff, 7 November 1916, inpatient case files of the NSH.

60. Nicholas Felson to Irwin Neff, 6 July 1916, inpatient case files of the NSH.

61. Charles Hanson Towne to Irwin Neff, 20 September 1917, inpatient case files of the NSH.

62. In 1911 "The Irish Players," an Abbey Theatre touring company led by Lady Gregory and W. B. Yeats, arrived in America to stage some of the new theatrical works that had developed in Ireland during the first decade of the twentieth century. These included works by Yeats, Gregory, and George Bernard Shaw. As a rule these works were well received both by audiences and by critics, but many Irish immigrants and Irish Americans found their portraits of Ireland and of the Irish people distasteful. Especially disturbing for those still struggling for middle-class status and respectability in the United States was J. M. Synge's *The Playboy of the Western World*, which was perceived as belittling Irish womanhood and maligning the nation of Ireland in general. Frequently, fraternal orders such as the Ancient Order of Hibernians would collaborate with Irish-language and literature interest groups to pressure local politicians and officials to prevent the plays' production or to insure their financial failure. See Gary Richardson, "The Irish in America: Long Journey Home," review in *New Hibernia Review* 2 (Summer 1998): 132–41.

63. Admission note, Robert Godson, inpatient case files of the NSH.

64. Robert Godson to Irwin Neff and Frank Carlisle, 18 September 1912, inpatient case files of the NSH.

65. Robert Godson to Frank Carlisle, 15 January 1915, inpatient case files of the NSH.

66. Robert Godson to Irwin Neff, 28 September 1912, inpatient case files of the NSH.

67. Robert Godson to Irwin Neff, 17 June 1914, inpatient case files of the NSH.

68. Robert Godson to Irwin Neff, 25 February 1914, inpatient case files of the NSH.

69. Irwin Neff to Robert Godson, 26 December 1912, inpatient case files of the NSH.

70. Robert Godson to Irwin Neff, 11 March 1913, inpatient case files of the NSH.

71. Robert Godson to Irwin Neff, 3 December 1915, inpatient case files of the NSH.

72. Irwin Neff to Robert Godson, 6 December 1915, inpatient case files of the NSH.

73. Robert Godson to Frank Carlisle, 6 April 1917, inpatient case files of the NSH.

74. Admissions forms, patient case file of Charles Winchester, inpatient case files of the NSH.

75. James Donald to Irwin Neff, 19 September 1911, inpatient case files of the NSH.

76. Charles Winchester to Irwin Neff, 18 February 1912, inpatient files of the NSH.

77. Charles Winchester to Irwin Neff, 10 March 1912, inpatient case files of the NSH.

78. Irwin Neff to Charles Winchester, 12 March 1912, inpatient case files of the NSH.

79. Charles Winchester to Irwin Neff, 7 April 1912, inpatient case files of the NSH.

80. Irwin Neff to Charles Winchester, 1 July 1912, inpatient case files of the NSH.

81. Admission notes, Charles Winchester, 27 August 1916, inpatient case files of the NSH.

82. Admissions forms, patient case file of Philip Rand, inpatient case files of the NSH.

83. Admissions forms, patient case file of Henry Rand, inpatient case files of the NSH.

84. Henry Rand to Irwin Neff, 5 August 1913, inpatient case files of the NSH.

85. Ibid.

86. Ibid.

87. Henry Rand to Irwin Neff, 28 May 1915, inpatient case files of the NSH.

88. Irwin Neff to Henry Rand, 2 June 1915, inpatient case files of the NSH.

89. Henry Rand to Irwin Neff, 9 June 1915, inpatient case files of the NSH.

90. Irwin Neff to Henry Rand, 19 July 1915, inpatient case files of the NSH.

91. "History of Present Illness," patient case file of Thomas Rand, 22 May 1912, inpatient case files of the NSH.

92. Admission note, Thomas Rand, 22 May 1912, inpatient case files of the NSH.

93. Thomas Rand to Irwin Neff, 2 July 1915, inpatient case files of the NSH.

94. Thomas Rand to Irwin Neff, September 1914, inpatient case files of the NSH.

95. Thomas Rand to Irwin Neff, 2 July 1915, inpatient case files of the NSH.

96. Ibid.

97. Irwin Neff to Thomas Rand, 21 September 1914, inpatient case files of the NSH.

98. Thomas Rand to Irwin Neff, 7 December 1914, inpatient case files of the NSH.

99. Irwin Neff to Thomas Rand, 28 April 1915, inpatient case files of the NSH.

100. Thomas Rand to Irwin Neff, 2 August 1915, inpatient case files of the NSH.

101. Irwin Neff to Thomas Rand, 5 August 1915, inpatient case files of the NSH.

102. Thomas Rand to Irwin Neff, 5 September 1915, inpatient case files of the NSH.

103. Thomas Rand to Irwin Neff, 2 February 1916, inpatient case files of the NSH.

104. Irwin Neff to Thomas Rand, 21 October 1916, inpatient case files of the NSH.

105. "Former Commitment," patient case file of Frank Casey, inpatient case files of the NSH.

106. Harriet Casey to Irwin Neff, 28 August 1908, Frank Casey, inpatient case files of the NSH.

107. Harriet Casey to Irwin Neff, 8 July 1909, Frank Casey, inpatient case files of the NSH.

108. Ibid. Emphasis in the original.

109. Georgia Casey to Irwin Neff, 18 September 1912, Frank Casey, inpatient case files of the NSH.

110. Irwin Neff to Georgia Casey, 28 September 1912, inpatient case files of NSH.

111. Clara Casey to Irwin Neff, 27 September 1912, Frank Casey, inpatient case files of the NSH.

112. Frank Casey to Irwin Neff, 10 April 1913, inpatient case files of the NSH.

113. Georgia Casey to Irwin Neff, 25 July 1913, Frank Casey, inpatient case files of the NSH.

114. Frank Casey to Irwin Neff, 3 August 1913, inpatient case files of the NSH.

115. Georgia and Frank Casey to Irwin Neff, 11 October 1913, inpatient case files of the NSH.

116. Harriet Casey to Irwin Neff, 29 July 1914, Frank Casey, inpatient case files of the NSH.

117. Irwin Neff to Georgia Casey, 8 May 1915, Frank Casey, inpatient case files of the NSH.

118. Georgia Casey to Irwin Neff, 21 May 1915, Frank Casey, inpatient case files of the NSH.

119. Georgia Casey to Irwin Neff, 15 May 1916, Frank Casey, inpatient case files of the NSH.

## Conclusion

1. House Document No. 2000, *Report of the Special Commision to Investigate the Problem of Drunkenness in Massachusetts,* March 7, 1945 (Boston: Wright and Potter, 1945), 10, 44.

2. Ibid., 11.

3. Ibid., 37–38.

4. Howard Haggard and E. M. Jellinek, *Alcohol Explored* (New York: Doubleday, 1942), 144.

5. Caroline J. Acker, "Stigma or Legitimation? A Historical Examination of the Social Potentials of Addiction Disease Models," *Journal of Psychoactive Drugs* 25 (July 1993): 193–205.

6. House Document No. 2000, 40.

7. Ibid., 41.

8. Chauncey Leake, Foreword, in *Alcohol—Basic Aspects and Treatment,* ed. Harold Himwich (Washington, DC: American Association for the Advancement of Science, 1957).

9. White, *Slaying the Dragon,* 178; see also Joseph Gusfield, *Contested Meanings: The Construction of Alcohol Problems* (Madison: University of Wisconsin Press, 1996); and Lender and Martin, *Drinking in America.*

10. Haggard and Jellinek, *Alcohol Explored.*

11. E. M. Jellinek, *The Disease Concept of Alcoholism* (New Haven, CT: Hillhouse Press, 1960), 19.

12. Ibid.

13. Ibid., 12.

14. For popular and medical references that discussed alcoholism as an allergy, see E. A. Strecker and F. T. Chambers, *One Man's Meat* (New York: Macmillan, 1938); Robert V. Seliger, "Working with the Alcoholic," *Medical Record of New York* 149 (1939): 147–50; and Robert S. Carroll, *What Price Alcohol?* (New York: Macmillan, 1941).

15. For the best treatment of the politics of the modern alcoholism treatment movement, see Ron Roizen, "How Does the Nation's 'Alcohol Problem' Change from Era to Era? Stalking the Social Logic of Problem-Definition Transformations since Repeal," in Tracy and Acker, *Altering American Consciousness,* 61–87.

16. Herbert Yahraes, *Alcoholism Is a Sickness* (New York: Public Affairs Committee, Inc., 1946), 2.

17. Robert V. Seliger, with Victoria Cranford and Harold S. Goodwin, *Alcoholics Are Sick People*, War Edition (Baltimore: Alcoholism Publications, 1945), frontispiece.

18. See E. M. Jellinek, ed., *Alcohol Addiction and Chronic Alcoholism* (New Haven, CT: Yale University Press, 1942); idem, *The Disease Concept of Alcoholism.*

19. E. M. Bluestone, Foreword, in E. L. Corwin and Elizabeth V. Cunningham, *Institutional Facilities for the Treatment of Alcoholism: A Report of the American Hospital Association*, Research Report No. 7 (New York: Research Council on Problems of Alcohol, 1944). This report was prepared for the Committee on Hospital Treatment of Alcoholism of the Council on Professional Practice of the American Hospital Association, based on a study conducted under a grant from the Research Council on Problems of Alcohol.

20. See White, *Slaying the Dragon*, 189–90, 300–301. See also Ralph Henderson and Seldon Bacon, "Problem Drinking: The Yale Plan for Business and Industry," *Quarterly Journal of Studies on Alcohol* 14 (1953): 247–62.

21. White, *Slaying the Dragon*, 190–92.

22. The alcoholic and the mentally ill patient were hardly distant cousins at midcentury, for most psychodynamically oriented psychiatrists regarded alcoholism as an expression of an underlying neurotic or psychotic state.

23. *AA Grapevine*, May 1947, 12.

24. White, *Slaying the Dragon*, 217. The report in question is R. Moore and T. Buchanan, "State Hospitals and Alcoholism: A National Survey of Treatment Techniques and Results," *Quarterly Journal of Studies on Alcohol* 27 (1966): 459–68.

25. G. Lolli, "Alcoholism as a Medical Problem," in *Bulletin of the New York Academy of Medicine* 31, no. 12 (December 1955): 882–83. Presented at the Stated Meeting of the New York Academy of Medicine, February 3, 1955, as part of the 30th Hermann M. Biggs Memorial Lecture program under the auspices of the Committee on Public Health.

26. White, *Slaying the Dragon*, 217. My discussion of the parameters of state institutional involvement post-Prohibition owes much to the work of William L. White.

27. See Jim Baumohl and Jerome Jaffe, "Treatment, History of, in the United States," in *Encyclopedia of Drugs, Alcohol and Addictive Behavior*, 2nd ed., ed. Rosalyn Carson-DeWitt (New York: Macmillan Reference, 2001), 1124.

28. Wooley, Foreword, in Seliger, *Alcoholics are Sick People.*

29. Haggard and Jellinek, *Alcohol Explored*, 160–61.

30. Baumohl and Jaffe, "Treatment, History of, in the United States," 1124–25.

31. White, *Slaying the Dragon*, 276.

32. Roizen, "How Does the Nation's 'Alcohol Problem' Change," 71–78.

33. Harry Tiebout, "Perspectives in Alcoholism," in *Selected Papers Delivered at the Sixth Annual Meeting National States' Conference on Alcoholism, Miami Beach, Florida, October 30-November 2, 1955* (Portland, OR: National States' Conference on Alcoholism), 1–7.

34. Institute of Medicine, *Broadening the Base of Treatment for Alcohol Problems* (Washington, DC: National Academy Press, 1990), 6–7.

35. Roizen, "How Does the Nation's 'Alcohol Problem' Change," 71–78.

36. Ibid.; see also Reinarman, "The Social Construction of an Alcohol Problem," *Theory and Society* 17 (1988): 91–120; David Wagner, *The New Temperance—The American Obsession with Sin and Vice* (Boulder, CO: Westview Press, 1997).

37. J. Madeleine Nash, "Addicted—Why Do People Get Hooked? Mounting Evidence Points to a Powerful Brain Chemical Called Dopamine," *Time*, 5 May 1997, 68–75.

38. Ibid.

39. Enoch Gordis, Introduction, in *Special Report to the U.S. Congress on Alcohol and Health—Highlights from Current Research* (Bethesda, MD: National Institutes of Health, 2000), xiii.

40. Ibid., xv.

41. Yahraes, *Alcoholism Is a Sickness*, 28–29.

# Index

AA. *See* Alcoholics Anonymous

AACI. *See* American Association for the Cure of Inebriates

Abbott, Lyman, 94–95

Abraham, Karl, 77

Acker, Caroline Jean, 40

Addams, Jane, 11, 309n9

addiction, as chronic disease, xii; competing theories of, 40; expansion of modern treatment facilities for, 287, 289; iatrogenic, 51, 303n54; psychological and environmental factors affecting, 293n11

aftercare: at Foxborough and Norfolk state hospitals, 186–87, 191, 245–72; Keeley Leagues, 86

AHA. *See* American Hospital Association

Al-Anon, 279

alcohol: consumption, 7–8, 288, 295n17; physiology of, 57, 79, 276, 303n47; poisoning, 42, 117; problems paradigm, xii, 24, 288–89, 294n12; stigmatized, 279; therapeutic use of, 51

alcoholics: as automatons, 66–67; personality of, 287; "repeaters," 137–38; stigmatized image of, 279. *See also* inebriates

Alcoholics Anonymous (AA), xiv, 4, 276, 279, 294n7, 319n85, 321n24

alcoholism: allergy model of, 4, 344n14; caused by alcohol poisoning, 117; as disease of consumption, 211; evolution of term, 41–45; first defined as a disease, 2, 41; free will and, ix, xiii, xiv, 19, 30, 40, 42, 66, 106; modern medical profession on, xi–xiv, 288; obesity and, 285; physiological vs. social explanations for, 276–79, 290; popular opinion on, 19, 23, 168, 176,

226–72, 279, 332n1; quest for mechanism in, 276–79, 290; as symptom of disease, 44; as syndrome, 287; tuberculosis and, 11, 57. *See also* dipsomania; inebriety; intemperance

*Alienist and Neurologist,* 97

allergy model of alcoholism, 4, 344n14

AMA. *See* American Medical Association

American Association for the Cure of Inebriates, x, 1–3, 12–14, 16, 26, 38–39, 53–55, 62, 70, 79–84, 88–91, 95, 97, 101–3, 106, 124, 156, 168, 201, 228

American Hospital Association (AHA), 282

American Medical Association (AMA), x, xiv, 25–26, 83, 84, 88, 91, 118, 292

American Medical Temperance Association (AMTA), 83

American Medico-Psychological Association, 178

American Psychiatric Association (APA), 288

American Psychoanalytic Association, 233

American Social Science Association, 56

American Society for the Promotion of Temperance, 6–7

American Society of Addiction Medicine, 282

American Temperance Society, 6–7

Ames, Oliver, 160

Anderson, Dwight, 280

Anderson, Lee, 201

Andrew, John, 153–54, 325n15

Anti-Saloon League, 41, 72

Appleton Temporary Home, 104, 326n27

Association of Medical Superintendents of American Institutions for the Insane, 15, 25–26, 96, 125–26, 128